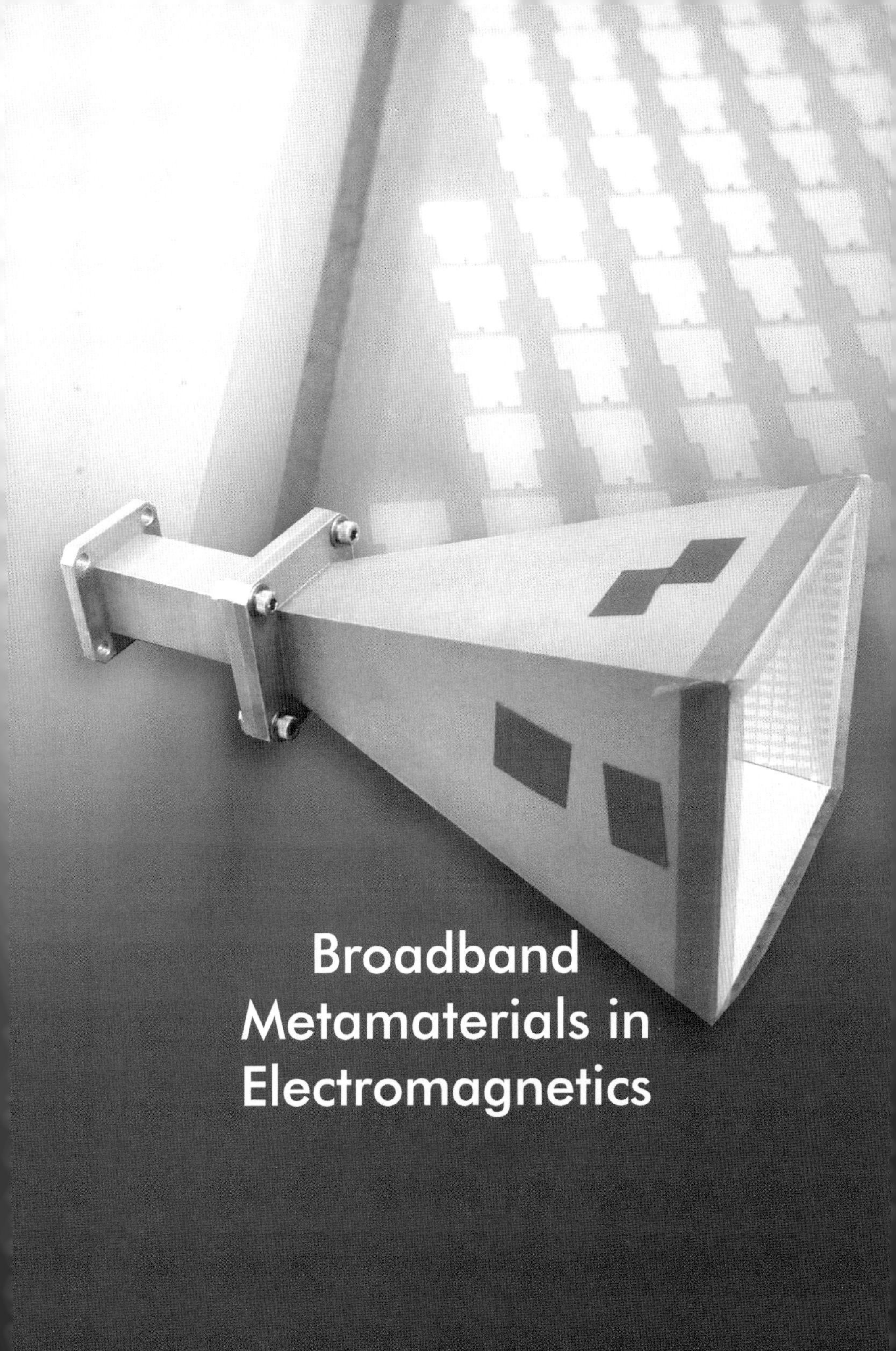

Broadband Metamaterials in Electromagnetics

Broadband Metamaterials in Electromagnetics

Technology and Applications

edited by
Douglas H. Werner

Published by

Pan Stanford Publishing Pte. Ltd.
Penthouse Level, Suntec Tower 3
8 Temasek Boulevard
Singapore 038988

Email: editorial@panstanford.com
Web: www.panstanford.com

British Library Cataloguing-in-Publication Data
A catalogue record for this book is available from the British Library.

Broadband Metamaterials in Electromagnetics: Technology and Applications

Copyright © 2017 by Pan Stanford Publishing Pte. Ltd.
All rights reserved. This book, or parts thereof, may not be reproduced in any form or by any means, electronic or mechanical, including photocopying, recording or any information storage and retrieval system now known or to be invented, without written permission from the publisher.

For photocopying of material in this volume, please pay a copying fee through the Copyright Clearance Center, Inc., 222 Rosewood Drive, Danvers, MA 01923, USA. In this case permission to photocopy is not required from the publisher.

ISBN 978-981-4745-68-0 (Hardcover)
ISBN 978-1-315-36443-8 (eBook)

Printed in Great Britain by Ashford Colour Press Ltd, Gosport, Hampshire

Contents

Preface

Metamaterials are artificial electromagnetic or optical media whose bulk (macroscopic) properties are controlled by engineering their structure on the subwavelength (microscopic) scale. The rapid development of technology based on metamaterials, coupled with the recent introduction of the transformation optics technique, provides an unprecedented ability for device designers to manipulate and control the behavior of electromagnetic wave phenomena. Many of the early metamaterial designs, such as negative index materials and electromagnetic bandgap surfaces, were limited to operation only over a very narrow bandwidth. However, recent groundbreaking work reported by several international research groups on the development of broadband metamaterials has opened up the doors to an exciting frontier in the creation of new devices for applications ranging from radio frequencies to visible wavelengths.

This book contains a collection of eight chapters, which cover recent cutting-edge contributions to the theoretical, numerical, and experimental aspects of broadband metamaterials. Chapter 1 provides an overview of recent advances in utilizing custom designed anisotropic metamaterials to enable broadband performance in microwave antennas. Several application examples are presented where anisotropic metamaterials with engineered dispersion and negligible loss have been employed to extend the impedance bandwidth and/or enhance the gain of antennas. These include the design of an ultrathin lightweight anisotropic metamaterial coating for extending the bandwidth of conventional monopole antennas, single-polarization and dual-polarization anisotropic zero-index metamaterial lenses for achieving high unidirectional radiation, transformation electromagnetics lenses with low-index properties for generating multiple highly directive beams along with extended impedance bandwidth, and a concept for realizing tunable versions of the low-index lenses capable of reconfigurable beam scanning.

One of the earliest documented examples of a broadband low-loss dispersion engineered metamaterials is highlighted in Chapter

2. This work represented a pivotal point in metamaterials' development by demonstrating the first practical broadband negligible-loss device (a horn antenna) to benefit from the application of metamaterials technology. By properly engineering the dispersive characteristics of a metamaterial liner inside a horn antenna, it is shown that the bandwidth of a conventional narrowband horn can be extended to operate over an octave or more with negligible intrinsic losses and low cross-polarization. Examples of broadband horn antenna prototypes that were fabricated and measured are presented for the C-band and K_u-band with wire-grid and printed circuit board type metamaterial liners, respectively. For maximum performance, it is shown that inhomogeneous metamaterials can be applied as horn liners to support broadband balanced hybrid modes and create polarization-independent patterns suitable for dual-polarized communication systems such as those commonly employed in satellite applications.

Chapter 3 provides, for the first time, an understanding of how exotic propagation phenomena can be explained and controlled using dispersion engineering. More importantly, these novel propagation modes are generated using a set of coupled transmission lines that can be printed on simple substrates. By controlling the propagation constant on each line, as well as their mutual capacitance and inductance, it is shown that these transmission line systems can be designed to support various exotic modes. These modes can be subsequently utilized to (i) improve the electronic efficiency of traveling wave tubes, (ii) miniaturize antennas, reaching their optimal limits, (iii) increase antenna directivity, and (iv) control antenna bandwidth. Several practical examples of this transformative design methodology are presented and discussed.

Chapter 4 presents a powerful and highly efficient synthesis technique for the custom design of multiband and broadband electromagnetic bandgap (EBG) devices, which are based on the well-known Sievenpiper surfaces. The synthesis approach optimizes the non-uniform capacitive loading of a periodic array of fixed metallic mushroom-type structures in order to meet a desired design objective. In this way, both the inhomogeneous and anisotropic properties of the metasurfaces can be engineered simultaneously. The approach exploits the fact that EBGs loaded with lumped capacitors can be conveniently represented as multi-port networks whose response is fully characterized by their *S*-matrix. Consequently, their analysis can be performed via simple and efficient circuit-based calculations rather than through computationally expensive full-wave simulations. It is shown that non-uniform capacitively loaded EBGs exhibit, in principle, wider bandgaps compared to those of the same EBG with uniform capacitive loads. Finally, it is demonstrated that this circuit-based analysis can be extended for the design of mushroom-type absorbers loaded with lumped tuning capacitors and resistors.

Chapter 5 establishes the conceptual foundations for performing ultrafast real-time spectrum analysis through the use of broadband metamaterials. The principle is based on the radio-analog signal processing (R-ASP) paradigm, which makes use of temporal and spatial phasers to discriminate the spectral components of a broadband signal in time and space, without resorting to digital computations. Three different systems for ultrafast real-time spectrum analysis applications are considered: (i) a 1D real-time spectrum analyzer (RTSA) based on a composite right/left-handed (CRLH) transmission line system to decompose a broadband signal along one dimension of space, (ii) a spatio-temporal 2D RTSA combining temporal and spatial phasers to perform spectral decomposition in two dimensions of space to achieve higher frequency resolution in the RTSA system, and (iii) a spatial 2D RTSA using metasurface phasers to spectrally decompose an incident pulsed wavefront onto two dimensions of space. These three systems are particularly well suited for mm-wave and terahertz high-speed applications, where digital computation is not readily available.

Chapter 6 provides a comprehensive treatment of the newly emerging and important topic concerned with broadband lens designs enabled or inspired by the quasi-conformal transformation optics (qTO) technique. Example lens designs are presented for both radio frequency and optical applications. The resulting lens designs are shown to possess desirable performance over a large field-of-view (FOV) and/or wide frequency range. First, the mathematics of transformation optics and the numerical algorithm qTO are described. Next, a series of broadband qTO-derived inhomogeneous metamaterial lenses are presented, which were created by transforming the geometry of classical designs, including the Luneburg and Fresnel lenses. Then, qTO is employed to transform wavefronts and design a broadband gradient-index (GRIN) multi-beam lens antenna. A systematic approach is introduced next for using qTO to design anti-reflective coatings. In addition, the wavefront matching (WFM) technique is discussed, which is a powerful design tool that can be used to correct for dispersion in complex multi-element optical lens systems. The final section looks at the effects of material dispersion on the performance of GRIN lenses, while a set of classical and qTO-inspired corrections are presented.

In Chapter 7, a distinct class of structures known as twisted metamaterials is explored. It is demonstrated that wave propagation through these metamaterials exhibits new phenomena, which are not available in conventional periodic metamaterials, such as a broadband chiral response. A detailed examination of the electromagnetic wave propagation in these unique metamaterial structures is provided. A generalized Floquet analysis is developed and applied to obtain rigorous modal solutions to lossless as well as lossy twisted metamaterials. The chapter concludes with a discussion of the wave polarization properties and potential applications of twisted metamaterials for broadband polarizer designs.

Chapter 8 summarizes recent breakthroughs in the development as well as applications of broadband optical metamaterials and metasurfaces. Dispersion engineering is introduced as a powerful design methodology for exploiting the resonant properties of metamaterials over broad wavelength ranges in order to enhance the performance of existing or realize new optical devices. Several examples are presented, including a demonstration of how metamaterial loss can be exploited for broadband absorption in the infrared regime. A robust genetic algorithm (GA) synthesis technique is employed to design super-octave and multi-octave metamaterial absorbers using only a single patterned metallic screen. Broadband optical metasurfaces that can control the phase and polarization of a reflected wave are also considered. Optical metasurface designs are presented along with measurement results that demonstrate broadband and wide-angle quarter-wave and half-wave plate functionalities. Another type of metasurface based on nanoantenna arrays that can artificially induce a phase gradient in the cross-polarized reflected or transmitted wave at an interface and, therefore, steer and focus light is also presented. Together, these nanostructured metamaterial and metasurface examples illustrate the potential for realizing practical optical devices with unique functionalities over broad bandwidths.

Douglas H. Werner

Spring 2017

Chapter 1

Broadband Anisotropic Metamaterials for Antenna Applications

Zhi Hao Jiang, Jeremiah P. Turpin, and Douglas H. Werner
Department of Electrical Engineering, Penn State University, University Park, PA 16802, USA
dhw@psu.edu

Artificially designed and structured electromagnetic materials, also known as metamaterials (MMs), allow the engineering of exotic electromagnetic properties, which in turn offers control over various wave–matter interactions, such as refraction, absorption, and radiation. Most reported MMs, both in the microwave and optical ranges, are limited either by a narrow bandwidth of operation and/or by significant absorption loss, both of which impede their integration into practical devices. By overcoming such limitations, broadband and low-loss MMs have the potential for providing great benefit to various engineering disciplines, especially for microwave antennas and circuits. This chapter presents the recent progress on advanced antenna designs by incorporating wideband and low-loss MMs with custom engineered anisotropic electromagnetic properties. First, an ultra-thin anisotropic MM coating is presented,

Broadband Metamaterials in Electromagnetics: Technology and Applications
Edited by Douglas H. Werner
Copyright © 2017 Pan Stanford Publishing Pte. Ltd.
ISBN 978-981-4745-68-0 (Hardcover), 978-1-315-36443-8 (eBook)
www.panstanford.com

which enhances the impedance bandwidth of a monopole to over an octave while simultaneously preserving the desirable radiation pattern properties. Next, static broadband anisotropic MM lenses for controlling the radiation patterns of antennas are discussed, including several demonstrated examples of MM lenses that produce a single or multiple highly directive beams. Finally, the concept for a tunable anisotropic MM lens for reconfigurable azimuthal beam scanning in the horizontal plane is presented, including experimental verification of the proposed unit cell configuration.

1.1 Introduction

The nascent field of MMs has enabled scientists and engineers to access a large range of refractive index and anisotropy that were previously unattainable [1–5]. These exotic electromagnetic properties further give rise to diverse new device functionalities for operation ranging from microwave frequencies to the terahertz regime and even at optical wavelengths. For instance, near- and far-field perfect lenses capable of obtaining sub-diffraction limited images [6], though with certain restrictions, have been implemented for operation at radio frequencies using two-/three-dimensional transmission line MMs [7] and in the visible spectrum by utilizing a slab of silver [8]. Electromagnetic free space invisibility cloaks [9], which can hide objects from being detected from their scattering fields, have also been proposed and/or experimentally demonstrated at microwave frequencies [10] as well as optical wavelengths [11]. Other more recently reported notable advances in this field include retro-directive reflectors [12], highly efficient absorbers [13], microwave illusion coatings [14, 15], terahertz phase modulators [16], electrically small antennas [17], asymmetric transmission surfaces [18], etc. However, the primary shortcomings of most current MM devices are their narrow operational bandwidth and high absorption loss originating from their resonant behavior. In order to produce more practical MM devices that can be readily integrated into existing systems or components, methods must be developed to broaden the operational bandwidth and reduce the intrinsic loss. Several efforts have been recently carried out to demonstrate such possibilities, including shifting the working frequencies of MMs outside of their resonant bands and tailoring their inherent dispersive properties to meet specific device requirements. Such approaches have enabled broadband carpet cloaks [19, 20], broadband polarization rotators [21], broadband flat Luneburg lenses [22, 23], broadband front-to-back end-fire leaky-wave antennas [24], broadband miniaturized couplers [25], octave bandwidth MM liners for horn antennas [26], etc.

In the field of antenna engineering, MMs have found widespread application for various radiating structures primarily in two aspects. First, the far-field pattern of an existing radiator can be shaped by incorporating MMs into the antenna structure due to the ability to tailor their refractive index or anisotropic properties. For example, MMs have been employed to manipulate the radiation properties of antennas, including MM-coated electrically small antennas [17], transmission line MM-enabled leaky-wave antennas [24], broadband negligible-loss metaliners for horn antennas that support low sidelobes and low cross-polarization [26], high-gain antennas based on electromagnetic bandgap structures [27], MM lenses for broadband highly directive multibeam antennas [28, 29], high-gain conformal antennas [30], and many others. Of particular interest in this research area is the ability to achieve highly directive radiated beam(s), which is beneficial for point-to-point communications, wireless power transfer, radar systems, base station systems, and various other wireless systems. Conventional approaches frequently employ Fabry–Pérot cavities [31] to obtain high directivity within a narrow bandwidth due to the high-quality factors of such resonating cavities. When they are constructed with a ground plane, the device thickness is usually on the order of one-half of the operating wavelength. Recent advanced Fabry–Pérot cavity designs employing artificial magnetic conductors (AMCs) allow the total device thickness to be reduced to around one-quarter of the wavelength [32]. Further reduction in the antenna profile can be achieved, e.g., down to $\lambda/9$, by using MM surfaces designed with customized reflection phases for both the top and bottom covers of the cavity [33]. Alternatively, isotropic zero-index MMs have been proposed to create narrow radiated beams based on Snell's law of refraction [34, 35]. In contrast to isotropic zero-index metamaterials (ZIMs), anisotropic ZIMs (AZIMs) [28, 29, 36, 37], which utilize their extreme anisotropy, can also generate highly directive radiation from an embedded isotropically radiating source. Second, the input impedance of an existing antenna can be tuned by properly loading

MMs in the antenna near-field so that the current distribution can be controllably modified. For instance, MMs have been utilized to enhance the impedance bandwidth of planar monopoles and microstrip antennas by proper loading with split-ring resonators (SRRs) [38] or using negative refractive index transmission lines [39]. They have also been exploited to create multiband antennas from a single-band radiating structure [40–42]. In most cases, the two aspects—radiation pattern shaping and impedance tuning—can be combined into an integrated design strategy. The MM can be tailored to account for its effect on both the far-field and near-field properties of antennas.

In this chapter, we present several application examples of broadband anisotropic MMs for improving the performance of microwave antennas. These include the design of an ultra-thin anisotropic MM coating for bandwidth extension of monopole antennas, single-polarization and dual-polarization AZIM lenses for achieving unidirectional radiation, multibeam AZIM lenses for generating multiple highly directive beams along with extended impedance bandwidth, and a tunable AZIM lens for reconfigurable beam steering.

1.2 MM Coatings for Monopole Bandwidth Extension

This section presents a type of novel MM coating that can extend the impedance bandwidth of a conventional quarter-wave monopole antenna to more than an octave. First, it presents the results of a parametric study of a monopole surrounded by a coating comprised of anisotropic material, which provide the guidelines for creating a practical broadband antenna design. Then it reports the MM unit cell designs targeting the *S*-band. Finally, it verifies the MM coating designs by both full-wave simulations and experimental results when integrated with a monopole. A *C*-band design is also included to further demonstrate the versatility of this approach.

1.2.1 Monopole with Anisotropic Material Coating

The configuration of the monopole antenna with an anisotropic material coating is shown in Fig. 1.1a. The monopole is 28.5 mm long and is designed to be resonant at 2.5 GHz. The cylindrical coating is composed of anisotropic dielectric material with a relative permittivity tensor $\boldsymbol{\varepsilon}_\mathbf{r} = [\varepsilon_{rt}, \varepsilon_{rt}, \varepsilon_{rz}]$, a diameter of d_c, and a height of h_c. The coating has a minimal impact on the monopole's omnidirectional radiation pattern in the H-plane. By fixing the value of ε_{rz} to be 8 and increasing the value of d_c, it can be seen that a second resonance occurs. When d_c = 14 mm, a wide impedance bandwidth can be achieved. Further increasing d_c will result in a single merged resonance at the center of the band. By fixing the value of d_c to be 14 mm and increasing the value of ε_{rz}, the S_{11} curve evolves from a single resonance at 2.3 GHz to dual resonances at 2.3 and 3.7 GHz and then to a single resonance at 3.4 GHz. The dual resonances give rise to the wideband matching. Increasing the value of ε_{rt} degrades the impedance bandwidth, so it should be maintained at a value near unity.

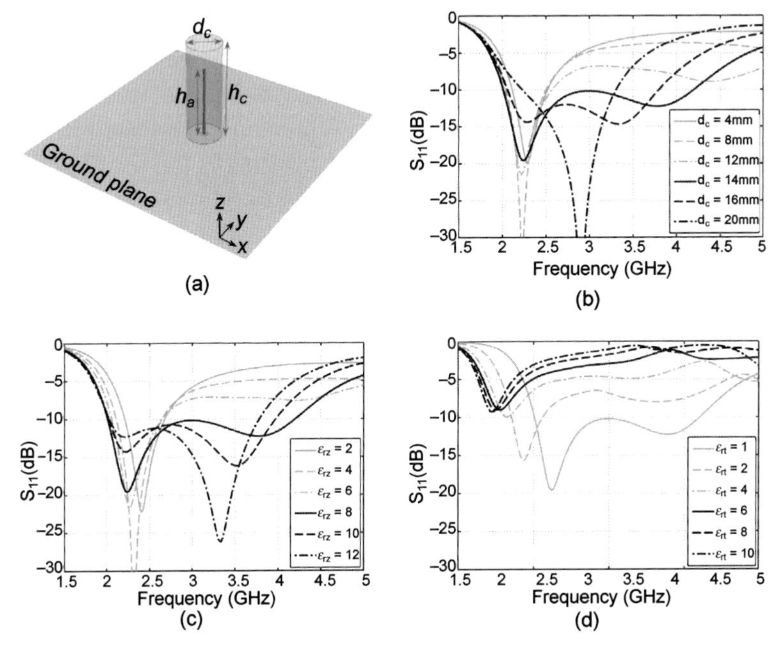

Figure 1.1 (a) Schematic of the monopole antenna (h_a = 28.5 mm) with an anisotropic dielectric coating, a diameter of d_c, and a height of h_c. (b) Simulated S_{11} of the coated monopole with h_c = 40 mm, ε_{rt} = 1, ε_{rz} = 8, and varying d_c. (c) Simulated S_{11} of the coated monopole with h_c = 40 mm, ε_{rt} = 1, d_c = 14 mm, and varying ε_{rz}. (d) Simulated S_{11} of the coated monopole with h_c = 40 mm, ε_{rz} = 8, d_c = 14 mm, and varying ε_{rt}.

1.2.2 Unit Cell Design and Full-wave Simulations

Based on the above parametric study, it can be concluded that in order to obtain a greatly extended impedance bandwidth of the coated monopole, the coating requires a high ε_{rz} with a value around 8 to 10, near-unity ε_{rx} and ε_{ry}, and a proper radius. However, no known natural materials have such high-contrast anisotropy with low loss at microwave frequencies. Hence, a custom-designed MM must be synthesized, which would respond to electromagnetic waves as if it were an effective homogeneous anisotropic material with the required properties. The unit cell of such an MM coating is illustrated in Fig. 1.2(a), which comprises two identical I-shaped copper patterns printed on both sides of a Rogers Ultralam 3850 substrate. The thicknesses of the substrate (d_s) and the copper (d_c) are 51 μm and 17 μm, respectively. Using this thin flexible substrate, the nominally planar MM structure can be easily rolled into a cylindrical configuration. In the simulation setup for the MM unit cell, periodic boundary conditions were assigned to the walls in the *y*- and *z*-directions. A *TE*/*TM*-polarized plane wave, with the *E*-field/*H*-field oriented along the *z*-direction, is assumed to be incident from the left half-space at an angle of φ ($0° \leq \varphi \leq 90°$) with respect to the *x*-axis. An anisotropic inversion technique presented in Ref. [43] was employed to extract the effective permittivity and permeability tensor quantities from the scattering parameters calculated at different angles of incidence using ANSYS HFSS (high frequency structure simulator).

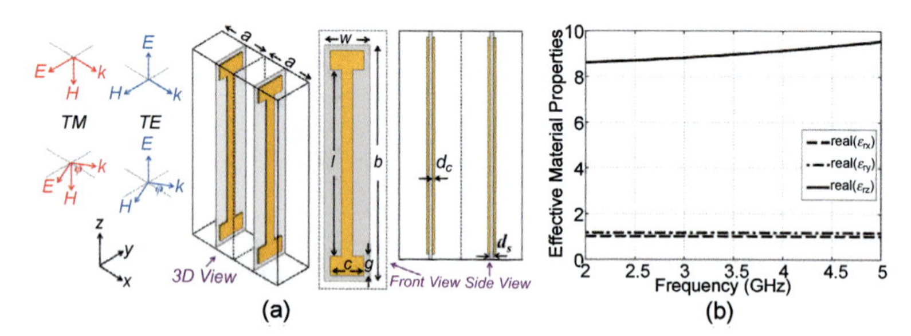

Figure 1.2 (a) Geometry and dimensions of the unit cells of the anisotropic MM coating. All dimensions are in millimeters: $a = 2.5$, $d_s = 0.051$, $d_c = 0.017$, $w = 2$, $b = 10$, $c = 1.5$, $g = 0.8$, and $l = 8$. (b) Real parts of the retrieved effective anisotropic permittivity tensor parameters (ε_{rx}, ε_{ry}, ε_{rz}). The imaginary parts are extremely small and are thus not shown.

The retrieved real parts of the effective permittivity tensor parameters are shown in Fig. 1.2b. Their imaginary parts are near zero (not shown) over the entire band of interest. It can be seen that none of the parameters exhibit a resonant response below 5 GHz, resulting from the subwavelength size of the I-shaped elements, which corroborates previously reported results on I-shaped unit cells utilized for broadband microwave transformation optics devices [19]. The retrieved effective ε_{rx} and ε_{ry} have values near unity throughout the band, whereas ε_{rz} exhibits a large value, which is attributed to the inductance provided by the central bar in the I-shaped elements and the capacitance associated with the gaps between the stubs of adjacent unit cells in the z-direction. The three effective permeability tensor parameters (not shown here) have unity values with very low loss, indicating that the MM does not have any effect on the radiated magnetic field.

When the actual MM coating is integrated with the monopole, two concentric layers are employed, each having four unit cells in the z-direction, as shown in Fig. 1.3(a). The inner and outer layers contain eight and sixteen unit cells along their circumference, respectively, in order to approximate a circular outer periphery. The outer radius of the MM coating is only 5 mm, *i.e.*, about $\lambda_0/24$ at 2.5 GHz, ensuring that the ultra-thin subwavelength coating is compact in the radial direction. The entire structure was simulated using HFSS. To examine the effect of the MM coating on the impedance bandwidth of the monopole and the efficacy of the anisotropic effective medium model, we compare the simulated S_{11} for three cases: monopole alone, monopole with the actual MM coating, and monopole with the anisotropic effective medium coating (see Fig. 1.3b).

The monopole alone has an $S_{11} < -10$ dB bandwidth of 0.4 GHz, from 2.3~2.7 GHz, with a resonance at 2.5 GHz. In contrast, with the actual MM coating present, the $S_{11} < -10$ dB bandwidth is remarkably broadened to 2.3 GHz, spanning a range from 2.1 to 4.4 GHz. The fundamental resonance shifts down slightly to 2.35 GHz, while the second resonance is located at 3.85 GHz. The current distribution is the same as that of the fundamental mode at this second resonance, leading to a well-maintained radiation pattern [44]. When a homogeneous anisotropic effective medium coating with the retrieved dispersive material parameters is employed, the S_{11} exhibits a similar behavior to that of the actual MM coating.

This indicates that the assumed homogeneous anisotropic effective medium model is a valid approximation for the actual curved MM, primarily due to the fact that a sufficient number of unit cells were used to form the cylindrical coating such that every MM unit cell still maintains local flatness.

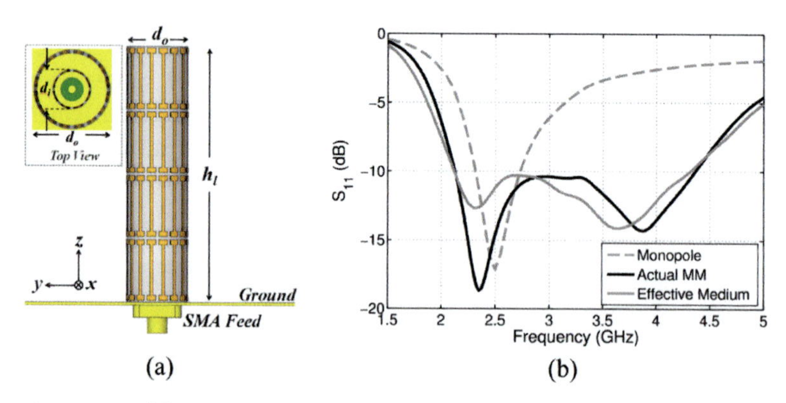

Figure 1.3 (a) Configuration of the quarter-wave monopole antenna with ultra-thin flexible anisotropic MM coating. Reprinted, with permission, from Ref. 44, Copyright 2011, IEEE. All dimensions are in millimeters: $d_i = 5$, $d_o = 2d_i$, and $h_1 = 40$. The dielectric is 51 μm thick Rogers Ultralam 3850 ($\varepsilon_r = 2.9$, $\delta_{tan} = 0.0025$). (b) Simulated S_{11} of the monopole alone, the monopole with the actual MM coating, and the monopole with the anisotropic effective medium coating.

1.2.3 Experimental Results

The MM coating and the monopole were fabricated and assembled, as shown in Fig. 1.4a. Four polypropylene washers were used as a frame to mechanically support the coating and to ensure correct inner and outer layer diameters. The impedance measurements were performed using a network analyzer. Figure 1.4b compares the simulated and measured voltage standing wave ratio (VSWR) curves of the monopole with and without the MM coating on a 32 cm × 32 cm ground plane. The measured S_{11} of the monopole alone is almost identical to the simulated results, having an $S_{11} < -10$ dB band from 2.3 to 2.7 GHz. With the MM coating present, a broad $S_{11} < -10$ dB band is achieved from 2.15 to 4.6 GHz, *i.e.*, a 2.14:1 ratio bandwidth. The radiation patterns of the monopole coated with the MM were measured in an anechoic chamber. Figure 1.4c presents the simulated and measured *E*-plane and *H*-plane patterns at the center frequency of 3.3 GHz. In the *H*-plane, an omnidirectional radiation pattern is observed, with a gain variation of around 0.5 dB and 1.2 dB for simulated and measured results, respectively. The increased measured gain variation as a function of the azimuthal angle is primarily caused by the imperfection of assembly and measurement noise, as well as the antenna rotation platform. In the *E*-plane, characteristic ear-shaped patterns are obtained, which are very similar to the patterns for the monopole alone, indicating that the added MM coating has negligible impact on the radiation properties of the monopole. The gain of the MM-coated monopole is about 4 dBi with the direction of the maximum at 26° from the horizon due to the employed finite-sized ground plane used in both simulation and measurement. The measured gain is 0.5 dB lower than the simulated values. The overall good agreement between the simulated and measured results confirms the expected performance of the proposed MM coating for monopole antennas.

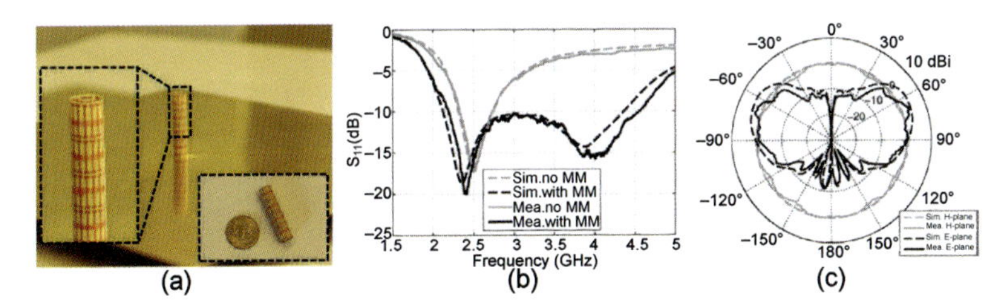

Figure 1.4 (a) Photographs of the fabricated MM coating placed over a monopole. (b) Simulated and measured S_{11} of the monopole with and without the MM coating. (c) Simulated and measured gain patterns in the *E*- and *H*-planes at 3.3 GHz.

1.2.4 C-Band Design

To show the flexibility of the proposed technique, an anisotropic MM coating was further designed for a *C*-band (4–8 GHz range) quarter-wave monopole antenna. The design procedure is similar to that already discussed for the S-band MM coating. The unit cell of the MM coating has the same shape and is printed on the same substrate material. The optimized geometrical dimensions are (all in millimeters): $a = 2.5$, $d_s = 0.051$, $d_c = 0.017$, $w = 1.9$, $b = 3.9$, $c = 1.1$, $g = 0.5$, and $l = 2.6$. The monopole was designed to resonate at 4.5 GHz, which corresponds to a length of 15 mm. The outer radius of the MM is 3.7 mm or about $\lambda_0/24$ at 4.5 GHz, ensuring that the ultra-thin subwavelength coating is compact in the radial direction. The monopole with and without the MM coating was simulated in HFSS. The simulated S_{11} curves are reported in Fig. 1.5. The monopole alone yields an $S_{11} < -10$ dB band of 0.5 GHz, namely from 4.5 to 5.0 GHz. When the actual MM coating is present, the $S_{11} < -10$ dB bandwidth is considerably broadened to 4.35 GHz, namely from 3.85 to 8.2 GHz. The fundamental resonance shifts down slightly to 4.55 GHz, while the second resonance occurs at 7.0 GHz. In all, this type of flexible MM coating is compact in size, extremely light weight, low in cost, and can be easily scaled to different frequency bands, thereby paving the way for widespread use as radiating elements in, for example, broadband arrays, wideband radios, and various portable wireless devices.

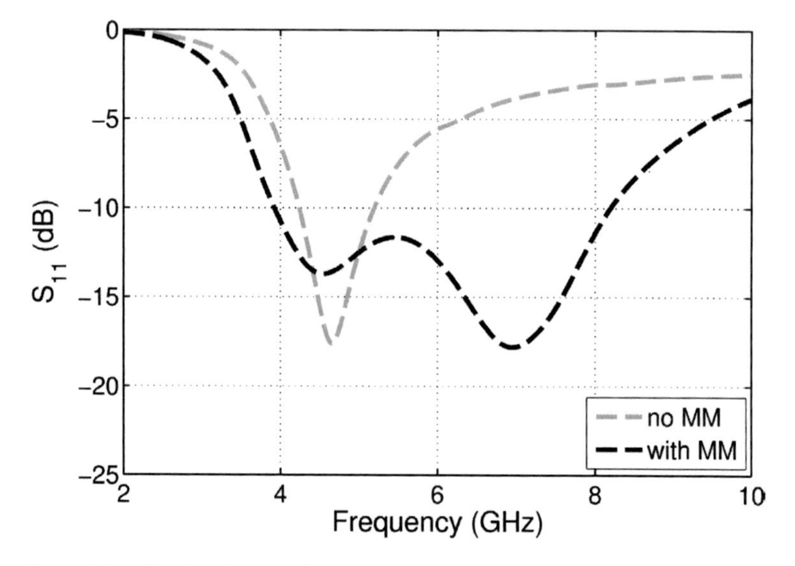

Figure 1.5 Simulated S_{11} of the monopole with and without the MM coating.

1.3 Anisotropic MM Lenses for Directive Radiation

In addition to broadening the impedance bandwidth of an omnidirectionally radiating wire antenna, MMs with certain anisotropy can also be utilized to change the radiation pattern of an antenna over a wide bandwidth. More specifically, the MMs are designed to change the angular distribution of electromagnetic energy radiated from an antenna to generate highly directive beams. Several examples are presented in this section, including AZIM lenses with engineered electric and/or magnetic properties for generating a single as well as multiple highly directive beams from a simple antenna feed.

1.3.1 Low-Profile AZIM Coating for Slot Antenna

1.3.1.1 Dispersion of grounded AZIM slab

First, we exploit the leaky modes supported by a grounded AZIM slab to produce a single unidirectional beam at broadside over a wide bandwidth. Let us consider a grounded slab that is infinite in the y-direction, as shown in Fig. 1.6a. The slab has a thickness of t and is composed of an anisotropic medium with its permittivity and permeability tensors expressed as [45]

$$\bar{\bar{\varepsilon}} = \left[\varepsilon_x, \varepsilon_y, \varepsilon_z\right] = \varepsilon_0 \left[\varepsilon_{rx}, \varepsilon_{ry}, \varepsilon_{rz}\right], \tag{1.1}$$

$$\bar{\bar{\mu}} = \left[\mu_x, \mu_y, \mu_z\right] = \mu_0 \left[\mu_{rx}, \mu_{ry}, \mu_{rz}\right], \tag{1.2}$$

The y-component of the complex propagation constant (k_y) of the leaky wave is studied under the following conditions. A time-harmonic dependence $e^{j\omega t}$ is assumed and used throughout the paper. Since no field variation is assumed in the x-direction, the two-dimensional nature of the configuration allows the problem to be separated into TE_z and TM_z modes. For the TE_z mode, the set of material tensor parameters (ε_{rx}, μ_{ry}, and μ_{rz}) are pertinent, while for the TM_z mode, the other set of material tensor parameters (μ_{rx}, ε_{ry}, and ε_{rz}) are responsible for the wave propagation properties. For simplicity, only the TM_z mode is considered here since a low-profile slot antenna will be employed, which represents a quasi-TM_z source. The dispersion equation can be obtained using the transverse equivalent network model, as shown in Fig. 1.6b [46], which is given by

$$\sqrt{k_0^2-(k_y^{TM})^2}+j\sqrt{\frac{k_0^2\varepsilon_{rz}-(k_y^{TM})^2}{\varepsilon_{ry}\varepsilon_{rz}}}\tan\left(\sqrt{k_0^2\varepsilon_{rz}-\left(k_y^{TM}\right)^2}\sqrt{\frac{\varepsilon_{ry}}{\varepsilon_{rz}}}t\right)=0. \tag{1.3}$$

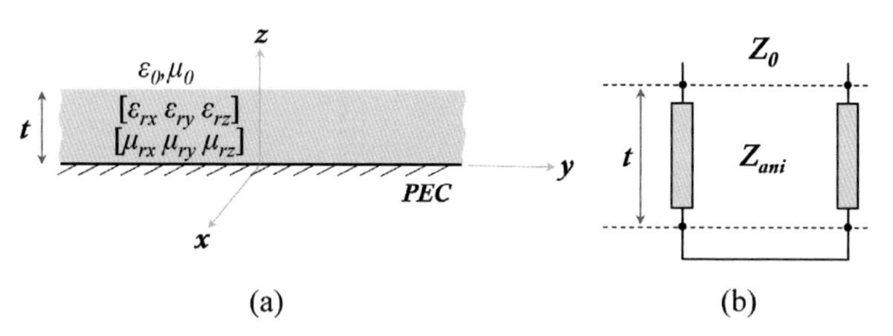

Figure 1.6 (a) Configuration of the grounded anisotropic slab with defined geometrical dimensions and material properties. (b) Equivalent transverse network for both TE and TM modes of the grounded anisotropic slab.

The propagation constant $k_y^{TM}=\beta_y^{TM}+j\alpha_y^{TM}$ of the leaky modes supported by the grounded anisotropic slab can be calculated numerically using Eq. (1.3). Figures 1.7a,b show the real and imaginary parts of k_y^{TM} (normalized to k_0), as a function of frequency for different ε_{rz} values. The values of ε_{ry} and μ_{rx} are fixed to be 2.4 and 1, respectively, while f_0 denotes the frequency at which the thickness of the slab (t) is equal to one wavelength. The ordinary isotropic case is also considered where $\varepsilon_{rz}=\varepsilon_{ry}=2.4$. Multiple solutions can be found, including the physical surface wave (SW), leaky wave, and improper nonphysical SW (IN), all of which are labeled in the figures. For the isotropic grounded slab, the leaky-wave band is narrow and highly dispersive. The value of β_y^{TM} is only slightly smaller than that of k_0. When the value of ε_{rz} decreases, the upper band improper SW mode disappears and the bandwidth of leaky-wave region is greatly extended. As the value of ε_{rz} is further decreased, the grounded anisotropic slab can be referred to as a grounded anisotropic low-index slab, where both β_y^{TM} and α_y^{TM} of the leaky modes become smaller and their profiles also get flatter. This is beneficial for maintaining a stable radiated beam near broadside across a wide frequency range. Theoretically, when ε_{rz} is extremely close to zero, k_y^{TM} is forced to be zero in order to satisfy the dispersion relation in Eq. (1.4), which possesses an elliptical isofrequency curve with a near-vanishing short axis [47].

$$\frac{(k_y^{TM})^2}{\varepsilon_{rz}} + \frac{(k_z^{TM})^2}{\varepsilon_{ry}} = k_0^2 \mu_{rx}, \tag{1.4}$$

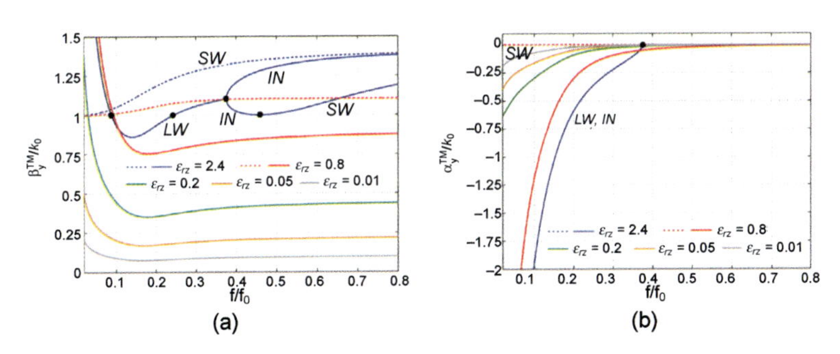

Figure 1.7 TM_z dispersion curves for a grounded anisotropic slab with ε_{ry} = 2.4, μ_{rx} = 1, and varying ε_{rz} as a function of frequency. (a) β_y^{TM}/k_0. (b) α_y^{TM}/k_0. Reprinted, with permission, from Ref. 48, Copyright 2014, IEEE.

1.3.1.2 Infinite TM$_z$ radiating source with realistic AZIM coating

To realize the anisotropic zero/low-index property mentioned above for the 5.8 GHz band, which is widely used for wireless local area network (WLAN) systems, periodic end-loaded dipole resonators (ELDRs) are employed [48]. Meandering end-loading is used to provide a degree of miniaturization. The unit cell geometry and dimensions are shown in the inset of Fig. 1.8a. An anisotropic parameter inversion algorithm was employed to retrieve the effective medium parameters (ε_{ry}, ε_{rz}, μ_{rx}) from the scattering parameters obtained from unit cell simulations [43]. The retrieved ε_{ry}, ε_{rz}, μ_{rx} are shown in Fig. 1.8b. While ε_{ry} and μ_{rx} are virtually non-dispersive with values around 2.4 and 1, respectively, the ε_{rz} curve crosses zero at 5.38 GHz. The value of ε_{rz} remains positive and below 0.15 within a broad frequency range extending from 5.4 to 6.1 GHz. In this frequency band, β_y^{TM} of the practical grounded dispersive MM varies from a near-zero value to 0.35. This geometry yields a broader low-index bandwidth compared to that of commonly used subwavelength electric LC (ELC) resonators [49] due to the lower quality factor provided by the large capacitance inherent in the meandered end-loadings.

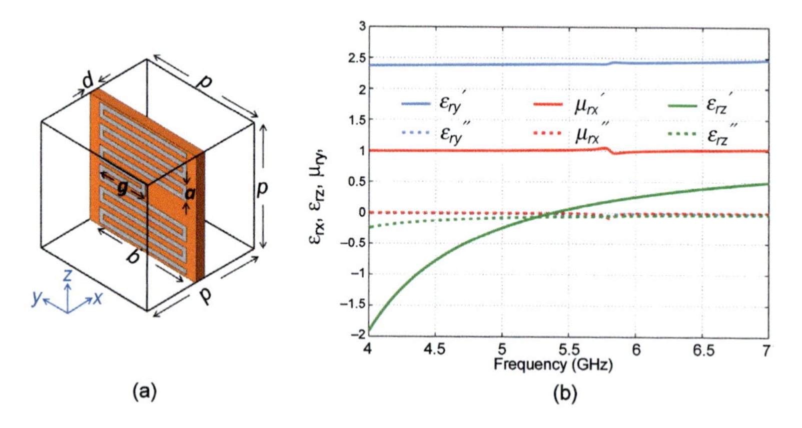

(a) (b)

Figure 1.8 Geometry of the ELDR unit cell. The dimensions are p = 6.5 mm, b = 5.35 mm, a = 0.7 mm, d = 0.508 mm, and g = 2.8 mm. The substrate material is Rogers RT/Duroid 5880 with a dielectric constant of 2.2 and a loss tangent of 0.009. (b) The retrieved effective medium parameters ε_{ry}, ε_{rz}, and μ_{rx}.

Full-wave simulations were then carried out to investigate the ability of a finite AZIM coating to control the radiation of an infinite slot. As illustrated in Fig. 1.9a, three cases were considered: (A) a 1 mm wide slot covered by a single-layer MM slab consisting of 14 cells in the x-direction, (B) a 1 mm wide slot covered by a dispersive effective medium slab with the retrieved relative permittivity and permeability tensors, and (C) a 1 mm wide slot alone. The structures are infinitely long i n the y-direction with only the field components H_y, E_x, and E_z existing in the far-zone, corresponding to the TM_z mode. The far-field patterns (normalized to the value of the slot alone at broadside) in the upper half-space are presented in Fig. 1.9b for the three cases at 5.4 GHz. It can be observed that for the slot alone, the radiated wave is maximum close to $\theta = \pm 40°$ due to the diffraction caused by the finite size of the ground plane in the x-direction. However, the actual MM or the effective medium slab causes the beam maximum to be located at broadside with about 9-fold enhancement. The radiated beam is sharpened and the radiation in other directions is greatly suppressed compared to that generated by the slot alone. The enhancement of radiated power at broadside as a function of frequency is shown in Fig. 1.9c. It can be seen that within a broad frequency range from 5.2 to 6.4 GHz, the broadside radiation enhancement remains above 4.5-fold. The drop in the enhancement factor is caused by the dispersive nature of the leaky MM slab, as previously discussed. However, compared to conventional directive leaky-wave antennas, stable unidirectional radiation at broadside is achieved over a much broader bandwidth. Overall, the numerical simulations show that the proposed subwavelength AZIM coating provides an efficient way of suppressing commonly encountered edge diffraction. In addition, the good agreement achieved between the simulation results using the actual MM and the effective medium slab justifies the homogenization approximation employed here, which is assumed valid due to the subwavelength size of the unit cells.

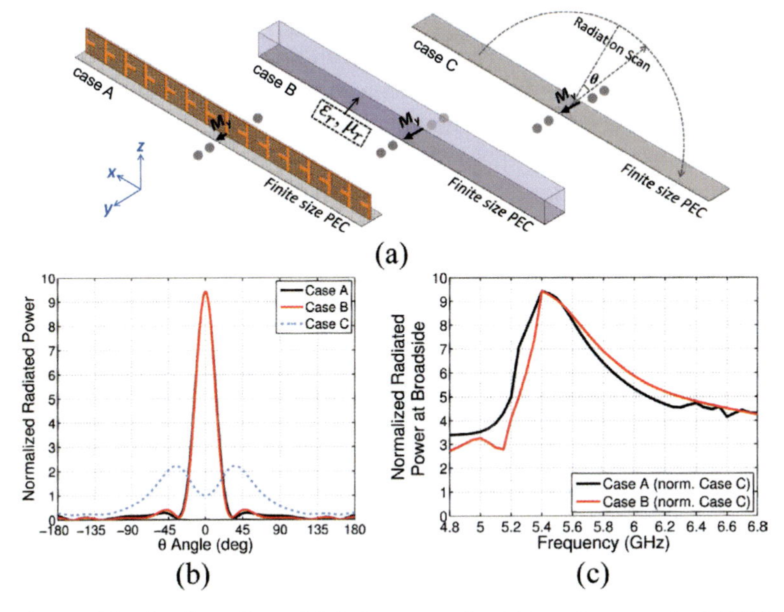

Figure 1.9 Configuration of infinite array simulations for an actual AZIM coating, a dispersive effective medium slab, and the slot alone. The structures are infinite in the *y*-direction with a periodicity of 6.5 mm. The finite-sized perfect electric conductor (PEC) plane is 92 mm long in the *x*-direction (located under an AZIM coating with 14 cells). A perfectly matched absorbing slab is placed underneath the slots in the simulations to absorb the radiation in the $-z$ half-space. (b) Normalized radiated power for the three cases at 5.4 GHz (*i.e.*, close to the effective plasma frequency of the metamaterial). All the curves are normalized to Case C at broadside (θ = 0°). (c) Normalized radiated power at broadside (θ = 0°) as a function of frequency.

1.3.1.3 High-gain SIW-fed slot antenna with realistic AZIM coating

In order to design a feeding antenna for the AZIM coating, a half-wave slot fed by a substrate-integrated waveguide (SIW) is adopted here due to its low profile compared to cavity backed or conventional waveguide-fed slots [50, 51]. The schematic view of the SIW-fed slot antenna is shown in Fig. 1.10a. It is composed of a 50 Ω microstrip and a shorted SIW with a longitudinal slot cut on its broad wall. A tapered microstrip is used for impedance matching between the 50 Ω feedline and the SIW. In order to achieve a magnetic dipole mode in the slot for efficient radiation, the length of the slot is chosen to be around $\lambda_g/2$ at 5.8 GHz. The distance between the center of the slot to the shorted wall of the SIW in the x-direction is set to be about $3\lambda_g/4$ at 5.8 GHz, which allows the standing wave peak to be located at the center of the slot. The geometrical dimensions of the SIW-fed slot antenna were optimized to achieve $S_{11} < -10$ dB in the 5.6–6.0 GHz band (see Fig. 1.11a). The E-plane (y–z plane) and H-plane (x–z plane) gain patterns at the center frequency 5.6, 5.8, and 6.0 GHz and are presented in Fig. 1.12. It can be seen that the E-plane has two peaks located at around 40–45° off broadside due to the diffraction of the finite-sized ground plane. The simulated broadside gain varies from 3 to 3.8 dBi.

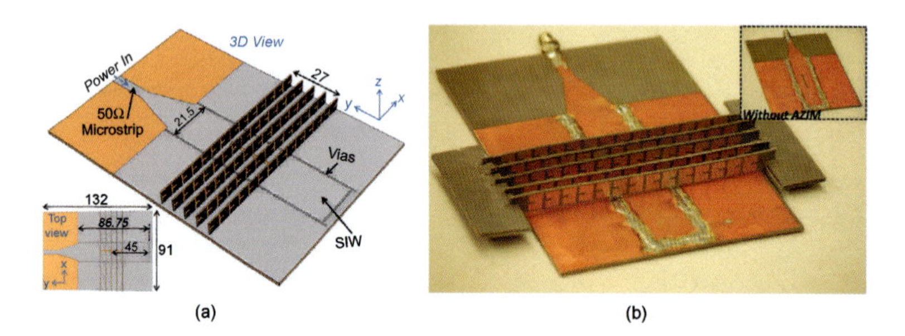

Figure 1.10 (a) Schematic view of the SIW-fed slot antenna with the actual AZIM coating. All the dimensions are in millimeters. (b) Photograph of the fabricated SIW-fed slot antenna covered by the AZIM coating. The inset shows the SIW-fed slot antenna alone.

Due to the fact that the AZIM coating has its near-zero permittivity tensor parameter only in the z-direction, the impedance of the slot is well maintained when it is covered by the AZIM (see Fig. 1.10b). As shown in Fig. 1.11b, the simulated S_{11} is below −10 dB from 5.6 to 6 GHz, which is very similar to that of the slot alone. This robust input impedance behavior ensures that the AZIM coating can be readily added onto or taken away from the slot to achieve different radiation properties without any additional modification to the slot antenna itself. The *E*-plane (*y*–*z* plane) and *H*-plane (*x*–*z* plane) gain patterns of the slot antenna with/without the AZIM at 5.6, 5.8, and 6.0 GHz and are presented in Fig. 1.12. Distinctly different from the two-peak patterns for the slot without the AZIM coating, a well-defined single beam at broadside is observed in the *E*-plane with a beam squint less than 2° off broadside and an half-power beam width (HPBW) of about 35° ~ 40°. Notably, with the presence of the AZIM coating, the broadside gain is significantly increased to 10.2~10.6 dBi, which indicates an improvement of about 7 dB. It should also be noted that the front-to-back ratio is greatly reduced by about 10 dB throughout the entire frequency range. Both the broadside-gain increase and the front-to-back ratio drop are primarily attributed to the reduction of fields at the edges of the ground plane. This is facilitated by the presence of the AZIM coating, resulting in much weaker diffracted fields at the back side of the antenna.

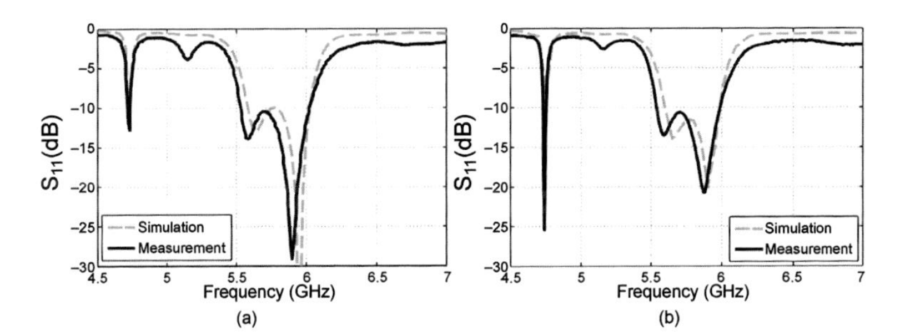

Figure 1.11 Simulated and measured S_{11} of the SIW-fed slot antenna (a) without and (b) with the AZIM coating.

The SIW-fed slot antenna and the AZIM coating structure were fabricated and assembled, as shown in Fig. 1.10b. The S_{11} of the slot with and without the AZIM coating was characterized by a network analyzer. As shown in Figs. 1.11a,b, good agreement can be found between simulations and measurements. The measured S_{11} of the slot with and without the AZIM coating has a −10 dB band from

5.52 to 6.03 GHz and from 5.54 to 6.01 GHz, respectively. They are both slightly broader than the simulations predict due to the minor frequency shift of the resonance at 5.6 GHz and the lower quality factors of both resonances within the −10 dB bandwidth. The radiation patterns and the gain of the slot antenna with and without the AZIM coating were characterized in an anechoic chamber, as shown in Fig. 1.12. Overall, the measured gain patterns in both the *E*-plane and *H*-plane agree well with the simulated results, confirming the proposed antenna design. Specifically, patterns exhibiting a double-peak can be seen for the slot alone, while a single sharp beam pointing at broadside can be observed for the slot with the AZIM coating. The measured HPBW in the *E*-plane is about 40–50°, which is slightly broader than the simulated beam width, especially in the high-frequency band. This is mainly attributed to fabrication and assembly imperfections, which result in a non-ideal symmetry in the actual MM structure. The gain at broadside for the slot with and without the AZIM coating is in the range of 9.8–10.4 dBi and 2.9–3.5 dBi, respectively, which indicates an improvement of about 6.9 dB.

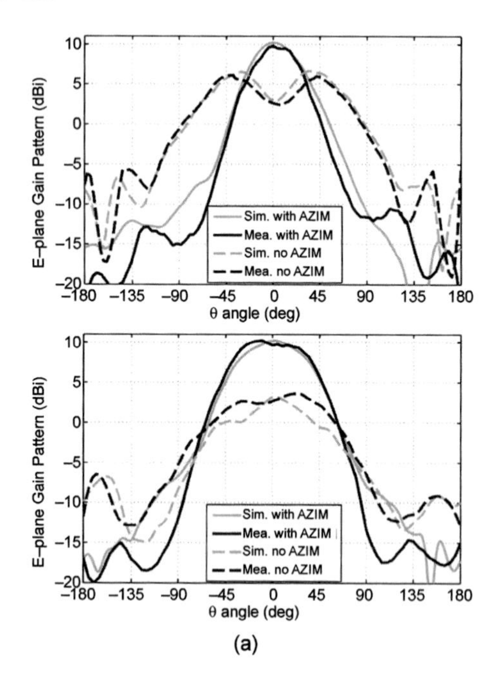

(a)

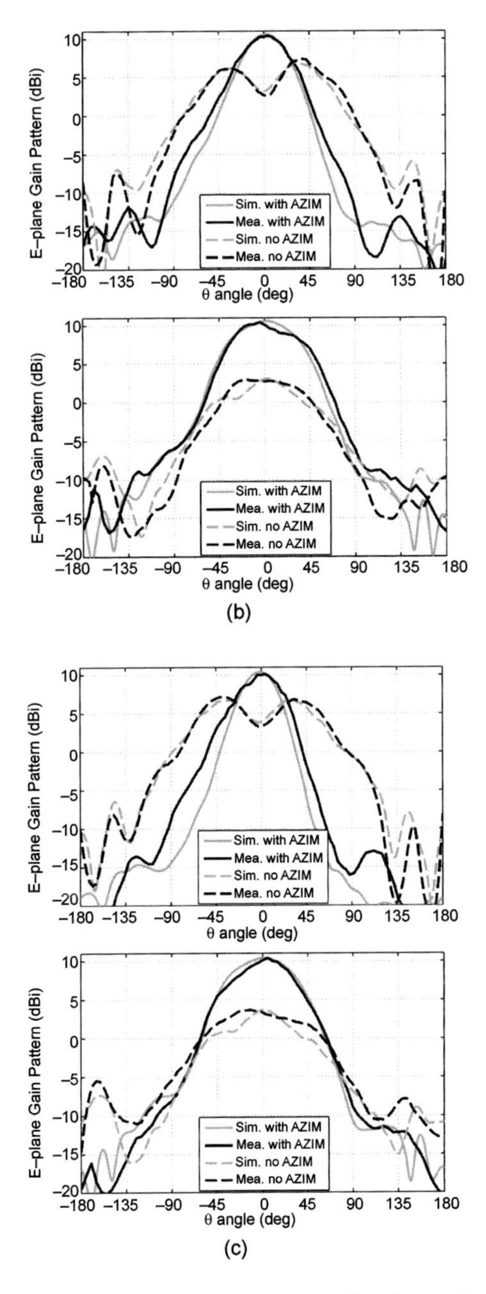

Figure 1.12 Simulated and measured *E*-plane and *H*-plane gain patterns of the SIW-fed slot antenna with and without the AZIM coating at (a) 5.6 GHz, (b) 5.8 GHz, and (c) 6.0 GHz.

1.3.2 Anisotropic MM Lens for Crossed-Dipole Antenna

High-gain circularly polarized (CP) and dual-polarized (DP) antennas have widespread applications as feeds for reflector antennas. Such a feed has specific gain requirements to maintain complete coverage over the entire reflector and maximize aperture efficiency. A typical solution for a large reflector uses horn antennas whose aperture size and mode distribution is selected to produce the required reflector illumination. These waveguide-fed horns represent substantial mass and volume investments, especially for multi-feed reflectors. An AZIM lens with both electric and magnetic resonances can be used for aperture enhancement while reducing the form factor to achieve a compact feed for reflector antennas.

1.3.2.1 Configuration and unit cell design

The collimating properties of AZIM lenses, as previously described for the leaky-wave AZIM antenna with embedded slot feed, can also be used for cases where the feed is not embedded into the MM slab. The MM slab, for example, may be used as a focusing lens in concert with a radiating source. A near-zero-index uniaxial lens acts as an angular-selective spatial filter that passes only plane wave components that are propagating within a small angular tolerance of zero degrees while all other plane wave components are reflected [52]. This property allows an AZIM lens placed in front of a low-gain radiating element to demonstrate very high aperture efficiencies. A corollary of this property is that by placing an AZIM lens over a highly collimated source, little or no improvement in beam characteristics will be produced. When used in isolation, such a lens may show a very uniformly illuminated aperture, but much of the incident energy would be reflected from the back surface and result in very low realized forward gain due to high reflection losses. As a partially reflective surface, the AZIM lens and feed arrangement may be backed with a conductive ground plane to form a Fabry–Pérot cavity antenna, which improves the realized gain and return loss of the system. An illustration of the MM slab placed over the feed antenna is shown in Fig. 1.13a.

Using an AZIM slab as the partially reflective surface of a Fabry–Pérot cavity produces improved behavior over either of the constituent cases of a lens illuminated by a feed-in free space, or a non-MM dielectric slab for the top surface of the cavity. Both the cavity and the AZIM MM act to distribute the energy from the sub-feed more evenly throughout the aperture to improve gain, while the combination of both effects improves the response even more [52, 53].

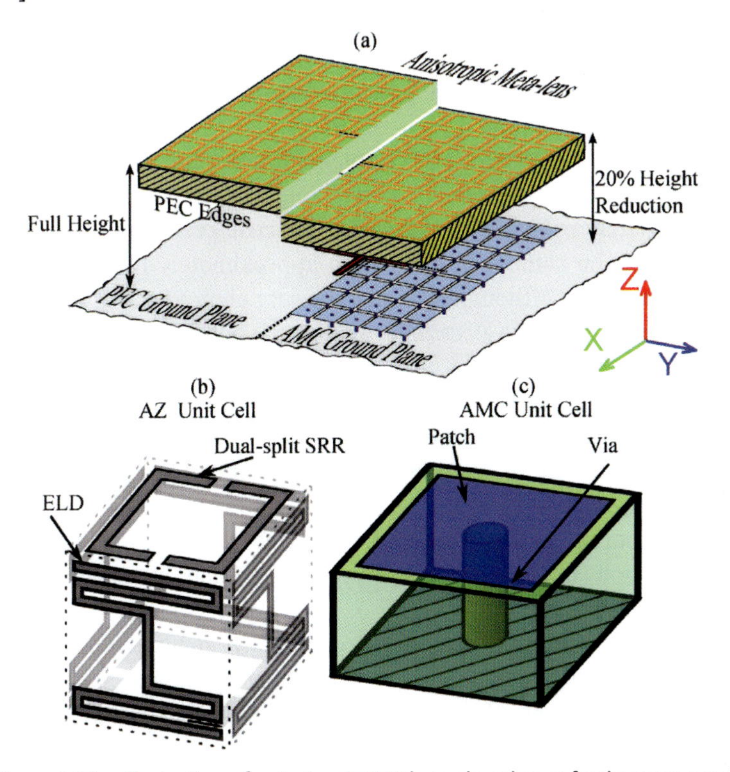

Figure 1.13 Illustration of anisotropic ZIM lens placed over feed antenna and ground plane to form a Fabry–Pérot-type cavity. The left side shows the lens constructed with a PEC ground plane, and the right shows the profile reduction possible when using an AMC ground plane. (b) Representative illustration of cubic AZIM unit cell. (c) Representative illustration of AMC unit cell. Reprinted, with permission, from Ref. 53, Copyright 2014, IEEE.

For a single-polarization configuration, either a near-zero-permittivity electric MM or a near-zero-permeability magnetic MM, oriented in the proper direction, can be sufficient to achieve substantial gain improvements. For a CP or DP antenna with requirements for a cylindrically symmetric radiation pattern, the MM requires both magnetic and electric properties. Although the behavior is identical at normal incidence for single-ZIM electric or magnetic lenses, off-normal responses and the resulting beam symmetry degrade if only one component is included. The AZIM lens for free-space beam collimation requires that the optical axis of the uniaxial material be parallel to the desired direction of propagation, which in turn requires, for the dual-polarization case, that the electric and magnetic properties be aligned on the same axis. For the following discussion, the optical axis is the *z*-axis, while the lens resides in the *x*–*y* plane.

A microwave-range MM implementation for the magneto-dielectric AZIM described above requires a combination of appropriately oriented magnetic and electric resonators tuned to operate at the same frequency with approximately the same unit cell dimensions. Although SRRs are the easy choice for the magnetic MM, there is no single electric resonator that has the same beneficial properties of the SRR across a wide frequency range. In general, an SRR operating at a given frequency can be smaller than an equivalent electric resonator for the same frequency. The properties desired in an electric resonator include matched resonance frequency to an equivalent-sized magnetic resonator, a square unit cell boundary, strong coupling efficiency to the z-oriented electric field, low bianisotropy (magneto-electric coupling), and minimal, symmetric response in the *x*–*y* plane.

Selections for electric MM resonators generally focus on variations of dipole configurations. However, dipole resonators are electrically large (half-wavelength) at resonance; without additional miniaturization, dipole elements alone would be much larger in comparison to their SRR counterparts. Several variations, including the ELC resonator and the ELDR elements, allow the length of the dipole to be reduced while remaining resonant. Planar electric resonators, such as the complimentary SRR (CSRR), which operate according to the Babinet principle for symmetrical electrical operation with the SRR [54], have limited bandwidth and strong bianisotropic tendencies. All other options for planar electric resonators require orientation parallel to the *z*-axis. The ELC uses lumped or distributed inductance and capacitance to promote resonance for electrically compact unit cells, but has limited field coupling area, which reduces the bandwidth of the resonant response and thus the operational ZIM band. The ELDR has a lower resonant frequency compared to an ELC of the same physical size due to a larger coupling region for the electric field, which helps to generate a stronger, broader resonance. The ELDRs function much better for the creation of a broadband magneto-dielectric response than other elements. For either unit cell, the electric resonators must be oriented vertically, parallel to the x–z or y–z planes. A set of four vertically oriented resonators can be arranged in a square to form a unit cell that will have a symmetric electric response in the x–y plane.

While the electric ELD resonators were reduced in size as far as possible relative to their resonant wavelength, it is necessary to increase the size of the SRRs such that the electric and magnetic unit cells are the same size. Using symmetric, square resonators with two splits in each resonator increases the electrical size to $\lambda_0/7.5$ to match the minimum electric resonator dimensions. Two copies of the SRR and four of the ELDRs are arranged to fit on the six interior faces of a hollow dielectric cube, with the SRR elements on the top and bottom and the ELD on the sides. This resonator arrangement within the unit cell creates a vertically oriented magnetic resonance and a vertically oriented electric resonance, which can be aligned at the same frequency through appropriate tuning of the resonator dimensions. Figure 1.13b illustrates the structure of a representative AZIM unit cell.

The anisotropic effective MM parameters were extracted during the design process and after finalizing the design using the scattering-parameter inversion method [55]. After simulating the scattering response from a periodic tiling of the unit cells, the effective response can be computed. The z-components of the permittivity and permeability show strong resonant behavior that produces a simultaneous zero-index band at 7.5 GHz. Although there are antiresonance responses in the tangential permeability [56], the tangential permittivity and permeability are not significantly dispersive over the zero-index band.

1.3.2.2 Numerical and experimental results

A slab of MM was constructed as a 12×12 array of unit cells to be placed over a horizontal crossed-dipole antenna, which is itself placed a quarter-wavelength above a metallic ground plane. The AZIM slab tends to spread out the radiating energy to fill the entire aperture, so in order to mitigate this effect, a perfect electric conductor (PEC) band was placed around the edge of the lens to contain the fields and prevent leakage from the sides of the lens. The feed element, ground plane, and lens together form a cavity antenna. The vertical spacing between the ground plane, feed antenna, and MM lens was selected to optimize the return loss and achievable gain.

The prototype antenna was modeled and measured using a single linear dipole in two separate orthogonal orientations, with no difference in response when the lens was rotated 90°. The combined antenna showed a 6~8 dB measured gain improvement across a 17% bandwidth over the dipole and ground plane alone, and also demonstrated low cross-polarized fields that are approximately 25 dB below the peak gain throughout the band. Figures 1.14a,b show the gain and S_{11} spectra for the AZIM lens with both PEC and AMC ground planes, showing the performance trade-offs for the reduced profile of the AMC design.

The MM lens antenna system has a relatively thick profile due to the required spacing between the ground plane and feed, and the feed and the lens. The thickness of the entire system can be reduced by replacing the PEC ground plane beneath the feed antenna by an AMC ground plane, as shown in Fig. 1.13a. Since the feed antenna may be placed arbitrarily close to the AMC ground plane, the ground–feed distance may be reduced to almost zero. This change does not affect the feed–lens thickness requirement however, and the AMC has a finite thickness, which limits the minimum height of the lens. Replacing the PEC with an AMC ground plane enabled about a 20% overall thickness reduction in the prototype lens. The AMC ground plane may be designed according to a number of possible design variations [32, 57, 58], but the simplest is the mushroom-type surface [59], whose template is illustrated in Fig. 1.13c.

The trade-off for thickness reduction is a decreased bandwidth. In the original lens, the dipole and its associated spacing requirements, the electric MM, and the magnetic MM were each independently tuned to operate at the same frequency. Of these effects, the electric permittivity was the limiting constraint on operational bandwidth. The addition of another strongly dispersive and narrow-band element to the structure limits the possible bandwidth of an optimized design compared to the original optimized antenna (10% versus 17%), and also decreases the peak gain of the antenna from 14 to 12 dBi. Although the achievable profile reduction is limited, such a design decision may still be worthwhile for selected applications.

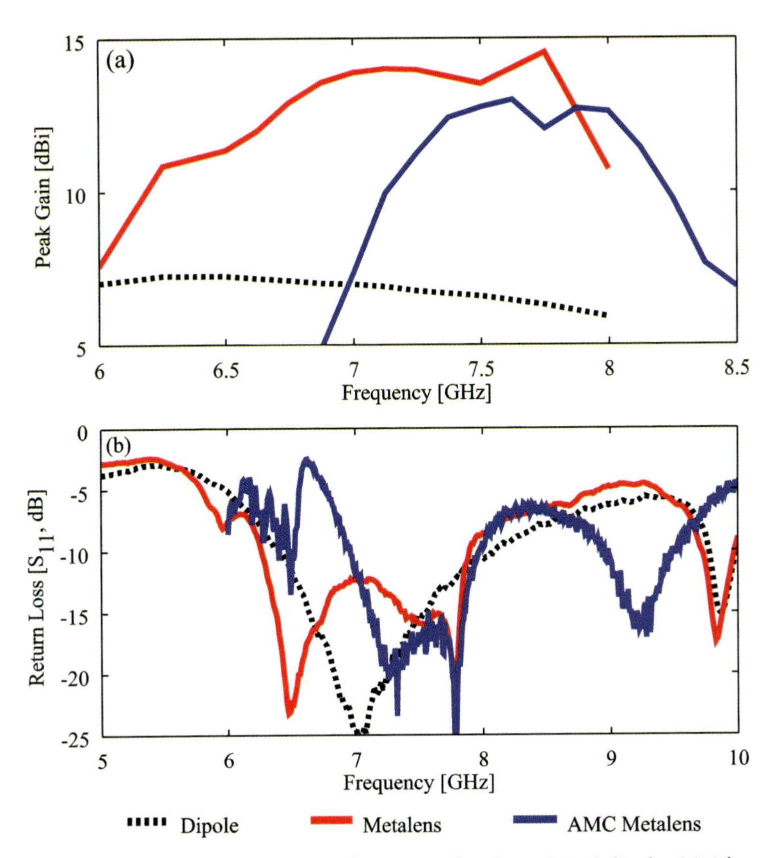

Figure 1.14 (a) Peak gain versus frequency for the original dipole, MM lens with PEC ground plane, and MM lens with AMC ground plane. (b) Return loss for the original dipole, MM lens with PEC ground plane, and MM lens with AMC ground plane. The MM lens with PEC ground plane shows better performance and bandwidth in both impedance and radiation properties but has a larger height profile.

Simulations of both antennas, with and without the AMC ground plane, were performed using HFSS, while measurements of the same structures confirmed the performance predictions. Each of the two designs demonstrated better than 10% impedance and pattern bandwidth, enabled through careful tuning of the geometric and electric parameters of the feed, electric MM, magnetic MM, and the AMC. While each additional dispersive and frequency-limited component increases the design constraints and reduces the possible operational bandwidth, these designs demonstrate that it is possible to create complex, coupled systems of resonant MMs while maintaining a useful operating bandwidth.

1.3.3 Anisotropic MM Multibeam Antenna Lens

In the previous two sub-sections, anisotropic MM slabs were used to generate a single highly directive beam. In this sub-section, we extend the idea to embed an isotropic source inside an AZIM lens with multiple lens segments for generating multiple highly directive beams over a wide operational bandwidth. Moreover, such an AZIM multibeam lens is experimentally demonstrated at microwave frequencies.

1.3.3.1 Two-dimensional/three-dimensional AZIM lens concept and numerical results

Two- and three-dimensional full-wave numerical simulations were performed to validate the concept of multiple highly directive beam generation via AZIM lenses. For the two-dimensional lens examples, only TM_z polarization with a z-directed E-field was used in the simulations for simplicity. The first example is a square lens with four collimated beams uniformly distributed in the x–y plane pointing at ϕ = {0°, 90°, 180°, 270°}, as shown in Fig. 1.15a, while the second example is a hexagonal lens with six beams uniformly distributed in the x–y plane pointing at ϕ = {30°, 90°, 150°, 210°, 270°, 330°}, as shown in Fig. 1.15b. To demonstrate the flexibility of controlling the radiated beams, a third lens is displayed in Fig. 1.15c having five customized collimated beams, each radiating at the desired angles of ϕ = {30°, 90°, 165°, 247.5°, 322.5°}. All three lenses have a low value for the $\overline{\overline{\mu_r}}'$ parameter with a magnitude of 0.01 in each of the segments. From the electric field distribution, it is observed that the waves radiated from the central isotropic source are well-collimated, even in close proximity to the source.

For the three-dimensional example, we surrounded a previously reported quasi-isotropic radiator [60], shown in Fig. 1.16 (left part), with six lens segments, each having low-value permittivity and permeability tensor parameters with a magnitude of 0.01 in the direction of the radiated beam. Figure 1.16 (right part) shows the three-dimensional radiation patterns with and without the presence of the transformation optics lens simulated by HFSS. It can be observed that the antenna alone has a near-isotropic radiation pattern, whereas with the lens present, the radiation pattern exhibits six highly directive beams. It should be noted that not all of the six directive beams have the same linear polarization, which is due to the quasi-isotropic source antenna employed here and not the AZIM lens since its response is polarization independent.

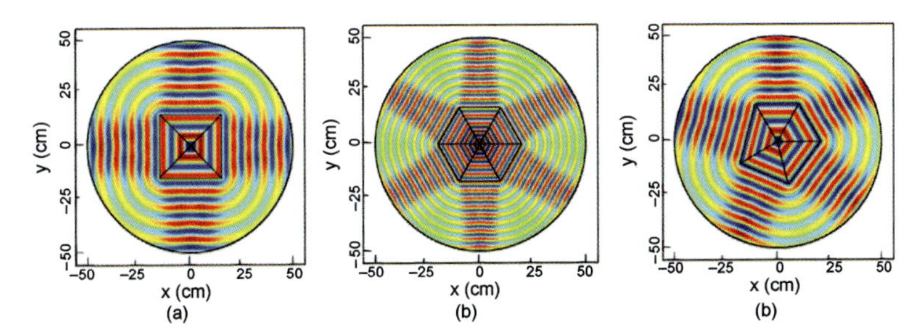

Figure 1.15 Snapshots of the *z*-directed near- and far-zone electric field determined via a two-dimensional COMSOL simulation of the transformation optics lens at 3 GHz with (a) four radiated beams uniformly distributed, (b) six radiated beams uniformly distributed, (c) five radiated beams non-uniformly distributed in the *x*–*y* plane.

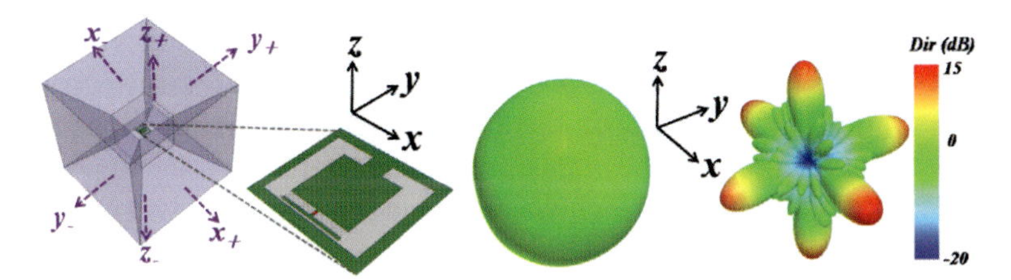

Figure 1.16 Three-dimensional coordinate transformation lens applied to a quasi-isotropic antenna proposed in Ref. 69. (a) The three-dimensional directive emission lens and the embedded quasi-isotropic antenna. The lens is designed to produce six highly directive beams: one normal to each face of the lens as indicated by the labels. (b) The simulated radiation pattern of the quasi-isotropic antenna without (left) and with (right) the lens. Reprinted, with permission, from Ref. 62, Copyright 2012, IEEE.

1.3.3.2 Realistic AZIM lens for monopole antenna

To experimentally verify the proposed multibeam AZIM lens, a quad-beam MM lens was designed that tailors the radiation of a *G*-band quarter-wave monopole, which nominally radiates omnidirectionally in the H-plane around 4–5 GHz. Because the radiated electric fields are nearly perpendicular to the ground plane in the H-plane, the lens needs only to work for the TM_z polarization. As presented in Fig. 1.17a, the monopole is surrounded with four triangular anisotropic MM lens segments. Segments 1 and 3 have a low value of effective μ_{rx}, and segments 2 and 4 have a low value of effective μ_{rx}. A 14 cm by 14 cm ($\sim 2\,\lambda_0 \times 2\,\lambda_0$) brass plate is used as the ground plane.

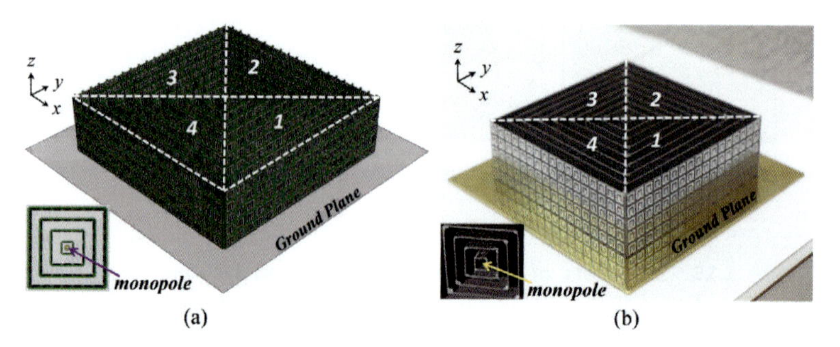

Figure 1.17 (a) Configuration of the multibeam MM lens with the monopole feed. (b) Photograph of the fabricated lens.

Broadside coupled capacitor loaded ring resonators (CLRRs) made of copper [61] are utilized as the building blocks (see Fig. 1.18a). The CLRRs are printed on each side of the dielectric substrate. The openings of the CLRRs are oriented in opposite directions to eliminate the bianisotropy due to structural asymmetry [54]. The MM building block was designed using HFSS, where the scattering parameters were calculated through the application of periodic boundary conditions assigned to the lateral walls in both x- and y-directions. A TE-polarized plane wave (contained in the y–z plane with the electric field along the x-direction) is assumed to be incident from the upper half-space at an angle θ_i ($0° \le \theta_i \le 90°$) with respect to $\hat{z}$, which is the axis of the MM. Three layers of unit cells were used in the z-direction in order to take into consideration the coupling between adjacent layers, thus enabling the acquisition of more accurate effective medium parameters. The three material tensor parameters active under TE-polarized illumination (μ_{ry}, μ_{rz}, and ε_{rx}), which can be retrieved using an anisotropic inversion algorithm [43], are shown plotted in Fig. 1.18b. The ε_{rx} and μ_{ry} are weakly dispersive and do not resonate in the band of interest. This provides a stable normalized impedance for waves propagating in the z-direction, *i.e.*, the direction of the outgoing beam in the directive emission coordinate transformation. Additionally, the lens is nearly matched to free space, ensuring low reflection at the lens–air interface. In contrast, the longitudinal tensor parameter μ_{rz} has a strong Lorentz-shaped resonance at 3.8 GHz and maintains a low value ($0 \leq |\mu_{rz}| \leq 0.4$) throughout the resonance tail over a broad bandwidth (4.2~5.3 GHz) with a weakly dispersive profile. This low μ_{rz} region, located away from the resonance band, is very low loss with the magnitude of Im{μ_{rz}} less than 0.04.

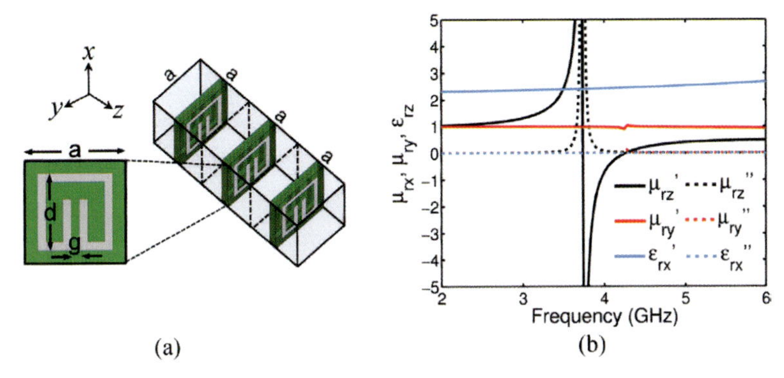

Figure 1.18 (a) Unit cell geometry. The dimensions are a = 6 mm, d = 4.5 mm, l = 4.5 mm, g = 0.5 mm, w = 0.5 mm, and h = 2.5 mm. Rogers RT/Duroid 5880 was used as the substrate. (b) Retrieved effective medium parameters (μ_{ry}, μ_{rz}, and ε_{rx}).

The simulated S_{11} of the monopole with and without the MM lens is presented in Fig. 1.19. The monopole alone has a single resonance at 4.8 GHz with a −10 dB bandwidth of around 0.78 GHz. With the MM lens present, the $S_{11} < -10$ dB bandwidth is increased to 1.35 GHz. This can be attributed primarily to the tuned near-field coupling between the monopole and the unit cells on the inner layers of the MM lens [62, 63]. The top parts of Fig. 1.20a–c show the simulated realized gain patterns in the H-plane of the monopole with and without the lens at 4.25 GHz, 4.85 GHz, and 5.30 GHz. Without the lens, the monopole exhibits an omnidirectional pattern in the horizontal plane with a variation of about 0.7 dB due to the finite ground plane size. The maximum realized gain is smaller than −0.35 dB. With the MM lens, in contrast, four highly directive beams can be observed in the ±x- and ±y-directions. Within the frequency band of

4.20–5.30 GHz, the peak realized gain varies from 4.3 dB to 5.8 dB, about 6 dB higher than that of the monopole alone. The HPBW of the four directive beams are approximately 35°, 32°, and 32° at the three frequencies. The realized gain patterns in the E-plane are reported in the bottom parts of Fig. 1.20a–c. Without the MM lens, the monopole has a maximum radiation at around 40° from the horizon due to the finite ground plane. The MM lens, on the other hand, redirects the maximum of the radiated beam in a direction, which is only about 8° above the horizon. The E-plane HPBW is also reduced from 51–60° to 35–41°. This effect shows that the MM lens not only effectively transforms a two-dimensional cylindrical wave into four collimated waves in the H-plane, but also pushes the radiated beam toward the horizon in the third dimension.

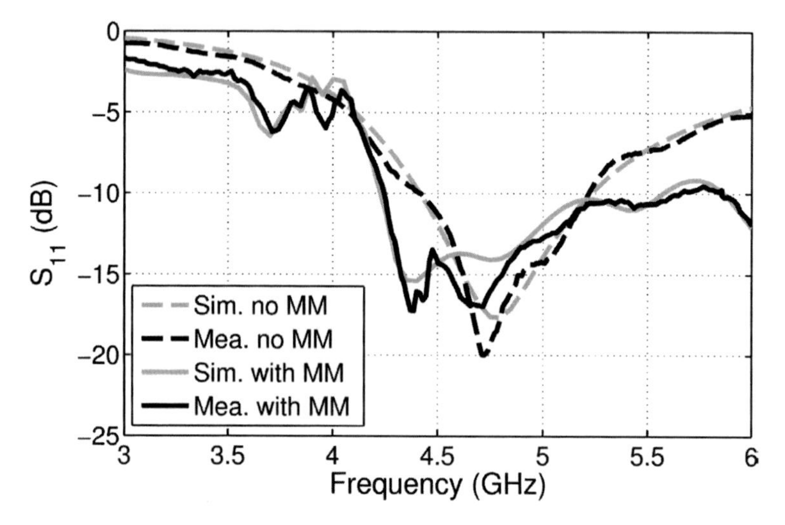

Figure 1.19 Simulated and measured S_{11} of the monopole antenna with and without the multibeam MM lens.

The MM lens and the monopole were fabricated and characterized to validate the proposed design (see Fig. 1.17b). The S_{11} of the monopole with and without the MM lens was measured using a network analyzer. As Fig. 1.19 shows, good agreement was achieved between simulations and measurements not only in terms of the −10 dB bandwidth but also the resonance positions. The measured S_{11} of the monopole alone has a resonance at 4.75 GHz, with a −10 dB bandwidth of 0.76 GHz. When the monopole is surrounded by the MM lens, the S_{11} is below −10 dB from 4.22 GHz to 5.60 GHz with three resonances located at 4.35 GHz, 4.7 GHz, and 5.45 GHz, exhibiting strong correspondence with the simulated results.

The realized gain patterns of the monopole antenna with and without the lens were measured in an anechoic chamber. In Fig. 1.20, the measured gain patterns in both the H- and E-plane at 4.25, 4.85, and 5.30 GHz are displayed. Again, good agreement is seen with the simulated results. The discrepancies are primarily due to minor inaccuracies in fabrication, non-ideal effects of the test setup, and noise in the measurements. For the case without the MM lens, the monopole has a nearly omnidirectional radiation pattern in this band. With the lens present, four high-gain beams can be observed with a measured realized gain varying in the range between 4.3 and 5.9 dB, indicating about 6 dB improvement. The measured average HPBWs of the four directive beams are approximately 36°, 30°, and 34° at the three frequencies considered. The measured E-plane patterns confirm the beam bending effect in the θ-direction. With the lens present, the beam maxima are maintained at approximately 10° from the horizon. In all, the experiment verifies the 3D collimating effect of the AZIM lens in reshaping the radiation of the embedded monopole antenna into multiple highly directive radiated beams.

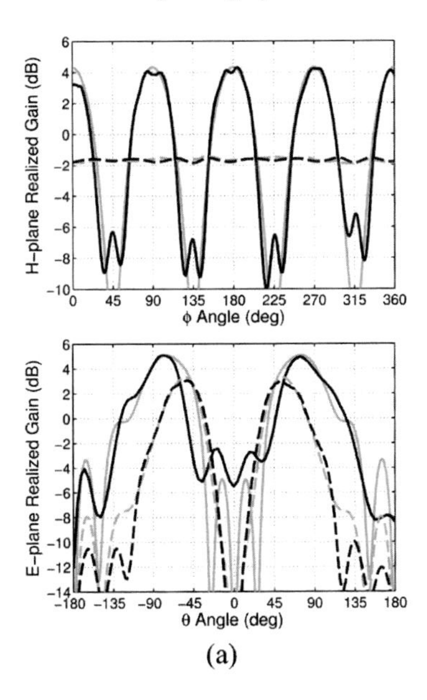

(a)

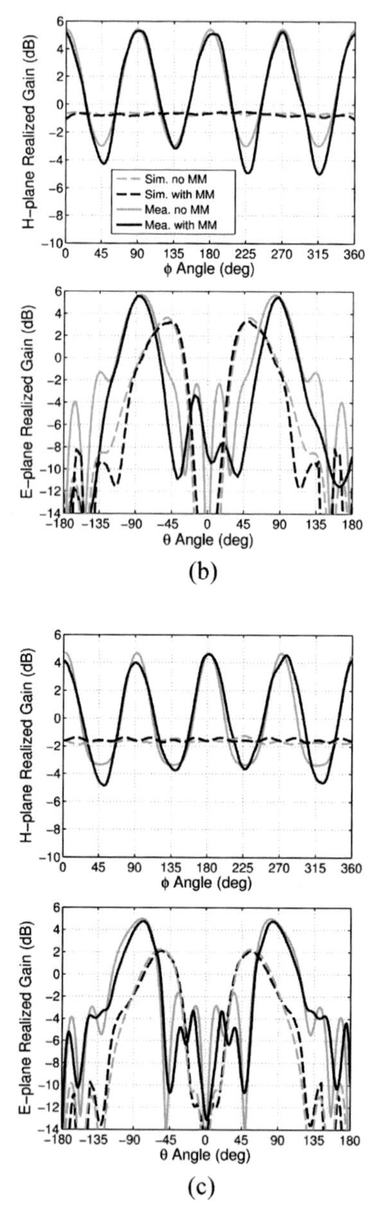

Figure 1.20 Simulated and measured *E*-plane and *H*-plane realized gain patterns of monopole with and without the multibeam MM lens at (a) 4.25 GHz, (b) 4.85 GHz, and (c) 5.30 GHz.

1.4 AZIM Lens for Reconfigurable Beam Steering

The properties of AZIMs, particularly their geometric shape-based control of radiation patterns, can be exploited for the implementation of reconfigurable MM lens antennas. Since the physical shape of the MM slab determines the radiation pattern, changing the shape allows tuning of the radiation pattern. Although changing the physical geometry of a structure without changing its electrical properties is a difficult task, changing the electrical properties within subregions of a reconfigurable MM slab will act to change the effective shape of the MM lens. Choosing a cylindrical or half-cylindrical slab of MM with an embedded feed, and allowing exterior regions of the slab to be disabled or switched from near-zero-index to near-free-space properties yields an antenna whose radiation pattern is governed by the shape of the effective near-zero-index subregion of the slab [64, 65]. An example is shown in Fig. 1.21. Such an antenna, with control over the azimuthal radiation pattern, serves the same purpose as phased array antennas. However, whereas a phased array requires many independent feeds with phase and magnitude control over each feed, the reconfigurable MM antenna uses only a single feed combined with the spatially reconfigurable lens to perform beam scanning. The MM lens may be manufactured using a printed circuit board approach, which allows rapid prototyping and inexpensive bulk production. The trade-off between the many feeds of a conventional phased array and the single-feed implementations of the lens allows designers to select the appropriate approach for a given problem.

Tunable MMs are designed such that their effective electromagnetic properties may be varied in response to a control signal or input [66, 67]. Control mechanisms can include electrical, mechanical, thermal, or optical inputs to vary the electrical properties of individual resonators or unit cells, and thus vary the effective medium properties of the MM structure as a whole. Tunable or reconfigurable MMs and devices are compelling as they yield additional degrees of control when combined with the new properties provided by the MMs themselves. The interest in tunable MMs, especially three-dimensional volumetric MMs, is growing but still limited by implementation challenges and complexities.

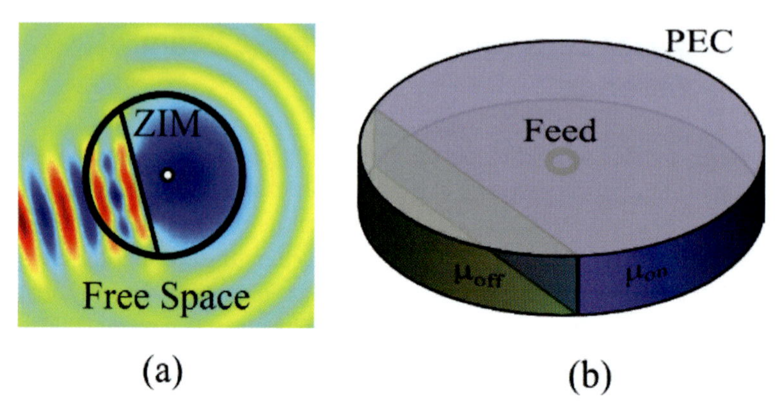

Figure 1.21 (a) Two-dimensional cylindrical reconfigurable lens, showing the directive beam due to the shape of the ZIM region. (b) A finite cylinder allows full azimuthal-plane scanning of one or more beams. The PEC plane on top and bottom and the thickness of the lens are selected to promote radiating modes from the dielectric resonator antenna.

The simplest form of tunable MMs involves a binary change in their properties, from an on to an off state, or high to low. Such a change could include a switch between resonant and non-resonant states, shifting an operating frequency between two values, or changing the effective index of the MM between high and low values. By only requiring a single control bit to specify the behavior of the MM, the complexity and scale of the control circuits necessary to affect the desired change are simplified and minimized.

After assembling an array of reconfigurable MM unit cells into a lens, there are several ways that their switching capability can be leveraged. If all unit cells are tuned together, which is the simplest arrangement, then the tuning behavior of the resulting lens will be limited, particularly with only two-state control over the MM properties. Effective bandwidth could be improved by shifting between two frequencies, or the operational mode of the antenna (beamwidth, polarization, gain) could be changed between two prescribed values. More interesting behavior is allowed, though, by expanding the constraints to allow either more tuning levels or spatial reconfigurability via independent control over the state of each unit cell, rather than tuning the entire lens synchronously.

Spatial reconfiguration is especially interesting as it allows for useful behaviors even when restricted to single-bit reconfigurable unit cells. The two key components for the design of any electromagnetic device are material and geometry. The geometrical shape, size, and relative orientation of a collection of one or more dielectric, metallic, or magnetic materials determine the behavior of electromagnetic fields and waves within the device. A spatially reconfigurable MM allows both geometry and materials to be electronically controlled—independent subregions of an array of MM resonators may be tuned to alter the effective geometry of the device, while the properties within each subregion are also varied by changing the MM state. The combination of the variable shape and material properties of the slab affects the electromagnetic properties of the structure as a whole. Some possible devices that could be implemented using this capability involve tunable dielectric resonators and antennas, dielectric lenses with variable focal points, and variable-phase reflectors. Due to the scale constraints introduced by manufacturing tolerances and the requirement of embedding control wires and circuitry in an EM-transparent fashion, designs for the lower microwave band are the most practical at the current time.

Given a device design that uses effective or bulk material parameters and ideal spatial tuning as described above for the near-zero-index uniaxial MM lens, it is necessary to determine an appropriate unit cell specification in order to construct a prototype antenna. For a single-polarization AZIM, either magnetic or electric resonators may be used. The planar nature of the magnetic SRR is convenient for fabrication and routing of control traces compared to the vertically oriented electric resonators. As a distributed LC circuit, the resonant frequency of SRR elements may be easily tuned by adjusting either the effective inductance or capacitance through the addition of a tunable lumped inductor or capacitor, respectively. Varactor diodes, acting as electrically tunable capacitors whose capacitance is controlled by the DC bias voltage across the junction, are a good choice for the active control element of the MM. Other choices include MEMS switches, transistors, or PIN diodes, but varactors possess advantages of extremely low power consumption, low cost, and compatible control voltages with standard CMOS logic levels. Limitations of varactors for reconfigurable MM devices include finite capacitance tuning ranges and the introduction of parasitic impedances when operated at microwave frequencies.

An additional challenge of a spatially reconfigurable MM over bulk reconfigurable MMs is the logistical challenge of distributing distinct control signals to each unit cell, rather than allowing a single or small number of signals to control the properties of the entire structure. In many cases, when tuning an entire panel of MMs at once, the bias or control wires can be integrated into the resonant unit cells with minimal effort, but the same is not true when each unit cell must be controlled independently. Once the number of unit cells is large enough to construct a structure that is an appreciable fraction of a wavelength, it becomes impractical to dedicate independent wires or traces to each unit cell. The control circuitry must be integrated into the unit cells at design time in order to ensure that the addition of the extraneous wires and elements do not change the effective material response of the final MM implementation. Instead, multiplexing the control signals onto a small number of wires becomes necessary. A serial-to-parallel distributed shift register, where each resonator is associated with a bit position, is one way to enable independent control over each resonator using a small number of wires [67]. Figure 1.22 demonstrates a sample implementation of a hexagonal unit cell with integrated shift register–based control circuits.

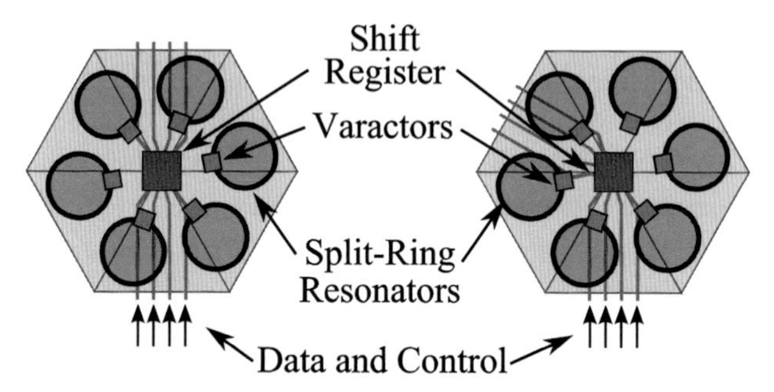

Figure 1.22 Hexagonal unit cells with integrated switching and shift register–based control circuitry may be tiled in a space-filling curve to create arbitrarily sized cylindrical lenses.

In order to treat a collection of electrically small resonators as an MM, the effective medium response of the structure must be well-defined. The common method requires a periodic MM, for which a single unit cell may be simulated with periodic boundary conditions to predict the behavior of an infinitely large sheet or block. This analysis technique then requires that the MM be constructed from a periodic array of identical unit cells, which applies constraints on the layout of the unit cells themselves. Specifically, the control circuitry must have the same or very similar geometric layout within each unit cell throughout the lens. A tiling of hexagonal unit cells allows a good approximation of a cylinder and facilitates interconnection between all unit cells of an arbitrarily sized lens with only two prototype elements: straight and angled [68]. With only two fundamental periodic elements, the effective properties of the MM slab can be conveniently determined and used for modeling as well as design purposes.

The construction of a spatially reconfigurable MM device with arbitrary control over the state of each resonator offers many advantages for electromagnetic designers and engineers. In the same way that field-programmable gate array (FPGA) and other programmable logic devices enable rapid prototyping of new computational circuits and applications, the presence of programmable materials enables similar results for transformation optics and gradient-index prototypes. Although still in the early stages of development and implementation, reconfigurable MMs offer significant benefits for future work in electromagnetics. This section of the chapter has described some of the design decisions and challenges associated with reconfigurable MMs from a device standpoint and should serve as a brief introduction to this rapidly growing field.

1.5 Conclusion

In summary, several concepts for realizing broadband negligible-loss MMs have been presented with application to antenna systems. Several examples were included, which demonstrate that MMs, with properly tailored dispersion and anisotropic properties, can be utilized to extend the impedance bandwidth and/or enhance the gain of antennas. Moreover, by incorporating active components, reconfigurable MM devices can be accomplished, which provide more degrees of freedom for MM-based antenna systems.

Acknowledgments

Portions of this work were supported by the Lockheed Martin University Research Initiative (LM URI) program and by the Penn State MRSEC, Center for Nanoscale Science, under the award NSF DMR-1420620.

References

1. Caloz, C., and Itoh, T. (2005). *Electromagnetic Metamaterials: Transmission Line Theory and Microwave Applications* (Wiley-IEEE Press, USA).
2. Eleftheriades, G. V., and Balmain, K. G. (2005). *Negative-Refraction Metamaterials: Fundamental Principles and Applications* (Wiley–IEEE Press, USA).
3. Engheta, N., and Ziolkowski, R. (2006). *Metamaterials: Physics and Engineering Explorations* (Wiley-IEEE Press, USA).
4. Cui, T. J., Smith, D. R., and Liu, R. (2009). *Metamaterials: Theory, Design, and Applications* (Springer, UK).
5. Werner, D. H., and Kwon, D.-H. (2014). *Transformation Electromagnetics and Metamaterials: Fundamental Principles and Applications* (Springer, UK).
6. Pendry, J. B. (2000). Negative refraction makes a perfect lens, *Phys. Rev. Lett.*, **85**, pp. 3966–3969.
7. Grbic, A., and Eleftheriades, G. V. (2004). Overcome the diffraction limit with a planar left-handed transmission-line lens, *Phys. Rev. Lett.*, **92**, pp. 117403(1)–(4).
8. Fang, N., Lee, H., Sun, C., and Zhang, X. (2005). Sub-diffraction-limited optical imaging with a silver superlens, *Science*, **308**, pp. 534–537.
9. Pendry, J. B., Schurig, D., and Smith, D. R. (2006). Controlling electromagnetic fields, *Science*, **312**, pp. 1780–1782.
10. Schurig, D., Mock, J. J., Justice, B. J., Cummer, S. A., Pendry, J. B., Starr, A. F., and Smith, D. R. (2006). Electromagnetic metamaterial cloaking at microwave frequencies, *Science*, **314**, pp. 977–980.
11. Cai, W., Chettiar, U. K., Kildishev, A. V., and Shalaev, V. M. (2007). Optical cloaking with metamaterials, *Nat. Photon.*, **1**, pp. 224–227.
12. Ma, Y. G., Ong, C. K., Tyc, T., and Leonhardt, U. (2009). An omnidirectional retroreflector based on the transmutation of dielectric singularities, *Nat. Mater.*, **8**, pp. 639–642.

13. Jiang, Z. H., Yun, S., Toor, F., Werner, D. H., and Mayer, T. S. (2011). Conformal dual-band near-perfectly absorbing mid-infrared metamaterial coating, *ACS Nano*, **5**, pp. 4641–4647.
14. Jiang, W. X., Qiu, C.-W., Han, T., Zhang, S., and Cui, T. J. (2013). Creation of ghost illusions using wave dynamics in metamaterials, *Adv. Funct. Mater.*, **23**, pp. 4028–4034.
15. Jiang, Z. H., and Werner, D. H. (2014). Quasi-three-dimensional angle-tolerant electromagnetic illusion using ultrathin metasurface coatings, *Adv. Funct. Mater.*, **24**, pp. 7728–7736.
16. Chen, H.-T., Padilla, W. J., Cich, M. J., Azad, A. K., Averitt, R. D., and Taylor, A. J. (2009). A metamaterial solid-state terahertz phase modulator, *Nat. Photon*, **3**, pp. 148–151.
17. Ziolkowski, R. W., Jin, P., and Lin, C.-C. (2011). Metamaterial-inspired engineering of antennas, *Proc. IEEE*, **99**, pp. 1720–1731.
18. Fedotov, V. A., Mladyonov, P. L., Prosvirnin, S. L., Rogacheva, A. V., Chen, Y., and Zheludev, N. I. (2006). Asymmetric propagation of electromagnetic waves through a planar chiral structure, *Phys. Rev. Lett.*, **97**, pp. 167401(1)–(4).
19. Liu, R., Ji, C., Mock, J. J., Chin, J. Y., Cui, T. J., and Smith, D. R. (2009). Broadband ground-plane cloak, *Science*, **323**, pp. 366–369.
20. Ma, H. F., and Cui, T. J. (2010). Three-dimensional broadband ground-plane cloak made of metamaterials, *Nat. Commun.*, **1**, pp. 21.
21. Chen, H., Hou, B., Chen, S., Ao, X., Wen, W., and Chan, C. T. (2009). Design and experimental realization of a broadband transformation media field rotator at microwave frequencies, *Phys. Rev. Lett.*, **102**, pp. 183903(1)–(4).
22. Kundtz, N., and Smith, D. R. (2010). Extreme-angle broadband metamaterial lens, *Nat. Mater.*, **9**, pp. 129–132.
23. Ma, H. F., and Cui, T. J. (2010). Three-dimensional broadband and broad-angle transformation-optics lens, *Nat. Commun.*, **1**, pp. 124.
24. Jackson, D. R., Caloz, C., and Itoh, T. (2012). Leaky-wave antennas, *Proc. IEEE*, **100**, pp. 2194–2206.
25. Okabe, H., Caloz, C., and Itoh, T. (2004). A compact enhanced-bandwidth hybrid ring using an artificial lumped-element left-handed transmission-line section, *IEEE Trans. Microw. Theory Tech.*, **52**, pp. 798–804.
26. Lier, E., Werner, D. H., Scarborough, C. P., Wu, Q., and Bossard, J. A. (2011). An octave-bandwidth negligible-loss radiofrequency metamaterial, *Nat. Mater.*, **10**, pp. 216–222.

27. Volakis, J. L., and Sertel, K. (2011). Narrowband and wideband metamaterial antennas based on degenerate band edge and magnetic photonic crystals, *Proc. IEEE*, **99**, pp. 1732–1745.
28. Zhou, R., Zhang, H., and Xin, H. (2010). Metallic wire array as low-effective index of refraction medium for directive antenna application, *IEEE Trans. Antennas Propag.*, **58**, pp. 79–87.
29. Jiang, Z. H., Gregory, M. D., and Werner, D. H. (2011). Experimental demonstration of a broadband transformation optics lens for highly directive multibeam emission, *Phys. Rev. B*, **84**, pp. 165111(1)–(6).
30. Ge, Y., Esselle, K. P., and Hao, Y. (2007). Design of low-profile high-gain EBG resonator antennas using a genetic algorithm, *IEEE Antennas Wireless Propag. Lett.*, **6**, pp. 480–483.
31. Guérin, N., Enoch, S., Tayeb, G., Sabouroux, P., Vincent, P., and Legay, H. (2006). A metallic Fabry–Pérot directive antenna, *IEEE Trans. Antennas Propag.*, **54**, pp. 220–224.
32. Feresidis, A. P., Goussetis, G., Wang, S., and Vardaxoglou, J. C. (2005). Artificial magnetic conductor surfaces and their application to low-profile high-gain planar antennas, *IEEE Trans. Antennas Propag.*, **53**, pp. 209–215.
33. Sun, Y., Chen, Z. N., Zhang, Y., Chen, H., and See, T. S. P. (2012). Subwavelength substrate-integrated Fabry–Pérot cavity antennas using artificial magnetic conductor, *IEEE Trans. Antennas Propag.*, **60**, pp. 30–35.
34. Enoch, S., Tayeb, G., Sabouroux, P., Guérin, N., and Vincent, P. (2002). A metamaterial for directive emission, *Phys. Rev. Lett.*, **89**, pp. 213902(1)–(4).
35. Ziolkowski, R. W. (2004). Propagation in and scattering from a matched metamaterial having a zero index of refraction, *Phys. Rev. E*, **70**, pp. 046608(1)–(12).
36. Turpin, J. P., Massoud, A. T., Jiang, Z. H., Werner, P. L., and Werner, D. H. (2010). Conformal mappings to achieve simple material parameters for transformation optics devices, *Opt. Express*, **18**, pp. 244–252.
37. Cheng, Q., and Cui, T. J. (2011). Multi-beam generations at pre-designed directions based on anisotropic zero-index metamaterials, *Appl. Phys. Lett.*, **99**, pp. 131913(1)–(3).
38. Palandoken, M., Grede, A., and Henke, H. (2009). Broadband microstrip antenna with left-handed metamaterials, *IEEE Trans. Antennas Propag.*, **57**, pp. 331–338.

39. Antoniades, M. A., and Eleftheriades, G. V. (2009). A broadband dual-mode monopole antenna using NRI-TL metamaterial loading, *IEEE Antennas Wireless Propag. Lett.*, **8**, pp. 258–261.

40. Antoniades, M. A., and Eleftheriades, G. V. (2012). Multiband compact printed dipole antennas using NRI-TL metamaterial loading, *IEEE Trans. Antennas Propag.*, **60**, pp. 5613–5626.

41. Mehdipour, A., Denidni, T. A., and Sebak, A.-R. (2014). Multi-band miniaturized antenna loaded by ZOR and CSRR metamaterial structures with monopolar radiation pattern, *IEEE Trans. Antennas Propag.*, **62**, pp. 555–562.

42. Kim, K., and Varadan, V. V. (2010). Electrically small, millimeter wave dual band meta-resonator antennas, *IEEE Trans. Antennas Propag.*, **58**, pp. 3458–3463.

43. Jiang, Z. H., Bossard, J. A., Wang, X., and Werner, D. H. (2011). Synthesizing metamaterials with angularly independent effective medium properties based on an anisotropic parameter retrieval technique coupled with a genetic algorithm, *J. Appl. Phys.*, **109**, pp. 1013515(1)–(11).

44. Jiang, Z. H., Gregory, M. D., and Werner, D. H. (2011). A broadband monopole antenna enabled by an ultrathin anisotropic metamaterial coating, *IEEE Antennas Wireless Propag. Lett.*, **10**, pp. 1543–1546.

45. Kong, J. A. (2000). *Electromagnetic Wave Theory* (EMW Cambridge, USA).

46. Volakis, J. (2007). *Antenna Engineering Handbook*, 4th ed. (McGraw-Hill Professional, USA).

47. Smith, D. R., and Schurig, D. (2003). Electromagnetic wave propagation in media with indefinite permittivity and permeability tensors, *Phys. Rev. Lett.*, **90**, pp. 077405(1)–(4).

48. Jiang, Z. H., Wu, Q., Brocker, D. E., Sieber, P. E., and Werner, D. H. (2014). A low-profile high-gain substrate-integrated waveguide slot antenna enabled by an ultrathin anisotropic zero-index metamaterial coating, *IEEE Trans. Antennas Propag.*, **62**, pp. 1173–1184.

49. Schurig, D., Mock, J. J., and Smith, D. R. (2006). Electric-field-coupled resonators for negative permittivity metamaterials, *Appl. Phys. Lett.*, **88**, pp. 041109(1)–(3).

50. Liu, B., Hong, W., Zhang, Y., Tang, H. J., Yin, X., and Wu, K. (2007). Half mode substrate integrated waveguide 180° 3-dB directional couplers, *IEEE Trans. Microw. Theory Tech.*, **55**, pp. 2586–2592.

51. Zhang, Y., Chen, Z. N., Qing, X., and Hong, W. (2011). Wideband millimeter-wave substrate integrated waveguide slotted narrow-wall fed cavity antennas, *IEEE Trans. Antennas Propag.*, **59**, pp. 1488–1496.
52. Turpin, J. P., Wu, Q., Werner, D. H., Martin, B., Bray, M., and Lier, E. (2012). Low cost and broadband dual-polarization metamaterial lens for directivity enhancement, *IEEE Trans. Antennas Propag.*, **60**, pp. 5717–5726.
53. Turpin, J. P., Wu, Q., Werner, D. H., Martin, B., Bray, M., and Lier, E. (2014). Near-zero-index metamaterial lens combined with AMC metasurface for high-directivity low-profile antennas, *IEEE Trans. Antennas Propag.*, **62**, pp. 1928–1936.
54. Falcone, F., Lopetegi, T., Laso, M. A. G., Baena, J. D., Bonache, J., Beruete, M., Marques, R., Martin, F., and Sorolla, M. (2004). Babinet principle applied to the design of metasurfaces and metamaterials, *Phys. Rev. Lett.*, **93**, pp. 197401(1)–(4).
55. Smith, D., Schultz, S., Markoš, P., and Soukoulis, C. (2002). Determination of effective permittivity and permeability of metamaterials from reflection and transmission coefficients, *Phys. Rev. B*, **65**, pp. 195104(1)–(5).
56. Koschny, T., Markoš, P., Economou, E. N., Smith, D. R., Vier, D. C., and Soukoulis, C. M. (2005). Impact of the inherent periodic structure on effective medium description of left-handed and related metamaterials, *Phys. Rev. B*, **71**, pp. 245105(1)–(22).
57. Bayraktar, Z., Turpin, J. P., and Werner, D. H. (2011). Nature-inspired optimization of high-impedance metasurfaces with ultrasmall interwoven unit cells, *IEEE Antennas Wireless Propag. Lett.*, **10**, pp. 1563–1566.
58. Xin, H., Matsugatani, K., Kim, M., Hacker, J., Higgins, J. A., Rosker, M., and Tanaka, M. (2002). Mutual coupling reduction of low-profile monopole antennas on high-impedance ground plane, *Electron. Lett.*, **38**, pp. 849–850.
59. Sievenpiper, D., Zhang, L., Broas, R. F. J., Alexopolous, N. G., and Yablonovitch, E. (1999). High-impedance electromagnetic surfaces with a forbidden frequency band, *IEEE Trans. Microw. Theory Tech.*, **47**, pp. 2059–2074.
60. Cho, C., Choo, H., and Park, I. (2008). Printed symmetric inverted-F antenna with a quasi-isotropic radiation pattern, *Microw. Opt. Technol. Lett.*, **50**, pp. 927–930.
61. Erenok, A., Luljak, P. L., and Ziolkowski, R. W. (2005). Characterization of a volumetric metamaterial realization of an artificial magnetic

conductor for antenna applications, *IEEE Trans. Antennas Propag.*, **53**, pp. 160–172.

62. Jiang, Z. H., Gregory, M. D., and Werner, D. H. (2012). Broadband highly directive multi-beam emission through transformation optics enabled metamaterial lenses, *IEEE Trans. Antennas Propag.*, **60**, pp. 5063–5074.
63. Wu, Q., Jiang, Z. H., Quevedo-Teruel, O., Turpin, J. P., Tang, W., Hao, Y., and Werner, D., H. (2013). Transformation optics inspired multibeam lens antennas for broadband directive radiation, *IEEE Trans. Antennas Propag.*, **61**, pp. 2910–5922.
64. Turpin, J. P., and Werner, D. H. (2014). Construction and measurements of a prototype near-zero-index reconfigurable metamaterial antenna, in *Proc. 2014 IEEE Ant. Propag. Int. Symp.*, Memphis, TN, USA.
65. Turpin, J. P., and Werner, D. H. (2012). Cylindrical metamaterial lens for single-feed adaptive beamforming, in *Proc. 2012 IEEE Ant. Propag. Int. Symp.*, Chicago, IL, USA.
66. Turpin, J. P., Bossard, J. A., Morgan, K. L., Werner, D. H., and Werner, P. L. (2014). Reconfigurable and tunable metamaterials: A review of the theory and applications, *Int. J. Antennas Propag.*, **2014**, pp. 1–18.
67. Turpin, J. P., Werner, D. H., and Wolfe, D. W. (2015). Design considerations for spatially reconfigurable metamaterials, *IEEE Trans. Antennas Propag.*, **63**, pp. 3513–3521.
68. Turpin, J. P., and Werner, D. H. (2013). Beam scanning antenna enabled by a spatially reconfigurable near-zero index metamaterial, in *Proc. 7th European Conf. Antennas Propag.*, Gothenburg, Sweden.
69. Cho, C., Choo, H., and Park, I. (2008). Printed symmetric inverted-F antenna with a quasi-isotropic radiation pattern, *Microw. Opt. Technol. Lett.*, **50**, pp. 927–930.

Chapter 2

Broadband Low-loss Metamaterial-Enabled Horn Antennas

Clinton P. Scarborough,[a] Qi Wu,[a] Douglas H. Werner,[a] Erik Lier,[b] and Bonnie G. Martin[b]

[a]*Department of Electrical Engineering, Penn State University, University Park, PA 16802, USA*

[b]*Lockheed Martin Commercial Space Systems, Littleton, CO 12257, USA*

dhw@psu.edu

Metamaterial surfaces have been shown to possess a wide range of uses in recent years, with many applications showing possible benefits from new ways to manipulate electromagnetic waves. While there are a variety of applications that would most benefit from broadband, non-dispersive material properties, some applications, including horn antennas for satellite reflector feeds, actually require careful engineering of dispersive properties over broad bandwidths for optimal operation. Metamaterial horn antennas have been created that take advantage of engineered dispersion and other properties to enable antenna operation over octave bandwidths with negligible intrinsic losses. These metasurface-enabled horn antennas promise to provide comparable or better operation than

Broadband Metamaterials in Electromagnetics: Technology and Applications
Edited by Douglas H. Werner
Copyright © 2017 Pan Stanford Publishing Pte. Ltd.
ISBN 978-981-4745-68-0 (Hardcover), 978-1-315-36443-8 (eBook)
www.panstanford.com

conventional alternatives, such as corrugated or dielectric core (dielcore) horns, while reducing weight, reducing cost, or improving operating bandwidths.

2.1 Introduction

2.1.1 Horn Antennas as Reflector Feeds

Horn antennas have served in numerous applications for much of the last century, but most notably for their role as feeds for reflector antennas. The simplest horn antennas (such as that shown in Fig. 2.1) have radiation patterns much like those of open-ended waveguides, with relatively high sidelobes in the *E*-plane and relatively low sidelobes in the *H*-plane. This behavior results from the aperture field distribution, which corresponds to the fundamental waveguide mode. For rectangular horns, the fundamental mode is typically TE_{10}, which has a uniform field in the *E*-plane and a smoothly tapered field in the *H*-plane. One of the important considerations for the design of a reflector antenna system is spillover loss, which can be mitigated by reducing the sidelobes in the feed's radiation pattern.

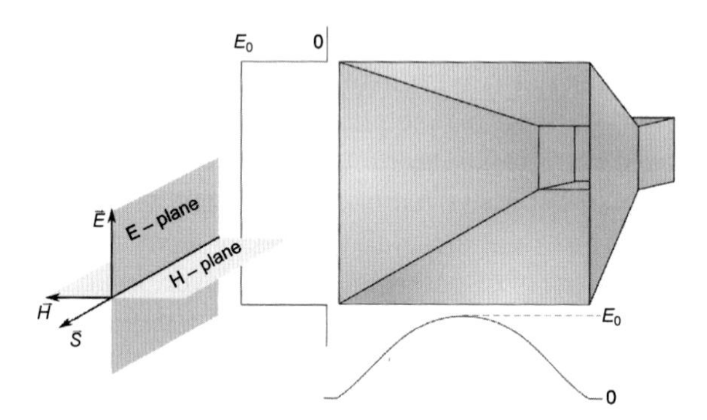

Figure 2.1 A simple pyramidal (rectangular) horn antenna, highlighting the principal planes and the typical corresponding aperture field distributions.

Reflector antennas are commonly used in satellite communication links, where the weight of the system is a critically important constraint. As a result, many satellite antennas reuse frequencies (and thus antennas) through dual-polarized communication links.

Antennas for dual-polarized systems must exhibit minimal cross-polarized radiation, as cross-pol from one polarization directly interferes with the alternative polarization.

Moreover, circular symmetry is often desirable for dual-polarized systems in order that the two communication channels will have similar characteristics. Simple circular (conical) feed horns exhibit high cross-pol as a result of the field line configurations of their associated mode distributions. To reduce cross-pol, hybrid-mode horns have been commonly employed for several decades, since the introduction of the corrugated horn [1], followed by the thorough description of hybrid-mode horns by Minnett and Thomas [2]. Ideal hybrid modes, which are linearly polarized modes characterized by perfectly straight field lines, can be supported by a waveguide or horn having an anisotropic boundary on its walls, such that it meets the balanced hybrid condition:

$$Z^{TE} Z^{TM} = \eta_0^2 \tag{2.1}$$

where Z^{TE} and Z^{TM} are the TE and TM boundary (surface) impedances, respectively, and η_0 is the free-space wave impedance. In the lossless case, the surface impedances become surface reactances, and the balanced hybrid condition may be expressed as:

$$Z^{TE} Z^{TM} = -X^{TE} X^{TM} = \eta_0^2 \tag{2.2}$$

2.1.2 Soft and Hard Horn Antennas

Hybrid-mode horns can be designed for tapered aperture distributions and thus low sidelobes (soft horns), uniform aperture distributions, and thus high sidelobes and maximum gain (hard horns), or for characteristics at nearly arbitrary points between the two extremes [3, 4]. Qualitatively, a soft surface forces an adjacent electric field to zero and is defined by:

$$X^{TE} = 0, X^{TM} = -\infty \tag{2.3}$$

where X^{TE} and X^{TM} denote surface reactances, while the surface resistances R^{TE} and R^{TM} are equal to zero. In contrast, a hard surface has no effect on an adjacent electric field and is defined by:

$$X^{TE} = \infty, X^{TM} = 0 \tag{2.4}$$

where, as for the ideal soft surface, the surface resistances are zero.

Traditionally, soft horns have been implemented using transverse corrugations in the horn walls [2, 5, 6] and later by dielectric loading [7–9], thin dielectrics covered with circular metallic strips [10], and finally by metamaterials [11–17]. Meanwhile, hard horns have been implemented using longitudinal, dielectric-filled corrugations, strips, and dielectric loading [3], as well as theoretically with metamaterials [13].

2.1.3 Metamaterial Horn Antennas

The first decade of the 2000s saw an abundance of research on electromagnetic metamaterials, including metamaterial surfaces, high-impedance surfaces [18], electromagnetic band-gap materials [19], and artificial magnetic conductors (AMCs) [20, 21], all of which have shown promise for a variety of antenna applications. As hybrid-mode horn antennas required boundaries characterized by their anisotropic surface impedances, the design techniques that had previously been developed for other antenna applications could be readily adapted to create metamaterial surface designs for hybrid-mode horns. Applying those surface designs to actual horn antennas requires numerous modifications and approximations, one of which is the fact that tapered horn walls do not permit the direct placement of a surface that was designed to be periodic in two dimensions. The following sections provide details and examples for many of the design considerations that are relevant to the development of metamaterial horn antennas, including successful metamaterial implementations of hybrid-mode horns.

2.2 Design and Modeling of Metamaterial Implementations for Soft and Hard Surfaces

2.2.1 Plane Wave Model of Metasurfaces

Any metasurface lining will necessarily be backed by the conducting horn walls. The generic metasurface structure of interest then takes on a form such as that shown in Fig. 2.2, with a patterned screen above a dielectric layer, which, in turn, is backed by a conductive ground plane. This structure can represent a variety of physical implementations, including a metallic frequency-selective-surface (FSS) type of screen [22, 23] or a wire-grid structure [24, 25]. We primarily consider FSS-like structures, as they are generally suitable to conventional printed circuit manufacturing techniques. In contrast with many FSS designs, however, the conducting vias are often critical to achieve the necessary surface impedance properties over a broad bandwidth. Moreover, they provide a conducting path that mitigates electrostatic discharge (ESD) concerns for satellite applications. Metamaterials also differ from FSS structures in that the unit cell sizes are typically a tenth of a wavelength or smaller, while FSS designs tend to have sizes on the order of half a wavelength.

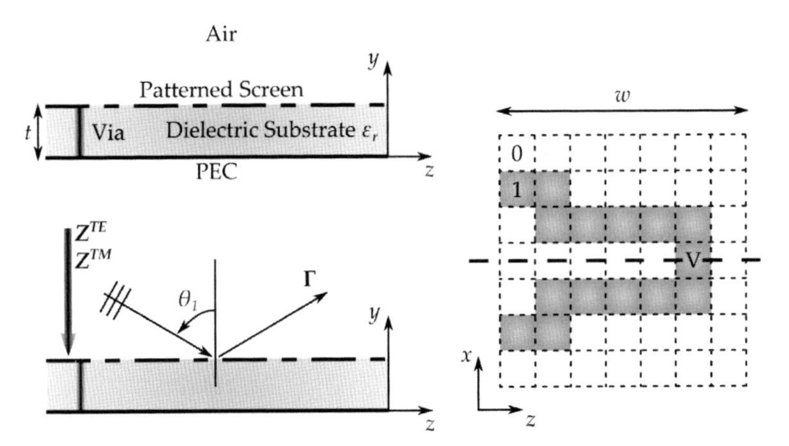

Figure 2.2 Cross sections of the PEC-backed metasurface, consisting of a patterned conducting layer atop a dielectric substrate (top left), with one or more conducting vias in each unit cell. Simulations are performed with plane waves at near-grazing incidence (lower left, $\theta_1 \to 90°$) to calculate both the TE and TM reflection coefficients, from which the surface impedances Z^{TE} and Z^{TM} are calculated. The square unit cell pattern (right) consists of an array of pixels, represented in the optimizer by either a "0" or a "1" to indicate the absence or presence of metal. The dashed line indicates a symmetry plane; the pattern on one half of the unit cell is optimized and then mirrored across the symmetry plane. The "V" indicates a typical via location.

As shown in Fig. 2.2, the metasurface has a patterned screen above a dielectric substrate with thickness t and dielectric constant ε_r. The pattern is based on square unit cells with a periodicity of w in both the x- and z-directions. Contrary to a conventional FSS, the periodicity w is restricted to be much smaller than the operating wavelength, and thus the metasurface structure can be approximated by its effective surface impedances [18]. The anisotropic surface impedances are defined by the ratios of the electric and magnetic field components tangential to the boundary. As such, they provide the boundary conditions for fields outside the metasurface while accounting for power dissipation and energy storage within the metasurface structure. Surface impedances provide convenient criteria for the design and optimization of metasurfaces for arbitrary electromagnetic responses.

The anisotropic surface impedances can be determined from the reflection coefficients for plane waves at near-grazing incidence upon the metasurface as follows:

$$Z^{TE} = R^{TE} + jX^{TE} = \eta^{TE} \frac{1+\Gamma^{TE}}{1-\Gamma^{TE}} = \frac{E_x}{H_z} \tag{2.5}$$

$$Z^{TM} = R^{TM} + jX^{TM} = \eta^{TM} \frac{1+\Gamma^{TM}}{1-\Gamma^{TM}} = -\frac{E_z}{H_x} \tag{2.6}$$

where Γ denotes the reflection coefficients, R^{TE} and R^{TM} are the surface resistances, X^{TE} and X^{TM} are the surface reactances, E and H are the electric and magnetic fields, respectively, at the surface of the metallic pattern, and η denotes the transverse wave impedance, defined as:

$$\eta^{TE} = \frac{\eta_0}{\cos\theta_1} \tag{2.7}$$

$$\eta^{TM} = \eta_0 \cos\theta_1 \tag{2.8}$$

for obliquely incident plane waves with an incident angle θ_1. Although the ideal metasurface would be designed with $\theta_1 = 90°$, it is generally necessary to keep $\theta_1 < 90°$ to make simulations possible, such that θ_1 can approach 90° as a limit.

2.2.2 Equivalent Homogeneous Metamaterial Model

In order to model a horn antenna lined with a metasurface, it is convenient to employ a homogeneous material model with effective permittivity and permeability values, rather than an infinitesimally thin boundary with the corresponding anisotropic surface impedances. Few, if any, simulation packages support anisotropic surface impedances, but many support a large range of effective material properties. This process begins by equating the composite metasurface structure to a homogeneous material backed by a conducting plane. The homogeneous material is assumed to have an effective permittivity ε_{eff} and an effective permeability μ_{eff}. The thickness of the homogeneous slab is set to be *t*, that of the composite metasurface. The effective parameters are chosen such that the homogeneous material gives rise to identical boundary conditions (*i.e.*, surface impedances) as the actual metasurface.

For a homogeneous slab with thickness *t*, backed by a conducting ground plane, the anisotropic surface impedances can be expressed as:

$$Z^{TE} = \frac{E_x}{H_z} = j\eta_0 \frac{\mu_{eff}}{\sqrt{\sin^2\theta_1 - z_{eff}\mu_{eff}}} \tanh\left(k_0 t\sqrt{\sin^2\theta_1 - \varepsilon_{eff}\mu_{eff}}\right) \tag{2.9}$$

$$Z^{TM} = -\frac{E_z}{H_x} = -j\eta_0 \frac{\sqrt{\sin^2\theta_1 - z_{eff}\mu_{eff}}}{z_{eff}} \tanh\left(k_0 t\sqrt{\sin^2\theta_1 - \varepsilon_{eff}\mu_{eff}}\right) \tag{2.10}$$

where k_0 and η_0 are the wave number and wave impedance of free space, respectively. Calculating the effective refractive index n_{eff} and wave impedance z_{eff} yields:

$$n_{\text{eff}} = \frac{\sin\theta_1}{\sqrt{1 - \dfrac{Z^{\text{TM}}}{Z^{\text{TE}}}}} \tag{2.11}$$

$$z_{\text{eff}} = \frac{\sqrt{Z^{\text{TE}} Z^{\text{TM}}}}{\tanh\left(j\, n_{\text{eff}} k_0 t\sqrt{1 - \dfrac{\sin\theta_1}{n_{\text{eff}}}}\right)} \tag{2.12}$$

from which we can calculate the permittivity and permeability as:

$$\varepsilon_{\text{eff}} = \frac{n_{\text{eff}}}{z_{\text{eff}}} \tag{2.13}$$

$$\mu_{\text{eff}} = n_{\text{eff}} \times z_{\text{eff}} \tag{2.14}$$

2.2.3 Design Goals and Optimization Methods

As mentioned previously, the typical primary requirement for an ideal horn liner is that it be a soft or hard surface. Certain periodic surfaces, such as corrugated metals, mushroom-like textured surfaces, and uniplanar compact band-gap structures, have been designed to exhibit soft and hard characteristics [19, 26]. Achieving a soft or hard boundary over a broad bandwidth is a significant challenge, particularly while maintaining the hybrid-mode condition. Global stochastic optimizers such as the genetic algorithm (GA) and more recent improvements [27, 28] provide a means for designing metamaterial surfaces that exhibit broadband and multi-band operation [21], or that achieve a specific frequency dispersion characteristic. The examples presented here were based on a GA optimization of the thickness t, the unit cell width (periodicity) w, and the metallic pixel pattern. The following cost functions were applied to the various optimizations:

$$\text{cost}_{\text{soft}} = \Sigma_{\text{frequencies}}\left(\left|X^{\text{TE}}\right| + \left|\frac{1}{X^{\text{TM}}}\right|\right) \tag{2.15}$$

$$\text{cost}_{\text{hard}} = \Sigma_{\text{frequencies}}\left(\left|\frac{1}{X^{\text{TE}}}\right| + \left|X^{\text{TM}}\right|\right) \tag{2.16}$$

$$\text{cost}_{\text{balanced}} = \Sigma_{\text{frequencies}}\left(Z^{\text{TE}}\,Z^{\text{TM}} - \eta_0^2\right) \tag{2.17}$$

The simulations and optimizations of unit cells were performed using a periodic finite-element boundary integral solver, which is now commercially available as *PFSS* from E x H, Inc. [29]. The following section presents several examples of metasurfaces that were designed using this optimization approach.

2.3 Metasurface Design Examples

2.3.1 Canonical Examples

To demonstrate the aforementioned modeling and optimization procedure, two canonical metasurface designs were created: a soft surface and a hard surface. For simplicity, both designs were optimized for a center frequency of 10 GHz, and neither design included a conductive via from the pattern to the ground plane beneath. Figure 2.3 shows the geometry and simulated responses for the soft and hard metasurface design examples.

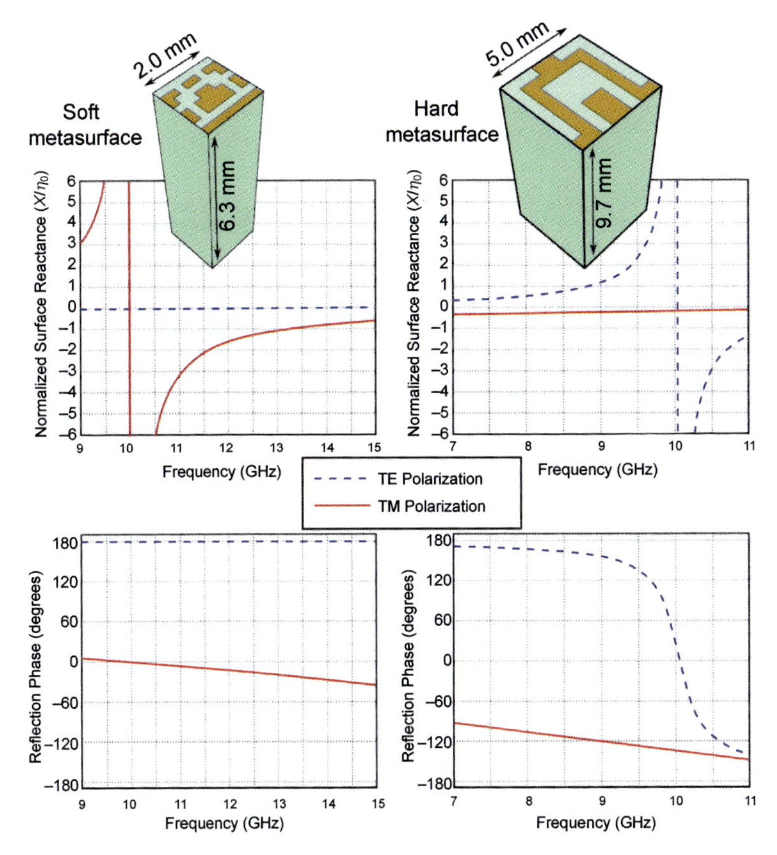

Figure 2.3 Pixilated soft (left) and hard (right) metasurface geometry (top), surface reactances (middle), and reflection phase (bottom).

Note that, consistent with the definitions in Eqs. (1.3) and (1.4), the soft metasurface exhibits a surface reactance X_{TE} that is essentially equal to zero, while X_{TM} has a magnitude much greater than zero at the center frequency, and it remains sufficiently below zero for many soft-surface applications across the band. In contrast, the hard metasurface has X_{TM} close to zero across the band, while X_{TE} shows the larger magnitude. An alternative way to interpret soft- and hard-surface operation is to consider the reflection phases of the metasurfaces in the bottom of Fig. 2.3. For the soft case, the metasurface acts as a perfect electric conductor (PEC) for TE-polarized incident waves, while it acts as an AMC for the TM polarization. Meanwhile, the hard design approximates a PEC for the TM-polarized case, and TE-polarized waves see an AMC near the center frequency. This anisotropic reflection behavior gives rise to the anisotropic surface reactances associated with the soft and hard surfaces.

2.3.2 Printed-Patch Balanced Hybrid Metasurface

Many horn antennas operate in the K_u-band, where wavelengths range from 1.7 to 2.5 cm. Metamaterial surfaces thus require features on the scale of 0.1 mm, which can easily and inexpensively be achieved using standard printed circuit board (PCB) technology. Moreover, the required substrate thicknesses also fall within typical PCB ranges. These properties make the K_u-band an ideal candidate for testing metasurfaces and metahorns.

The soft- and hard-surface examples in the previous section satisfy the balanced hybrid condition only over a narrow bandwidth. This means that, although they can be useful over a broadband when lining a single-polarization horn, they would only allow low cross-polarization over a narrow band when lining a dual-polarized horn. In order to achieve broadband operation in a dual-polarized communication system, a metasurface was optimized using the combination of Eqs. (2.15) and (2.17), resulting in an approximately soft metasurface that also met the balanced hybrid condition. The optimization incorporated a single conducting via in each unit cell, as well as several constraints, including symmetry across the x–z plane to reduce cross-polarized reflections from the metasurface and restrictions on diagonally connected pixels for reasonable fabrication. The substrate material was chosen to be Rogers 5880LZ, which has a low dielectric constant of 1.96 and a loss tangent of 0.002. A lower substrate dielectric constant leads to larger operating bandwidths. Removing some unnecessary pixels from the optimized structure to simplify manufacturing yielded the geometry shown in Fig. 2.4, which also shows its simulated properties for a unit cell periodicity $p = 0.134\lambda$, thickness $t = 0.134\lambda$, $w_1 = 0.074\lambda$, $w_2 = 0.045\lambda$, $s_1 = 0.060\lambda$, and $s_2 = 0.060\lambda$, where λ is the wavelength at 12 GHz.

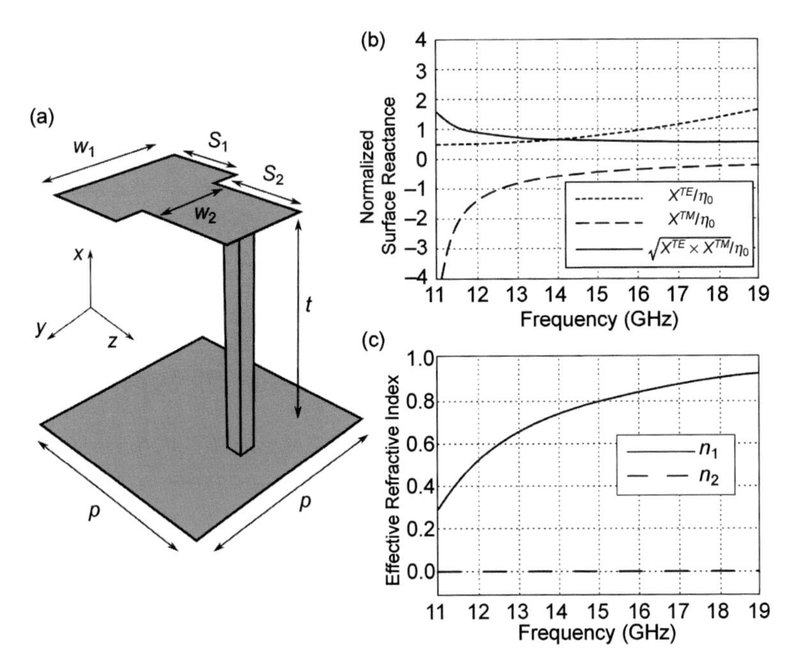

Figure 2.4 Printed-patch metasurface geometry (a), surface reactances (b), effective index of refraction (c). Although it is not shown, the unit cell geometry includes a substrate material between the top patch and the ground plane beneath. Note that X^{TM} is significantly less than zero across the band, and that the hybrid-mode condition is approximately satisfied across the band. Also note that the imaginary component of the effective index is essentially zero, indicating minimal intrinsic loss in the metasurface.

The simulated properties in Fig. 2.4 correspond to an obliquely incident plane wave at an angle approaching 90°. TM-polarized waves at these oblique angles interact with the conducting via, leading to an effective surface reactance appropriate for soft operation. Although both the TE and TM surface reactances are dispersive across the band, their product remains in the vicinity of one, approximately satisfying the balanced hybrid condition across the K_u-band. The remaining plot in Fig. 2.4 shows the effective refractive index achieved by the metasurface. Note that across the band of interest, the refractive index is below unity; *i.e.*, the metasurface acts as a low-index metamaterial. Moreover, the imaginary component of the refractive index remains on the order of 0.001, indicating minimal intrinsic loss across the K_u-band.

2.3.3 Wire-Grid Metasurface

Although printed circuit technology is ideally suited to K-band horns and nearby frequency bands, lower frequency bands such as C-band may be better served by other structures. One such structure is a wire grid, not unlike Rotman's simulated plasma [24], but operating on different principles. While Rotman's goal was to simulate the volumetric effects of a plasma, we are interested instead in creating a soft (or hard) boundary condition at a surface. Figure 2.5a shows four unit cells of a wire-grid structure that results in a soft boundary condition, analogous to the basic structure of the pixilated soft-surface designs. The unit cells connect in the x-direction, forming a long continuous wire, while the wires are cut in the z-direction. The y-directed wire is 0.19λ and primarily couples with the perpendicular electric fields. The long wire in the x-direction prevents the appearance of electric fields parallel to the surface in that direction. The cut wire in the z-direction is 0.12λ and controls the shape of the dispersion curve, as well as provides end-loading to reduce the thickness of the metamaterial surface. The distance between vertical wires represents the unit cell periodicity and is equal to 0.14λ. The final dimensions were chosen by sweeping the three wire dimensions over a range of values to find the optimum wire lengths. Note that simple metal pins or a "bed of nails," which are commonly used as AMCs in gap waveguides [30, 31], cannot be directly applied to the metamaterial liner as they do not provide sufficient control over the surface impedances and thickness requirements.

Figure 2.5b shows the dispersive surface reactances achieved by the metamaterial structure. Note that the TM surface reactance is significantly lower than zero; this signifies that the metamaterial will force the electric field to zero at its surface, creating the desired tapered field distribution in the horn aperture. Figure 2.5c shows the effective dispersive index of refraction and wave impedance of the metamaterial liner. Figure 2.5d shows the dispersive effective permittivity and permeability of the surface, which are the values that were used for simulations of a horn with a homogeneous isotropic liner. The imaginary components are essentially zero, being on the order of 10^{-4}. This implies that the metamaterial has negligible intrinsic loss. All the results shown in Fig. 2.5 are for a unit cell composed of realistic copper wire elements. The following section details a metamaterial horn antenna based on this wire-grid structure and yielding excellent broadband, low-loss performance.

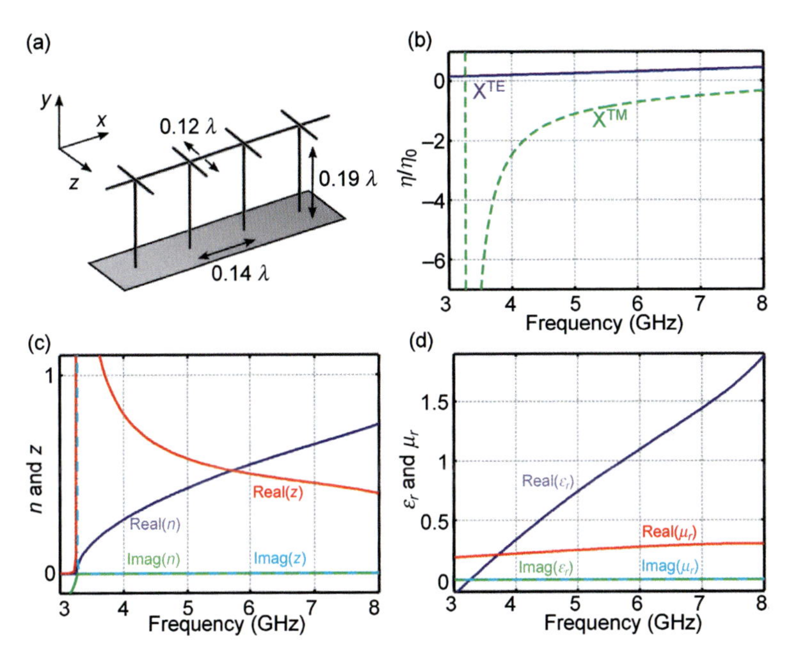

Figure 2.5 Wire-grid metasurface geometry (a), surface reactances (b), effective index of refraction and wave impedance (c), and effective relative permittivity and permeability (d). Note that X^{TM} is significantly less than zero across the band, and that the imaginary components of the effective parameters are essentially zero, being on the order of 10^{-4}, indicating minimal intrinsic loss in the metasurface. Reprinted, with permission, from Ref. 14, Copyright 2013, IEEE.

2.4 Octave-Bandwidth Single-Polarization Horn Antenna with Negligible Loss

2.4.1 Application Background

The super-extended *C*-band presents one of the greatest challenges to traditional low-sidelobe horns, with its comparatively long wavelength requiring large, massive structures. The trifurcated horn does not require excessive mass at *C*-band compared to a simple horn, but its performance is less than optimal. Here, we consider a metamaterial-based rectangular horn that simultaneously maintains both the low-sidelobe and low-backlobe performance of the corrugated horn and the low weight of the trifurcated horn.

Figure 2.6 compares the aperture distributions and three-dimensional radiation patterns for several rectangular horns: an unlined horn, a trifurcated horn, and a metamaterial horn. For an unlined rectangular horn, the significant sidelobes are in the *E*-plane, resulting from the abrupt change in the aperture field magnitude at the *E*-plane walls of the horn, shown in Fig. 2.6a. The *H*-plane sidelobes are very low as a result of the smoothly tapered aperture field magnitude in the *H*-plane. The trifurcated horn creates a stepped aperture distribution in the *E*-plane, resulting in lower primary sidelobes, but comparable backlobes to an unlined horn, shown in Fig. 2.6b. Placing the appropriate metamaterial liner on the *E*-plane walls of the horn creates a smoothly tapered aperture field distribution in the *E*-plane, shown in Fig. 2.6c. With the aperture field magnitude tapered in all planes, the metamaterial horn radiates a pattern with very low sidelobes and backlobes, unlike horns with more uniform field distributions, such as the plain unlined horn or the trifurcated horn.

The horn considered here has a $2.8\lambda \times 2.6\lambda$ (25 × 23 cm) aperture, and a total length of 5.7λ (50 cm), where λ is calculated at the low end of the band. The feed waveguide is WR177. These values lead to a flare angle of 23°. The *H*-plane dimension was chosen to give approximately equal *E*- and *H*-plane patterns, maximizing the efficiency when illuminating a circularly symmetric reflector. The waveguide feed was chosen to be close to cutoff at the low end of the band in order to maximize the operating bandwidth. These dimensions lead to a peak gain ranging from about 16 to 24 dB across the super-extended *C*-band.

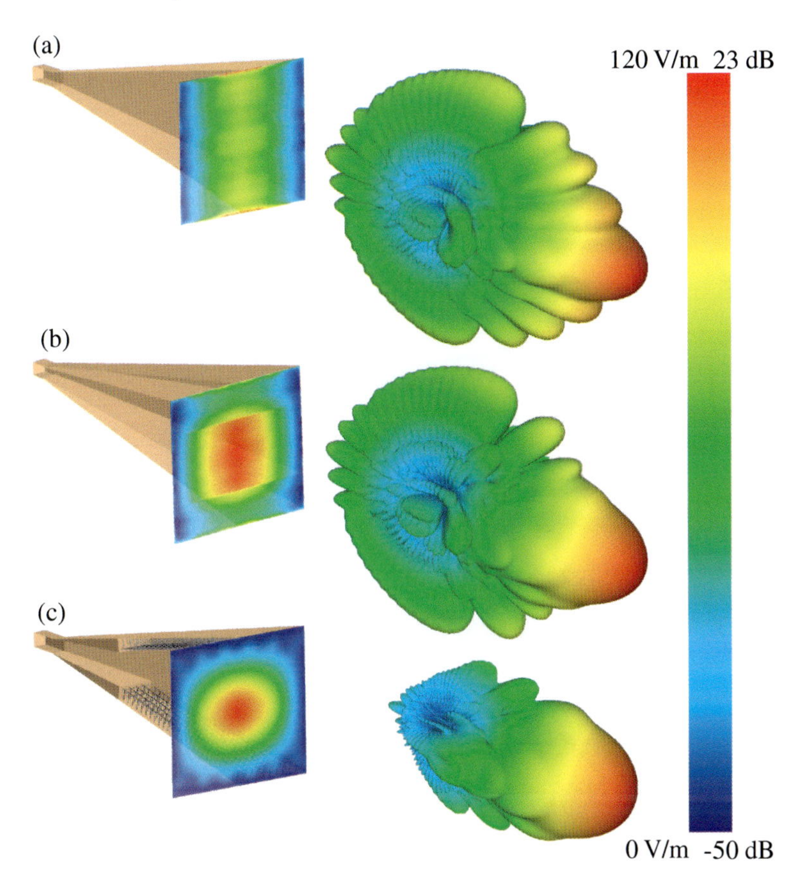

Figure 2.6 Aperture electric field distributions and 3D radiation patterns for the (a) unlined horn, (b) trifurcated horn, and (c) metahorn. Note that the metahorn outperforms both of the alternatives in sidelobe levels and especially in backlobe levels. This performance is comparable to that available from a corrugated horn, but at a fraction of the weight. Reprinted, with permission, from Ref. 14, Copyright 2013, IEEE.

2.4.2 Modeling and Simulation

Preliminary simulations calculated the horn's performance when lined with a homogeneous material having the effective electromagnetic properties of the wire-grid metasurface, as shown in Section 2.3 previously. Figure 2.7 illustrates the horn geometry with the effective homogeneous liner and with the final wire-grid metasurface. The preliminary simulations promised low-sidelobe performance across the band, validating the basic design before undertaking the more complex modeling required for the actual metasurface structure.

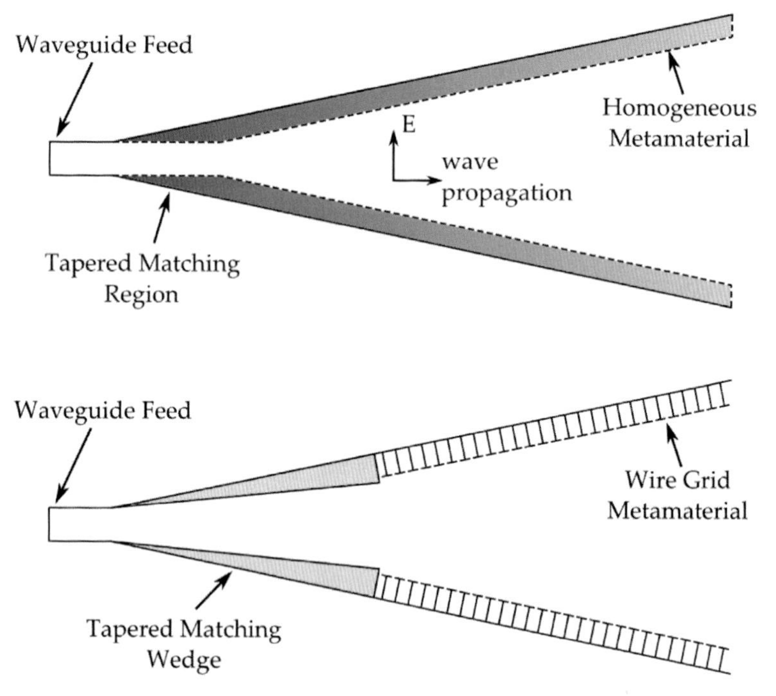

Figure 2.7 Top: Horn geometry as modeled with an effective homogeneous metamaterial liner. Reprinted, with permission, from Ref. 14, Copyright 2013, IEEE. Bottom: Final horn geometry with the actual wire-grid structure and a conductive matching wedge at the horn throat.

Modeling the actual wire grid involved several hurdles that had to be overcome before the metamaterial-lined horn could approach the performance of the horn with the effective homogeneous liner. Each of these iterations gave valuable insights into the performance of the metamaterial inside the horn, so they are included in the discussion here.

The first attempt at placing the wire grid in the horn resulted in very poor antenna performance. To keep from obstructing the waveguide feed (which resulted in all the power being reflected back into the feeding waveguide), the liner was truncated partway up the horn wall. Moreover, the horizontal wires connecting all the unit cells in each row were not connected to the horn walls, a significant change from the assumption that those wires were infinitely long. Both of these factors led to undesired modes and thus high cross-polarization and irregular radiation patterns at some frequencies, as well as unacceptable levels of return loss.

Through successions of later attempts, the horizontal wires were connected to the horn walls. The conducting walls approximate image planes, effectively enabling the wires to appear infinite in length. This resulted in dramatic improvements to the radiation performance. A conducting wedge that functions as an effective matching section was placed between the waveguide and the start of the liner, resulting in a significant improvement in return loss and a moderate improvement in radiation patterns. The return loss becomes a challenge due to the fact that the low-index metamaterial liner increases the cutoff frequency compared to an empty horn, in contrast with regular dielectric loading that decreases the cutoff frequency.

The length of the wedge determined a tradeoff between return loss and radiation performance. To an extent, a longer wedge improved return loss, but raised sidelobe levels because of the shortened liner, reducing the tapering effect on the aperture field distribution. The final wedge was 0.24λ (2.1 cm) tall at the metamaterial end (slightly taller than the wire-grid liner itself). The entire wedge was about 2λ (18 cm) in length. This wedge length, approximately one-third of the total length of the horn, led to reasonable return loss while maintaining good low-sidelobe performance.

The final wire-grid simulations with the horizontal wires connected to the side walls, and the matching wedge placed between the waveguide and the liner produced comparable performance to that predicted by the homogeneous liner simulations. Relative cross-polarization increased compared to the ideal homogeneous liner, but remained below 30 dB across the band. Figure 2.8 shows typical simulated radiation patterns for the horn with the final wire-grid liner at 4.0 and 6.2 GHz.

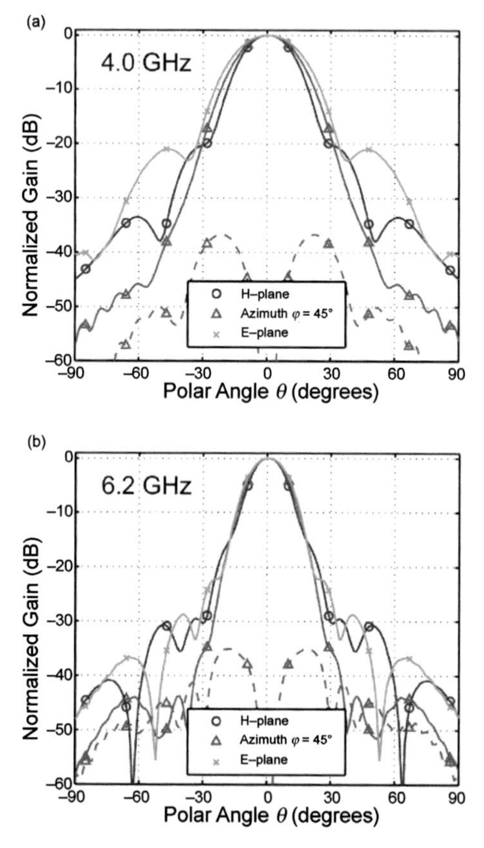

Figure 2.8 Simulated normalized radiation patterns for the metahorn with the final, full wire-grid structure lining its *E*-plane walls. Note that the sidelobes are significantly lower than the −10 dB levels of an unlined horn. Reprinted, with permission, from Ref. 14, Copyright 2013, IEEE.

2.4.3 Prototype and Measurements

Based on the promising results afforded by the simulations, a prototype antenna was constructed. Each row of the wire-grid metasurface was formed from AWG20 copper wire and soldered into two large copper plates. The copper plates were then placed inside a rectangular horn antenna, along with the two matching wedges in the horn throat. Figure 2.9 shows photographs of the prototype. The wire grid exhibited dimensional inaccuracies of several percent, but the overall horn performance is highly tolerant of slight geometrical changes, as evidenced by the measurements that follow.

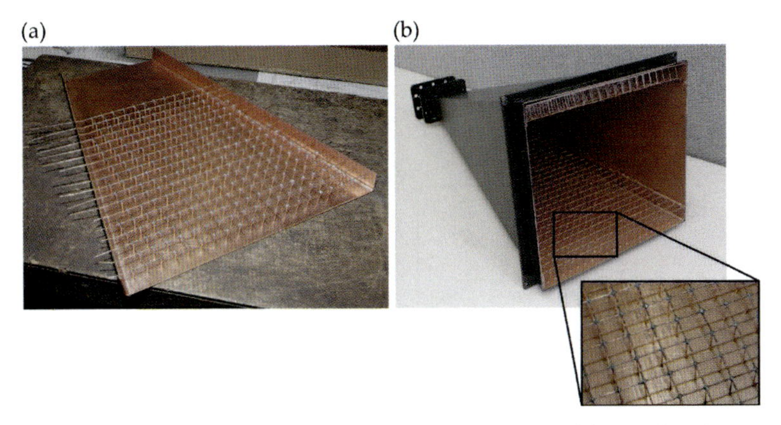

Figure 2.9 Photographs showing (a) fabrication of and (b) completed wire-grid metahorn antenna prototype. Reprinted, with permission, from Ref. 14, Copyright 2013, IEEE.

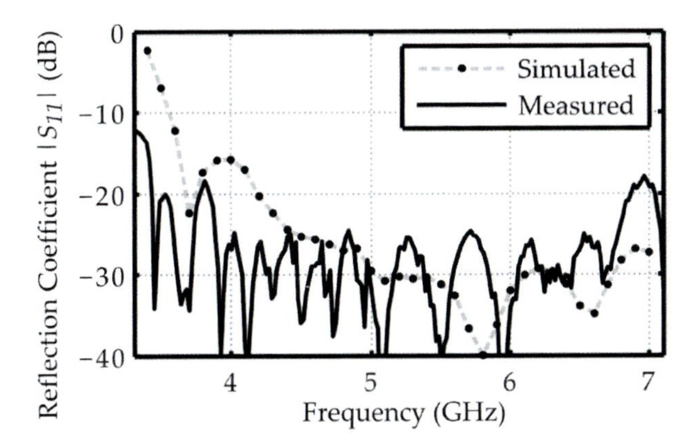

Figure 2.10 Measured and simulated metahorn input reflection coefficient. While the measured data differ from the simulated data as a result of standing wave/measurement artifacts, both show that the horn maintains a low reflection coefficient better than −15 dB across most of the band.

The metahorn's reflection coefficient remained below −15 dB across its operating band, as shown by the measured and simulated data in Fig. 2.10. Pattern measurements, shown in Fig. 2.11, confirmed the excellent performance predicted by the simulations. The measured patterns show some slight asymmetries; these are a result of imperfect manufacturing (hand-soldering) and asymmetries in the antenna measurement range. The antenna was rotated 180° between two sets of pattern measurements to help isolate the asymmetries. Inconsistencies appearing on the solid and dashed curves at the same θ value correspond to asymmetries in the range, while inconsistencies appearing on opposite values of θ correspond to asymmetries in the metahorn itself.

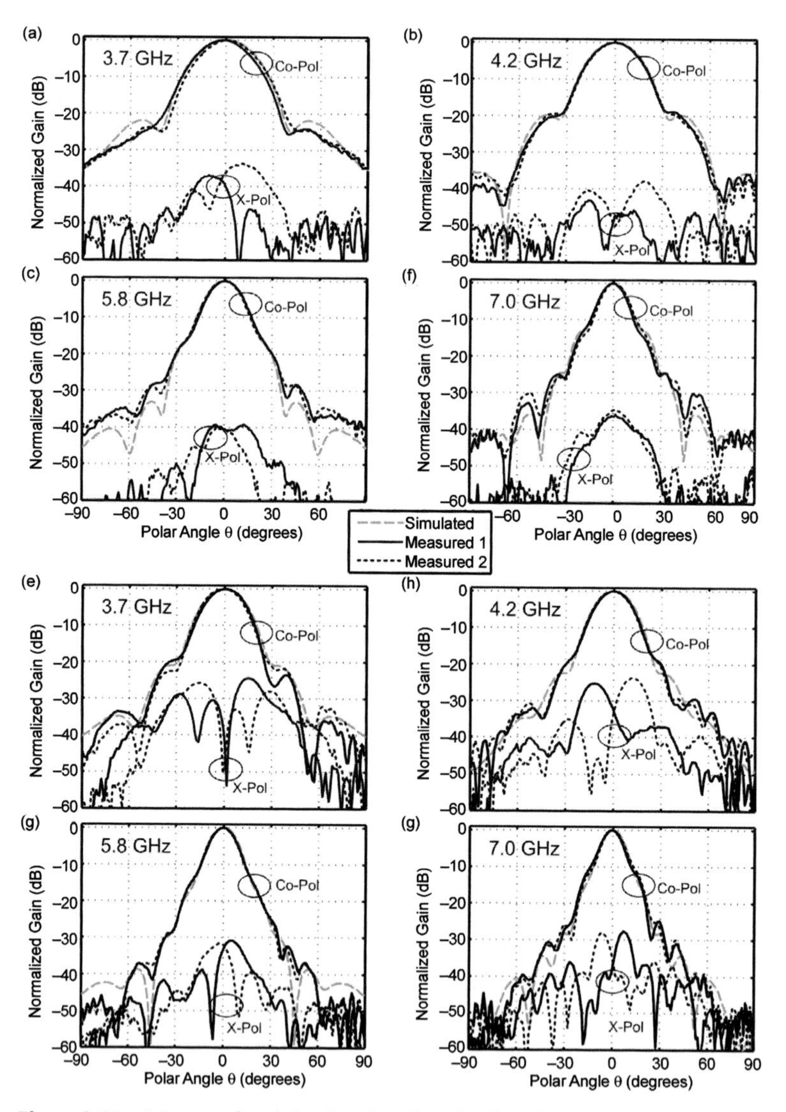

Figure 2.11 Measured and simulated *E*-plane (a–d) and *H*-plane (e–h) radiation patterns for the wire-grid metahorn. The two measurements represent two orientations (rotated 180° relative to each other) of the metahorn to help separate errors in the testing range from errors in the horn itself.

Measurements of the cross-polarized radiation produced values that were somewhat higher than those predicted by simulations, but this is to be expected from an imperfectly manufactured horn. In spite of the imperfections, the peak relative cross-polarization remained around −25 to −30 dB across the operating band. Higher-quality manufacturing would improve these levels further.

The wire-grid metahorn exhibited successful operation with negligible intrinsic loss across more than an octave bandwidth. Compared to the trifurcated horn, which is the state-of-the-art feed horn for single linear polarization at *C*-band, the metahorn yields lower sidelobes and lower backlobes across the operating band. Moreover, this demonstration shows promise for lightweight metamaterial horns to replace circular corrugated feed horns for dual-polarization over the super-extended *C*-band, where such horns are heavy and expensive.

2.5 Dual-Polarization K_u-Band Metamaterial Horn

2.5.1 Application Background

Many satellite communication systems employ dual-polarized K_u-band antennas, including corrugated horns as reflector feeds. K_u-band corrugated horns, while providing excellent performance over bandwidths approaching an octave, are relatively expensive to manufacture with sufficient tolerances, and they have a high weight. Metasurfaces based on PCB manufacturing, however, promise a less expensive and lightweight option while providing comparable sidelobe levels and cross-polarization to corrugated horns.

2.5.2 Modeling and Simulation

Figure 2.12 shows the geometry of the metahorn. The horn is fed by a WR62 waveguide, which connects via a tapered adapter to a 15.8 mm square waveguide, followed by a straight section of square waveguide before connecting to the throat of the metahorn itself. The chosen horn dimensions lead to a gain in the vicinity of 20 dB across the operating band. The metamaterial liner is the printed-patch balanced hybrid metasurface presented earlier.

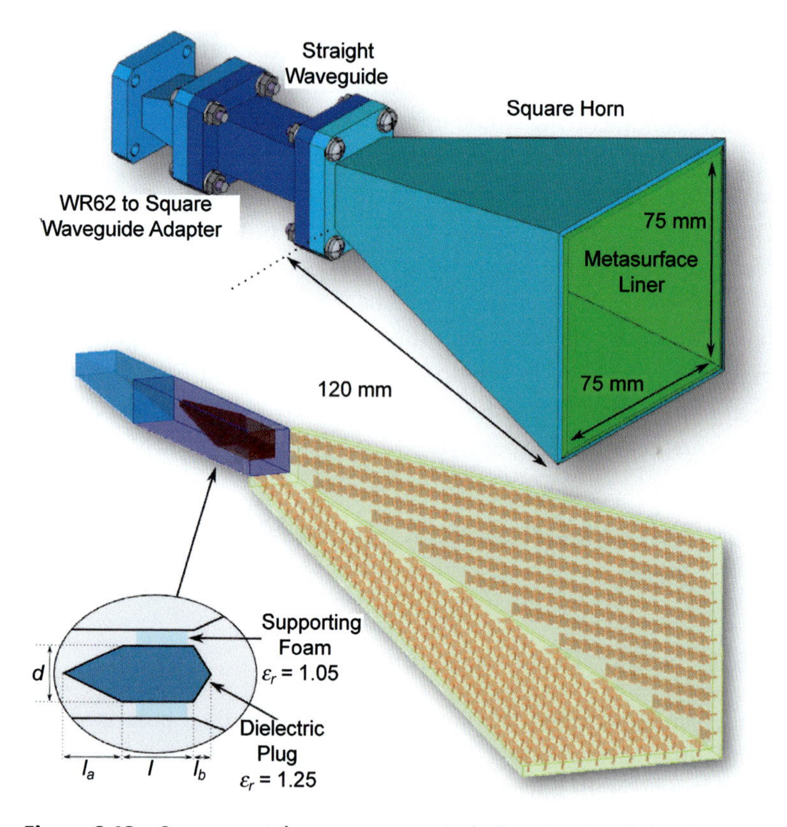

Figure 2.12 Square metahorn geometry, including details of the dielectric mode-matching plug located in the horn throat.

Unlike the wire-grid horn, the metasurface in the dual-polarized square horn must extend all the way to the throat of the antenna to avoid increased cross-polarization levels from the higher-order modes that would be excited. Similar to corrugated horns, the square metahorn then required a mode converter at the horn throat to convert from the TE_{10} fundamental waveguide mode to the desired hybrid mode. Corrugated horns achieve mode conversion through a sequence of grooves (corrugations) of varying depth in the waveguide and horn throat. As corrugations would negate the potential cost and weight savings of the metasurfaces, an alternative approach had to be found. Analogous to dielectric core horns, a dielectric plug was created, but terminated at the horn throat, as shown in Fig. 2.12. Typical dimensions of the plug were $l = 0.6\lambda$, $d = 0.4\lambda$, $l_a = 0.6\lambda$, and $l_b = 0.16\lambda$. As evidenced by later measurements, this dielectric plug proved to provide an effective conversion mechanism for the waveguide modes, leading to low cross-polarization across the entire K_u-band.

Because of the dielectric materials involved, a high-frequency structure simulator (HFSS) proved to be the suitable choice for simulating the horn's performance. As with the wire-grid horn, symmetry allowed the size of the structure to be reduced to a quarter of the full horn. The quarter-metahorn was simulated, lined with the metamaterial structure, as shown in Fig. 2.12. The metasurface liner consisted of four trapezoidal panels fitting the walls of the horn. Partial unit cells were removed from the edges for manufacturability; their absence did not significantly affect the horn's performance because the electromagnetic fields are minimal in the corners of a soft horn.

2.5.3 Prototype and Measurements

Figure 2.13 shows photographs of the fabricated prototype. The horn and metasurfaces were constructed separately, and then the four trapezoidal sections of metasurfaces were placed inside the horn. The metasurfaces were fabricated using standard PCB techniques, including vias. The dielectric mode-matching plug was mounted in the section of straight waveguide behind the horn throat. Alternatively, a metahorn could be constructed almost solely of the metasurface PCBs, as the ground plane layer of the PCBs could serve as the horn walls. The only potential issues are the mounting flange and the connections between the four PCB walls, depending on the requirements of the specific application. This approach would further reduce the weight and cost of the metahorn antenna.

Figure 2.14 shows the measured radiation patterns for the metahorn at 12, 14, and 17 GHz, compared with the patterns for an unlined horn. The low sidelobes and negligible backlobes indicate that the metasurface indeed tailors the horn's field distributions appropriately for a soft surface, in contrast with the high sidelobes and backlobes exhibited by the unlined horn. Moreover, the main beams from the metahorn are nearly rotationally symmetric, *i.e.*, the metahorn exhibits (approximately) a polarization-independent pattern—an ideal characteristic for dual-polarized communication systems with circularly symmetric reflectors.

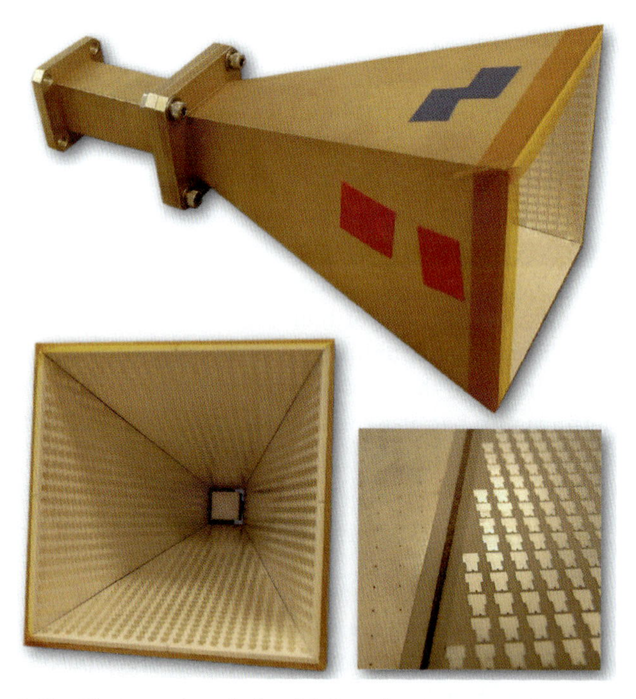

Figure 2.13 Photographs of the fabricated square metahorn prototype, together with a close-up of the front and back of its metasurface.

Figure 2.15 shows detailed comparisons between the measured and simulated metahorn radiation patterns. Again we see the nearly identical main beams regardless of the ϕ-plane. The patterns show some slight asymmetries, but these are likely artifacts of the measurement range rather than due to higher-order modes arising from imperfections in the horn. Cross-polarization is only apparent for the $\phi = 45°$ plane, as expected. At 12 and 14 GHz, the cross-polarization remains below approximately −30 dB, but it creeps up around −25 dB at 17 GHz. This increase results from the excitation of undesired higher-order modes in the waveguide and/or horn. An improved mode converter would likely lower these levels.

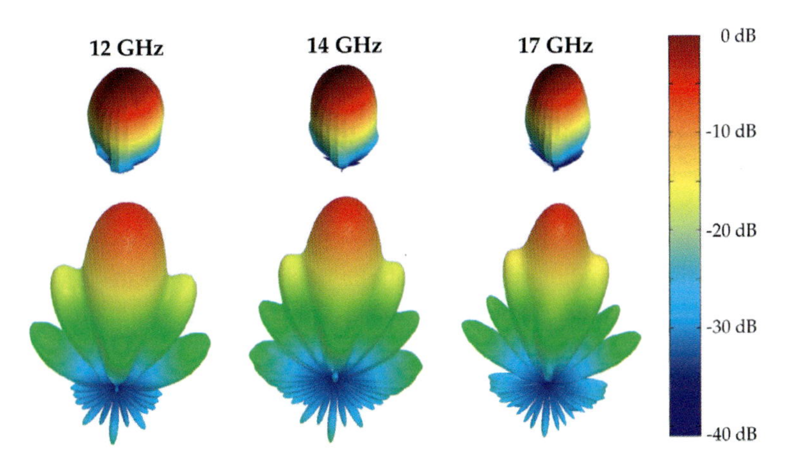

Figure 2.14 Three-dimensional normalized radiation patterns for the measured metahorn (top) and simulated unlined horn (bottom).

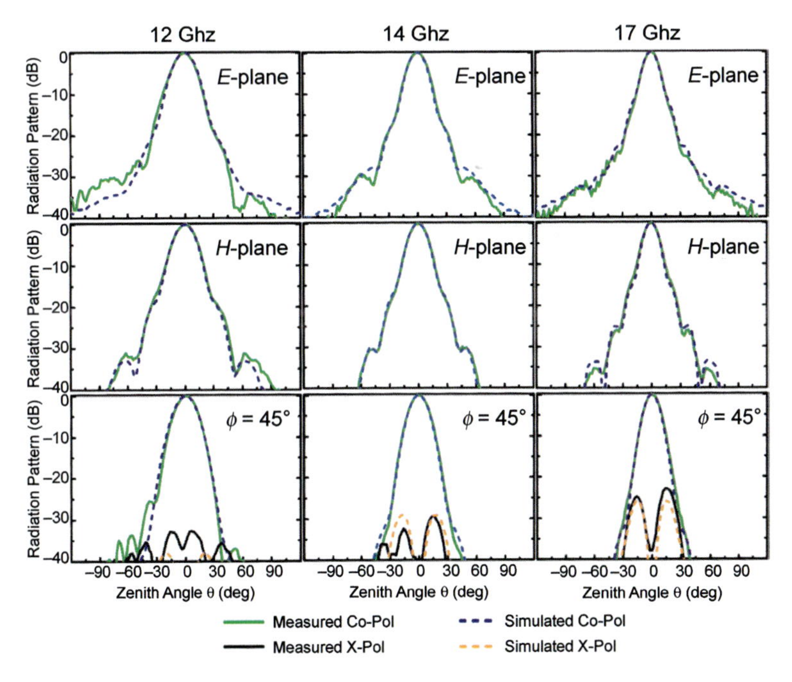

Figure 2.15 Simulated and measured metahorn radiation patterns in principal planes at 12, 14, and 17 GHz. Note that the main beam patterns are approximately identical for all planes at a given frequency, producing radiation patterns that are nearly independent of polarization.

Figure 2.16 shows simulated and measured metahorn performance across the K_u-band. Measurements and simulations agree reasonably well. The metasurface exhibits a resonance just below 11.5 GHz, which leads to a corresponding peak in the S_{11} curves. This resonance frequency is most likely the cutoff frequency in the horn throat, since the low-index metamaterial increases the cutoff frequency compared to that of an empty horn. Above that point and throughout the K_u-band, the metasurface operates on the tail of its resonance, which leads to better return losses and minimal absorption loss.

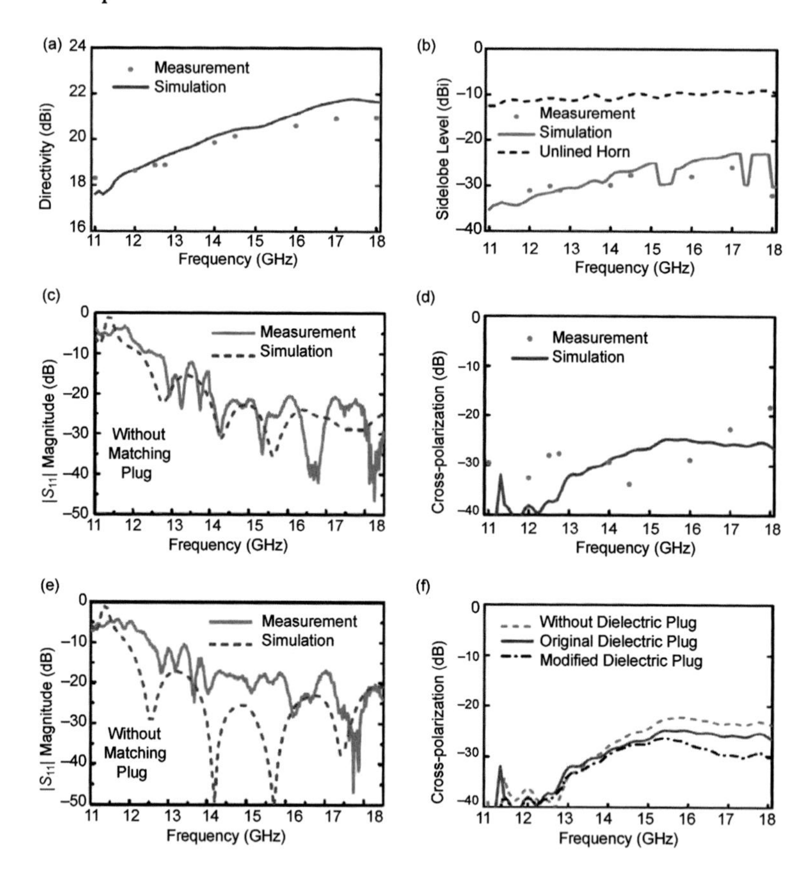

Figure 2.16 Summary simulated and measured performance of the square metahorn: directivity (a), sidelobe level (b), input reflection coefficient without (c) and with (e) the dielectric plug, cross-polarization measurements versus simulations (d), and simulated dielectric plug comparisons (f).

It is apparent from Fig. 2.16 that the dielectric mode converter is necessary to reduce cross-polarization levels. Moreover, modifying the plug design from its initial geometry to improve its mode conversion operation significantly improved cross-polarization levels at the higher frequencies, visible by the black dot-dashed curve in Fig. 2.16f, where the cross-polarization remains on the order of −30 dB across the K_u-band. Conveniently, the region around 15–16 GHz where the cross-polarization is highest lies outside the typical transmit and receive bands used for satellite applications.

2.6 Improved-Performance Horn Enabled by Inhomogeneous Metasurfaces

2.6.1 Motivation and Rationale

The dual-polarized square horn in the previous section is eminently practical for fabrication, but a metahorn with a circular profile could lead to even broader operating bandwidths. This behavior results from the corner areas of rectangular horns where the balanced hybrid condition cannot be met, causing differences in the higher-order modes supported by square versus circular waveguides. This section, while remaining as practical as possible, focuses on what would be necessary to maximize the potential of metahorn technology.

The unit cell geometry employed for this section is related to the wire-grid structure shown previously, but it is intended for printed circuit manufacturing like the pixilated unit cell designs. Figure 2.17a shows the unit cell structure, with a patch serving as an end-load above a conducting via and a continuous strip of copper connecting all unit cells in the x-direction. The rectangular patch width and length provide a simple means for adjusting the metasurface properties depending on position within the horn, analogously to corrugations of varying depth acting as mode converters in conventional corrugated horns. The unit cell periodicity and thickness were fixed to be p = 3 mm and t = 5.2 mm, respectively, leading to operation between approximately 10 GHz and 20 GHz, including the K_u-band. The transverse connecting wire was set at a width of 0.4 mm. Fixing these parameters allows for more practical manufacturing and application to a horn antenna's interior.

2.6.2 Effects of Parameter Variations on Metasurface Characteristics

Figures 2.17b,c show how the anisotropic surface reactances change as the patch length and width change. In general, the frequency response of the surface reactances shifts to lower frequencies when either the patch width or length is increased, with length having a stronger effect. In all cases, X^{TE} remains nearly zero, while X^{TM} varies from a large negative value toward zero. These characteristic curves provide insight for the design of metasurfaces optimized for a specific frequency band.

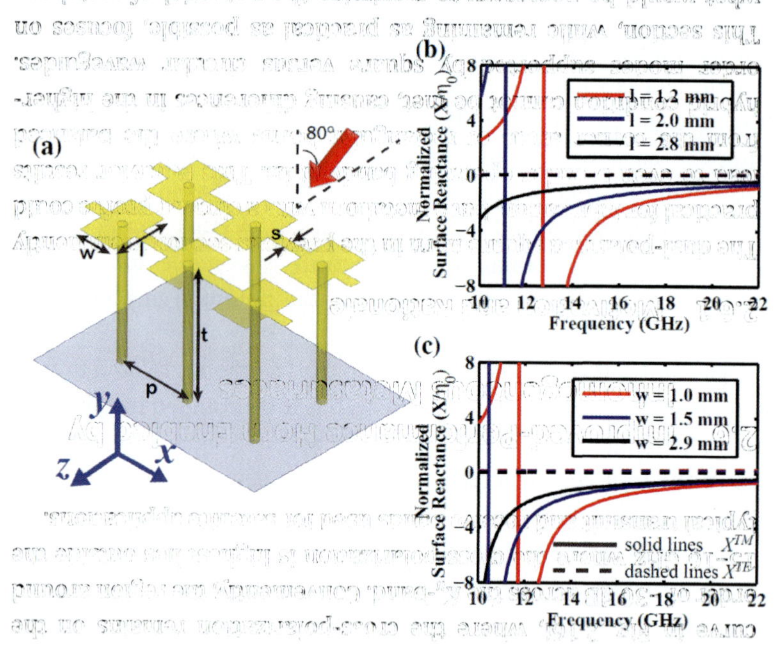

Figure 2.17 Unit cell geometry (a) and surface reactances for varying patch length (b) and width (c). For (b), the width was fixed at 2.0 mm, and for (c), the length was fixed at 2.3 mm. In all cases, $t = 5.2$ mm, $s = 0.4$ mm, and $p = 3.0$ mm. Reprinted, with permission, from Ref. 16, Copyright 2013, IEEE.

2.6.3 Metasurfaces in Cylindrical Waveguides

Figure 2.18a shows a section of cylindrical waveguide lined with the metasurface under consideration. Importantly, the field distributions in a hybrid-mode cylindrical waveguide are determined by the anisotropic surface impedances on the waveguide walls [32]. Of interest here is the fact that the aperture field magnitude becomes independent of azimuth angle and no cross-polarized field exists when the following relationship is satisfied:

$$\frac{X^{\mathrm{TE}}}{\eta_0}+\frac{\eta_0}{X^{\mathrm{TM}}}=0 \tag{2.18}$$

where η_0 is the wave impedance in a vacuum. This relationship can be satisfied either by $X^{\mathrm{TE}} \times X^{\mathrm{TM}} = -\eta_0^2$, or with both terms on the left of Eq. (2.18) being zero. The former is the balanced hybrid condition presented earlier, which can be satisfied by dielectric-loaded horns [7, 8, 31]. The latter is the definition of a soft surface and can be realized by appropriately sized corrugations or by many of the metasurfaces presented in this chapter. The metasurface of Fig. 2.17 provides a soft surface with an easily designed operating frequency band, and proved an ideal candidate for creating hybrid modes in a metasurface-lined circular waveguide. The connecting strip of width s becomes a circular ring, while the vias connect from the patches radially outward to the waveguide wall. The eigenmode solver in HFSS was used to create the dispersion diagrams and mode patterns shown in Figs. 2.18b,c. These characteristics of the metasurface provide more insight into how the surface behaves in the context of a cylindrical waveguide or horn than is possible with plane wave models used for the previously presented horn designs. The calculated mode patterns verified the operation of the metasurface as a soft surface, exhibiting tapered field distributions and nearly perfect linearly polarized electric fields. Also of note is the fact that metasurfaces with smaller patches lead to a higher cutoff frequency and more confined mode patterns, while larger patches lower the cutoff frequency further.

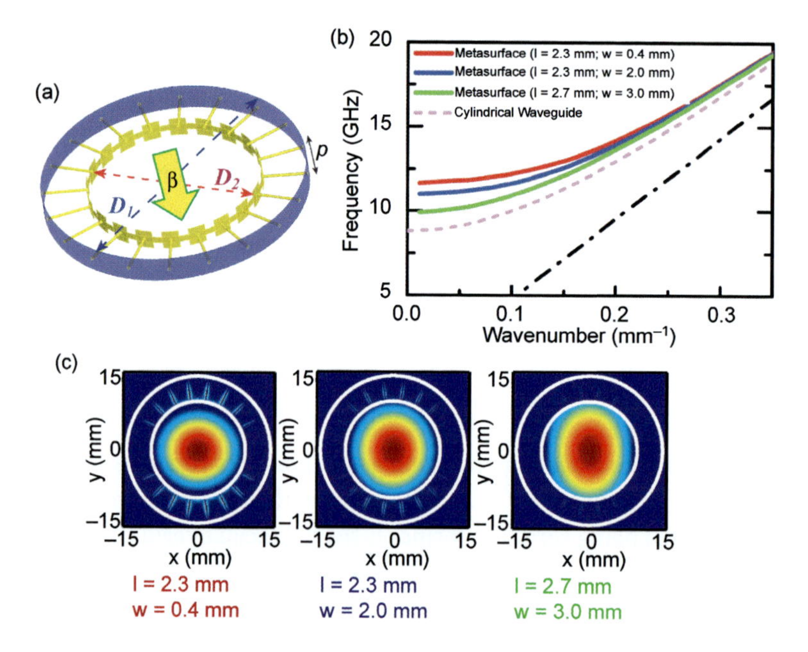

Figure 2.18 (a) Schematic of a section of metasurface lining a cylindrical waveguide. D_1 = 30.4 mm and D_2 = 20.0 mm. (b) Dispersion diagram for metasurfaces with varying dimensions. The dispersion of a cylindrical waveguide with an inner diameter of 20.0 mm is shown for reference, and the plane wave behavior is indicated by the dash-dotted line. (c) Electric field mode patterns for the three metasurface-lined waveguides at 12 GHz. Reprinted, with permission, from Ref. 16, Copyright 2013, IEEE.

2.6.4 Comparison of Metahorns with Homogeneous and Inhomogeneous Metasurfaces

Lining a cylindrical horn antenna with a homogeneous metasurface yielded the geometry and simulated performance shown in Figs. 2.19a–c. The horn's bandwidth was limited by the cutoff frequency at the low end and by the excitation of higher-order modes and thus undesired cross-polarization at the upper end of the band, both depending on the exact metasurface patch dimensions. Larger patches lowered the cutoff frequency, but also lowered the frequency at which the higher-order modes appeared and rose above the tolerated level of −30 dB.

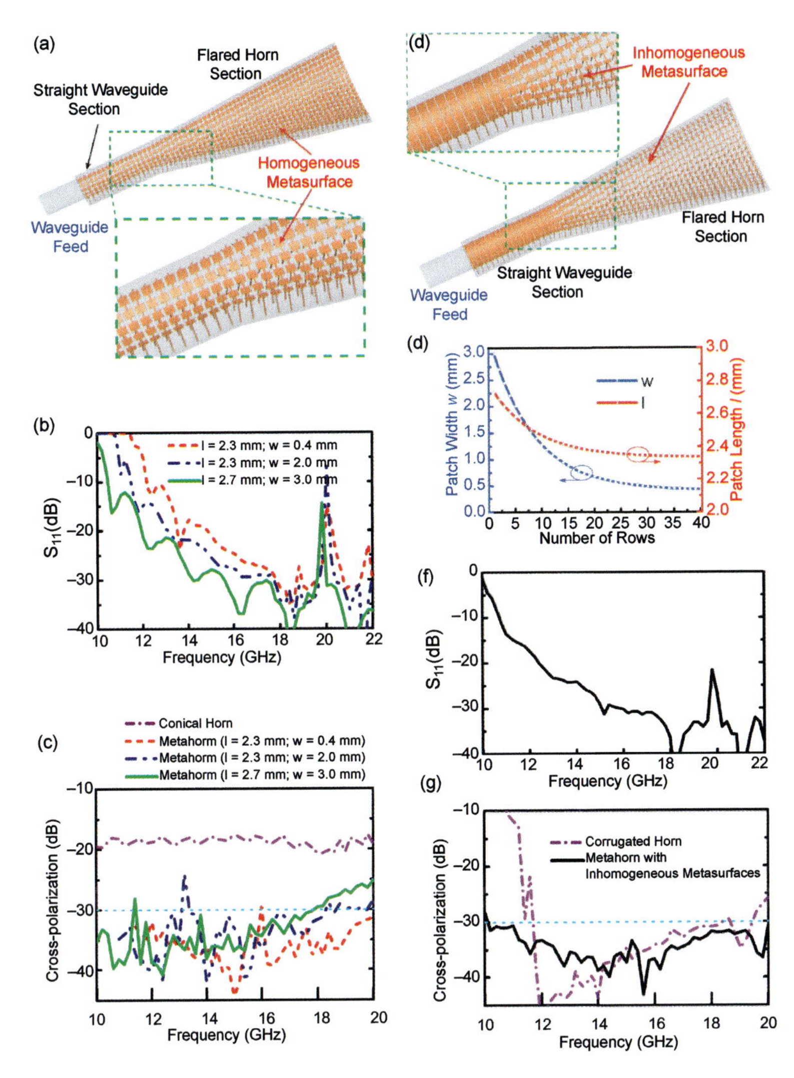

Figure 2.19 (a) Rendering of horn with a homogeneous metasurface liner and simulated S_{11} (b) and cross-polarization (c) for various metasurface dimensions. (d) Rendering of horn with an inhomogeneous metasurface liner. (e) Exponential curves defining the metasurface patch dimensions' dependence on position in the horn. Simulated S_{11} (f) and cross-polarization (g) for the inhomogeneous metahorn. Reprinted, with permission, from Ref. 16, Copyright 2013, IEEE.

Creating a horn with an inhomogeneous metasurface, as shown in Fig. 2.19d, allowed larger patches in the waveguide and horn throat to lower the waveguide cutoff frequency below 10 GHz, while smaller patches further out on the horn walls prevented higher-order modes from being excited significantly below 20 GHz, as visible in Figs. 2.19f–g. Rather than having a discrete transition between patch sizes, the exponential curves in Fig. 2.19e smoothly tapered the patch sizes to minimize any higher-order modes that would be excited by a metasurface transition.

For comparison, the cross-polarization of a comparable corrugated horn is also shown in Fig. 2.19g. Although the corrugated horn has the best cross-polarization level over a narrow frequency range at the low end of the K_u-band, the metahorn exhibits comparable or better cross-polarization over a much broader frequency range, including the entirety of the K_u-band.

Similar to corrugated horns, other important considerations for the design of metasurface-based hybrid-mode horns include the horn flare angle and the tapering profile of the metasurface unit cells. Numerical studies of these parameters and their effects on the horn performance showed that, like the corrugated horn, the flare angle has a significant effect on the horn's reflection and radiation properties, as shown in Figs. 2.20a,b. The tapered profile of the inhomogeneous metasurface patch dimensions also has a significant effect. Rapid tapering from the large patches to the small patches creates a stronger discontinuity that can excite higher-order modes and thus increase cross-polarization levels, while tapering that is too slow fails to suppress higher-order modes properly. In spite of these effects, it is readily apparent that the proper choice of flare angles and tapering profiles can lead to a metasurface-lined horn that performs comparably or better than the corrugated horn across a very broad bandwidth. As these metasurfaces operate on the tails of resonances and exhibit minimal intrinsic losses, the metahorn antennas have negligible added losses due to the metasurface liners.

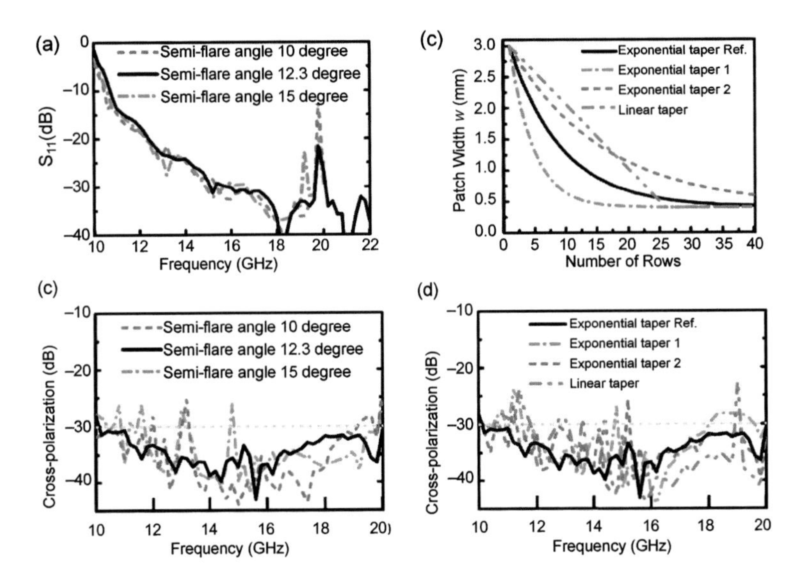

Figure 2.20 Input reflection coefficient (a) and relative cross-polarization (b) for metahorns with various semi-flare angles. Metasurface patch tapering profiles (c) and their corresponding effects on the horn's cross-polarization (d). Reprinted, with permission, from Ref. 16, Copyright 2013, IEEE.

2.7 Summary and Conclusions

Broadband, low-loss metamaterial surfaces provide an excellent alternative to conventional enhancements for horn feeds. Depending on the frequency band and exact application requirements, metasurface-enabled horns can reduce weight, increase bandwidth, or lower manufacturing costs. Properly engineering the dispersive characteristics of the metasurfaces can enable horns that operate over an octave bandwidth or more with negligible intrinsic losses, all while providing the benefits of conventional soft horns. For optimal performance, inhomogeneous metasurfaces can be applied as horn liners to support balanced hybrid modes and create polarization-independent patterns suitable for dual-polarized communication systems such as those commonly found on satellites.

Acknowledgments

Several others besides the authors helped to make this work possible. Notably, Micah Gregory and Xiande Wang provided support for the early metamaterial simulations, and Robert Shaw assisted in the development of the metamaterial horn concepts.

References

1. A. J. Simmons and A. F. Kay, "The scalar feed-A high performance feed for large paraboloid reflectors," *Design and Construction of Large Steerable Aerials*, pp. 213–217, 1966.
2. H. Minnett and B. Thomas, "A method of synthesizing radiation patterns with axial symmetry," *IEEE Transactions on Antennas and Propagation*, vol. 14, no. 5, pp. 654–656, 1966.
3. E. Lier and P.-S. Kildal, "Soft and hard horn antennas," *IEEE Transactions on Antennas and Propagation*, vol. 36, no. 8, pp. 1152–1157, 1988.
4. E. Lier and A. Kishk, "A new class of dielectric-loaded hybrid-mode horn antennas with selective gain: design and analysis by single mode model and method of moments," *IEEE Transactions on Antennas and Propagation*, vol. 53, no. 1, pp. 125–138, 2005 (Special issue on soft and hard surfaces).
5. A. F. Kay, *The Scalar Feed*. AFCRL Rep. 64–347, AD601609, 1964.
6. V. Rumsey, "Horn antennas with uniform power patterns around their axes," *IEEE Transactions on Antennas and Propagation*, vol. 14, no. 5, pp. 656–658, 1966.
7. H. E. Bartlett and R. E. Moseley, "Dielguides—Highly efficient low noise antenna feeds," *Microwave Journal*, vol. 9, pp. 53–58, 1966.
8. P. J. B. Clarricoats, A. D. Olver, and M. Rizk, "A dielectric loaded conical feed with low cross-polar radiation," *Proceedings of URSI Symposium on Electromagnetic Theory, Spain*, pp. 351–354, 1983.
9. E. Lier and J. A. Aas, "Simple hybrid mode horn feed loaded with a dielectric cone," *Electronics Letters*, vol. 21, no. 13, pp. 563–564, 1985.
10. E. Lier and T. Schaug-Pettersen, "The strip-loaded hybrid-mode feed horn," *IEEE Transactions on Antennas and Propagation*, vol. 35, no. 9, pp. 1086–1089, 1987.
11. E. Lier and R. K. Shaw, "Design and simulation of metamaterial-based hybrid-mode horn antennas," *Electronics Letters*, vol. 44, no. 25, pp. 1444–1445, 2008.

12. E. Lier, D. H. Werner, C. P. Scarborough, Q. Wu, and J. A. Bossard, "An octave-bandwidth negligible-loss radiofrequency metamaterial," *Nature Materials*, vol. 10, no. 3, pp. 216–222, 2011.

13. Q. Wu, C. P. Scarborough, D. H. Werner, E. Lier, and X. Wang, "Design synthesis of metasurfaces for broadband hybrid-mode horn antennas with enhanced radiation pattern and polarization characteristics," *IEEE Transactions on Antennas and Propagation*, vol. 60, no. 8, pp. 3594–3604, 2012.

14. C. P. Scarborough, Q. Wu, D. H. Werner, E. Lier, R. K. Shaw, and B. G. Martin, "Demonstration of an octave-bandwidth negligible-loss metamaterial horn antenna for satellite applications," *IEEE Transactions on Antennas and Propagation*, vol. 61, no. 3, pp. 1081–1088, 2013.

15. Q. Wu, C. P. Scarborough, B. G. Martin, R. K. Shaw, D. H. Werner, E. Lier, and X. Wang, "A K_u-band dual polarization hybrid-mode horn antenna enabled by printed-circuit-board metasurfaces," *IEEE Transactions on Antennas and Propagation*, vol. 61, no. 3, pp. 1089–1098, 2013.

16. Q. Wu, C. P. Scarborough, D. H. Werner, E. Lier, and R. K. Shaw, "Inhomogeneous metasurfaces with engineered dispersion for broadband hybrid-mode horn antennas," *IEEE Transactions on Antennas and Propagation*, vol. 61, no. 10, pp. 4947–4956, 2013.

17. E. Lier, "Soft and hard horn antennas," Chapter 6 (*Feed Systems*), *Handbook of Reflector Antennas and Feed Systems*, vol. 2, Artech House, 2013.

18. D. Sievenpiper, L. Zhang, R. F. J. Broas, N. G. Alexopolous, and E. Yablonovitch, "High-impedance electromagnetic surfaces with a forbidden frequency band," *IEEE Transactions on Microwave Theory and Techniques*, vol. 47, no. 11, pp. 2059–2074, 1999.

19. F. Yang and Y. Rahmat-Samii, *Electromagnetic Band Gap Structures in Antenna Engineering*. Cambridge University Press, Cambridge, 2009.

20. P.-S. Kildal, A. A. Kishk, and S. Maci, "Special issue on artificial magnetic conductors, soft/hard surfaces, and other complex surfaces," *IEEE Transactions on Antennas and Propagation*, vol. 53, no. 1, pp. 2–7, 2005.

21. D. J. Kern, D. H. Werner, A. Monorchio, L. Lanuzza, and M. J. Wilhelm, "The design synthesis of multiband artificial magnetic conductors using high impedance frequency selective surfaces," *IEEE Transactions on Antennas and Propagation*, vol. 53, no. 1, pp. 8– 17, 2005.

22. B. A. Munk, *Frequency Selective Surfaces: Theory and Design*. John Wiley & Sons, Inc., 2000.

23. A. Fallahi, M. Mishrikey, C. Hafner, and R. Vahldieck, "Efficient procedures for the optimization of frequency selective surfaces," *IEEE Transactions on Antennas and Propagation*, vol. 56, no. 5, pp. 1340–1349, 2008.
24. W. Rotman, "Plasma simulation by artificial dielectrics and parallel-plate media," *IRE Transactions on Antennas and Propagation*, vol. 10, no. 1, pp. 82–95, 1962.
25. J. B. Pendry, A. J. Holden, D. J. Robbins, and W. J. Stewart, "Low frequency plasmons in thin-wire structures," *Journal of Physics: Condensed Matter*, vol. 10, no. 22, p. 4785, 1998.
26. N. Engheta and R. W. Ziolkowski, *Metamaterials: Physics and Engineering Explorations.* John Wiley & Sons, 2006.
27. R. L. Haupt and D. H. Werner, *Genetic Algorithms in Electromagnetics.* John Wiley & Sons, 2007.
28. D. Hadka and P. Reed, "Borg: An auto-adaptive many-objective evolutionary computing framework," *Evolutionary Computation*, vol. 21, no. 2, pp. 231–259, 2013.
29. *PFSS.* E x H, Inc. (www.exhengineering.com), 2015.
30. P.-S. Kildal, A. U. Zaman, E. Rajo-Iglesias, E. Alfonso, and A. Valero-Nogueira, "Design and experimental verification of ridge gap waveguide in bed of nails for parallel-plate mode suppression," *IET Microwaves, Antennas Propagation*, vol. 5, no. 3, pp. 262–270, 2011.
31. A. Polemi, S. Maci, and P.-S. Kildal, "Dispersion characteristics of a metamaterial-based parallel-plate ridge gap waveguide realized by bed of nails," *IEEE Transactions on Antennas and Propagation*, vol. 59, no. 3, pp. 904–913, 2011.
32. P. J. B. Clarricoats and A. D. Olver, *Corrugated Horns for Microwave Antennas.* IET, 1984.

Chapter 3

Realization of Slow Wave Phenomena Using Coupled Transmission Lines and Their Application to Antennas and Vacuum Electronics

Md. R. Zuboraj and John L. Volakis
ElectroScience Laboratory, Department of Electrical and Computer Engineering, The Ohio State University, Columbus, OH 43212, USA
zuboraj.1@buckeyemail.osu.edu; volakis.1@osu.edu

Over the past decade, slow waves, backward traveling waves, and other exotic propagation phenomena has been extensively used to modify and control the radiation and coupling properties of microwave and RF devices. Example applications include the development of miniature antennas, microwave filers and couplers, traveling wave tubes (TWTs), and backward wave oscillators. Typically, designs are carried out using a variety of computational tools, but with limited reference to the dispersion diagrams that control the radiation and propagation properties of these devices. For the first time,

Broadband Metamaterials in Electromagnetics: Technology and Applications
Edited by Douglas H. Werner
Copyright © 2017 Pan Stanford Publishing Pte. Ltd.
ISBN 978-981-4745-68-0 (Hardcover), 978-1-315-36443-8 (eBook)
www.panstanford.com

in this chapter, we endeavor to provide an understanding of how exotic propagation phenomena can be explained and controlled using dispersion engineering. More importantly, these novel propagation modes are generated using a set of coupled transmission lines that can be printed on simple substrates. By controlling the propagation constant on each line, as well as their mutual capacitance and inductance, it is shown that dispersion diagrams of second, third, and fourth order can be realized. In turn, this leads to transmission lines that support various exotic modes. These modes are subsequently used to (i) improve the electronic efficiency of TWTs, (ii) miniaturize antennas, reaching their optimal limits, (iii) increase antenna directivity, and (iv) control antenna bandwidth. Examples of the latter are provided.

3.1 Introduction

Slow wave propagation has been of strong interest to electromagnetics because of its exotic properties that can lead to (i) small antennas, filters, and RF circuits; (ii) novel functionalities, and (iii) realize high-power RF and optical devices. Among some already demonstrated applications are (i) intense light source [1, 2], (ii) miniaturized antennas [3] with improved directivity, and (iii) high-power microwaves such as oscillators and amplifiers [4]. The slow wave properties are realized using periodic layered media that exhibit effective anisotropic structure. Alternately, slow waves can be formed using coupled transmission lines (CTLs). In this case, one of the transmission lines propagates a strong forward wave, whereas the other adjacent line supports a weak backward wave [5].

To understand the physics of slow waves, this chapter provides an analysis of CTLs (Fig. 3.1) that are loaded with (L,C) elements in a periodic manner. As is well-known, any periodic loading supports Bloch waves [6] and exhibits passband and stopband behaviors. In this chapter, it is shown that the usual dispersion relation of these Bloch waves is of second order and can achieve nearly zero group velocity at the band edge. However, by changing the material properties or loading of the CTLs, as in Fig. 3.1a, the order of the dispersion curve can be altered, a process often called dispersion engineering. Specifically, higher-order dispersion curves (third or fourth) can be attained. Indeed, the use of anisotropic material in periodic stacks, instead of isotropic ones, can lead to third- or fourth-order dispersion relation [7] (Fig. 3.2). The third-order dispersion is achievable with magnetic photonic crystals (MPC). On the other hand, fourth-order dispersion relation leads to maximally flat curves. But this maximally flat behavior is difficult to achieve and leads to degenerate band edge (DBE) modes that are inherently narrowband phenomena. Of importance is the fact that these slow wave modes provide for large field enhancement as the group velocity nearly drops to zero. This reduction in group velocity also implies device miniaturization [8]. Specifically, the DBE and MPC properties were applied to enhance the directivity of antennas [3, 9, 11, 12]. But as noted, DBE resonances are associated with very narrow bandwidth. Wider bandwidth is typically desired, and can be achieved by introducing magnetic anisotropy into the periodic layers. By doing so, a third-order dispersion relation was achieved [7], which is associated with an inflection point within the Bloch diagrams. Such a third-order relation provides for more bandwidth [10]. The associated modes are referred to as MPCs [11, 12] and have led to highly directive, wideband, and miniature antennas. DBE and MPC modes were also realized and demonstrated by fabricating periodic volumetric stacks of dielectrics [13]. But these material layers are bulks. Therefore, there is interest to develop printed or wire-type versions of MPC structures for more practical applications.

An important development in realizing the DBE and MPC modes came from Locker et al. [14], who proposed an equivalent printed form of DBE and MPC using printed microstrip lines. The microstrip lines are CTLs and depended on the substrate to emulate the behavior of volumetric DBE and MPC [14, 15]. In this chapter, a simple theoretical analysis of coupled transmission lines is presented. This is done first for the second-order dispersion curves and is then expanded to higher-order dispersion engineering. The chapter closes with the description of several applications.

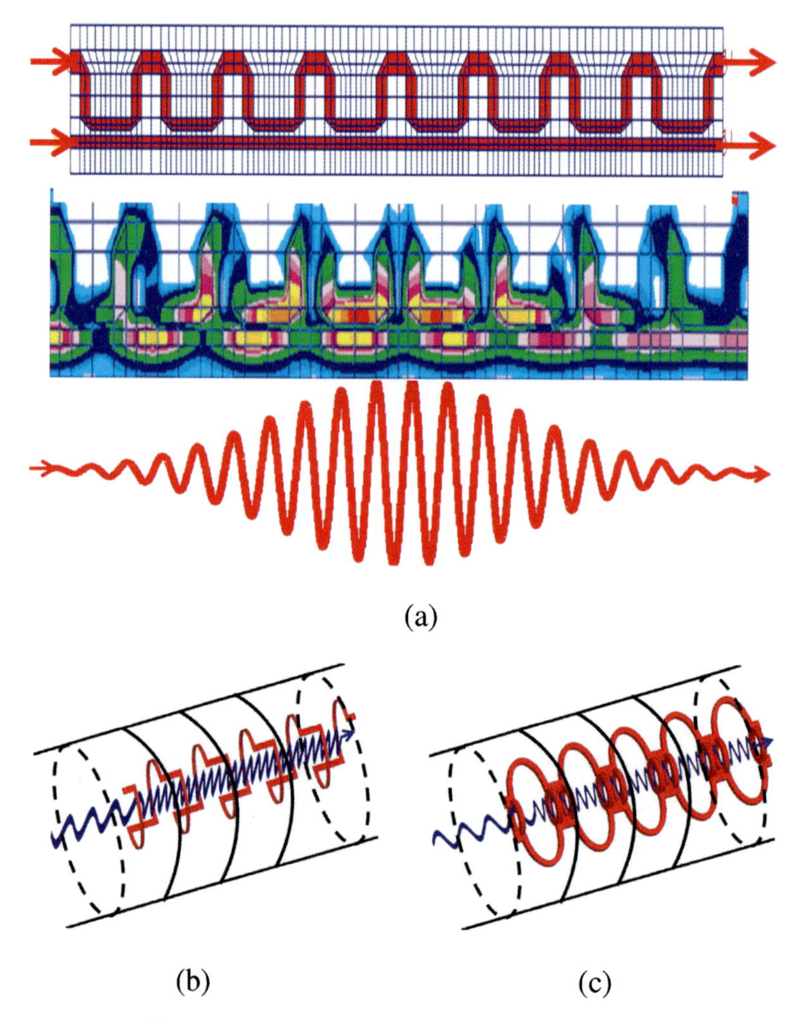

Figure 3.1 (a) Top: printed coupled microstrip lines to realize the frozen mode (finite length); middle: field strength within the coupled transmission lines showing strong field intensities in the middle due to wave velocity slowdown; bottom: wave amplitude representing field growth. (b) Slow wave propagation within a helical wire structure placed in a waveguide. (c) Slow wave propagation within a curved ring-bar structure placed in a waveguide, the latter refers to traveling wave tube applications.

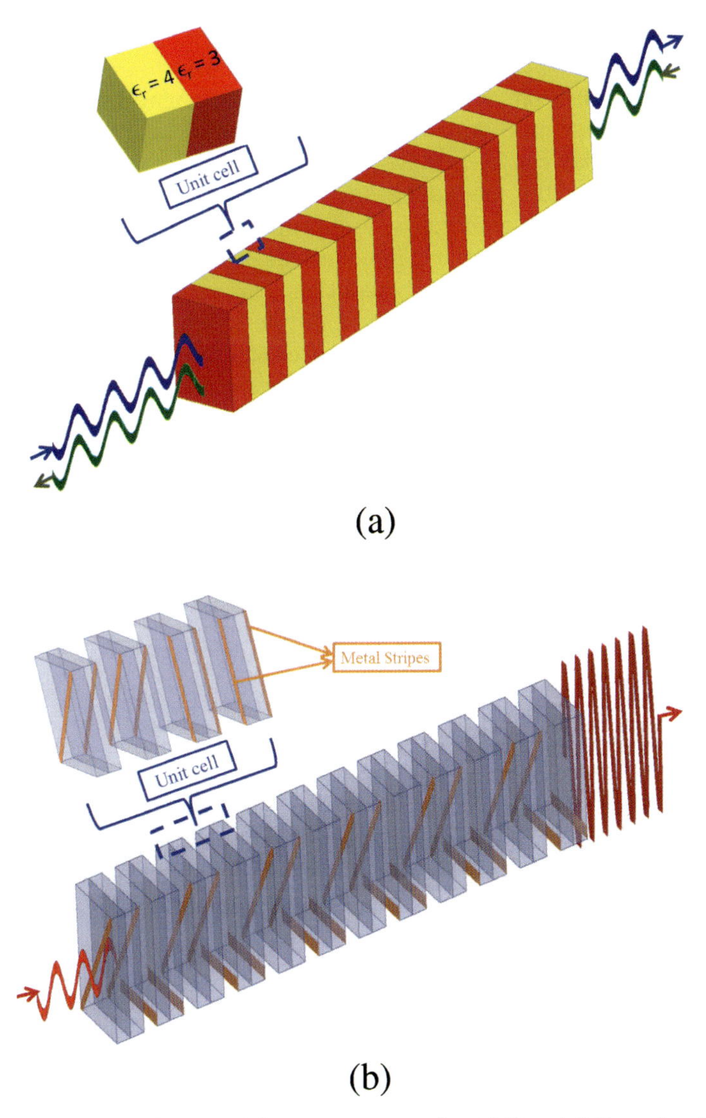

Figure 3.2 Periodic material structures to realize MPC and DBE modes: (a) photonic crystals with simple dielectric stacking. Forward and backward waves are depicted by blue and green arrows. (b) Magnetic photonic crystals using volumetric stack of anisotropic materials created using metal strips printed on the dielectric layers [3].

3.2 Slow Wave Theory

Slow waves refer to the propagation of electromagnetic waves with phase velocity much smaller than the speed of light, c [16]. This wave behavior can be achieved by changing the constitutive (ε,μ) parameters of the medium properties. Of course, the traditional way is to load the medium with appropriate dielectric properties. But, this is an expensive and cumbersome process and often implies use of exotic and low loss materials, not available in practice. Thus, engineering techniques and synthetic materials need to be considered to realize slow wave phenomena. These techniques often involve the construction of periodic stacks of dielectric layers for the goal to emulate the behavior of artificial dielectrics. The concept of metamaterial is based on periodic structures and leads to engineered media with unusual properties. Indeed, periodic media have received considerable attention due to their usefulness in dispersion engineering. In this section, we introduce the concept of slow wave propagation using periodic loading of CTLs followed by experimental demonstration.

3.2.1 Periodic Structures

Periodic stacks of dielectric material arrangements are often called photonic crystals. They support waves that exhibit multiple reflections as they propagate through photonic crystals. The superposition of all these reflections leads to various modes that can be exploited. Propagation through periodic media is typically modeled by invoking the well-known Floquet theorem [17]. As stated by Collin [17], "in a periodic system, if a given mode of propagation exists, fields inside a given cross-section of the system differs from fields that are a period distance away (or integer multiple of period distance away) by a complex constant." The fields $(\bar{E},\bar{H})$ in such a periodic system can be described by the function $F(x,y,z)$. $F(x,y,z)$ is periodic along z, the direction of material periodicity. Using $F(x,y,z)$, the $\bar{E}$ field can be written as:

$$E(x, y, z) = F(x, y, z)e^{-j\beta_0 z} \tag{3.1}$$

In this, β_0 is the propagation constant of the mode being described, and because $F(x,y,z)$ is periodic, it can be written as the sum of all possible supported modes. We can write E as,

$$E(x,y,z)=\sum_{n=-\infty}^{+\infty} E_n(x,y)e^{-j\beta_n z} \tag{3.2}$$

where p refers to the periodicity of the medium. In the above, each summation represents the nth mode or harmonic and is associated with the propagation constant:

$$\beta_n = \beta_0 + \frac{2\pi n}{p} \tag{3.3}$$

The associated phase and group velocity for each of the modes in Eq. (3.2) are given by:

$$v_{pn} = \frac{\omega}{\beta_n} = \frac{\omega}{\beta_0 + \frac{2\pi n}{p}} = \frac{1}{\frac{1}{v_{p0}} + \frac{2\pi n}{\omega p}} \tag{3.4}$$

$$v_{gn} = \frac{d\omega}{d\beta_n} = \frac{d\omega}{d\left(\beta_0 + \frac{2\pi n}{p}\right)} = \frac{d\omega}{d\beta_0} = v_{g0} \tag{3.5}$$

Thus, depending on the value of n, the phase velocity of the nth harmonic, v_{pn}, can be greater than or less than the phase velocity, v_{p0}, of the non-periodic medium. A typical dispersion diagram is shown in Fig. 3.3, a curve referred to as Brillouin's diagram.

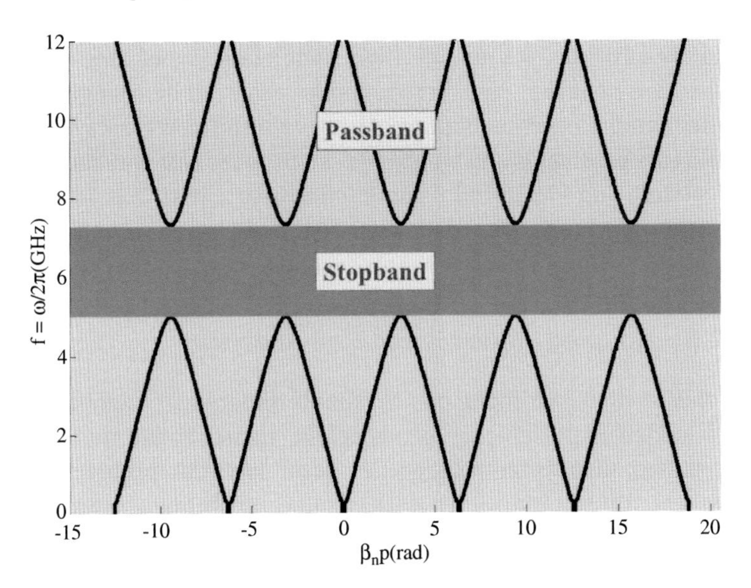

Figure 3.3 Typical dispersion diagram of the propagation frequency versus $K = \beta_n p$ in a periodic medium. When $\beta_n p$ is not given for a band, this band is referred to as "stopband."

3.2.2 Second-Order Dispersion

Three consequences can be identified when propagating with a linear, homogenous, and isotropic medium:

1. Periodic structures slow down waves in a similar manner as high-contrast dielectrics.
2. Backward waves (associated with negative n) are a consequence of periodicity.
3. Coupling of forward and backward wave leads to passbands and stopbands in a manner similar to Bloch waves in semiconductors.

Photonic crystals (see Fig. 3.2a) formed by simple isotropic dielectric stacks are the simplest example of these properties. Dielectric stacking causes reflections from the interfaces between two different materials that contribute to form backward wave [18]. The resulting dispersion is only of second order but clearly shows the formation of passbands and stopbands. The stacking of dielectrics can be simulated by a circuit model using inductance and/or capacitance (L,C) lumped elements to emulate the relative dielectric constants (ε_r, μ_r) forming the periodic cell. Such circuit analysis provides for an intuitive understanding of the overall dispersion behavior. Also, design can be achieved by changing the circuit parameters. The coupling of backward and forward waves that can create passbands and stopbands can be more easily understood, calculated, and illustrated using circuit theory. With this in mind, the next section derives a circuit model of CTLs and discusses its dispersion behavior using a simple second-order dispersion.

3.2.3 Coupled Transmission Line Analysis

In this section, we focus on the simple theoretical analysis of a CTL loaded with inductive or capacitive elements and derive the associated dispersion relation. To formulate the problem, two identical transmission lines with similar constitutive parameters (L,C) are considered, as illustrated in Fig. 3.4. For the moment, only inductive coupling is considered. The specific CTL is illustrated in Fig. 3.4.

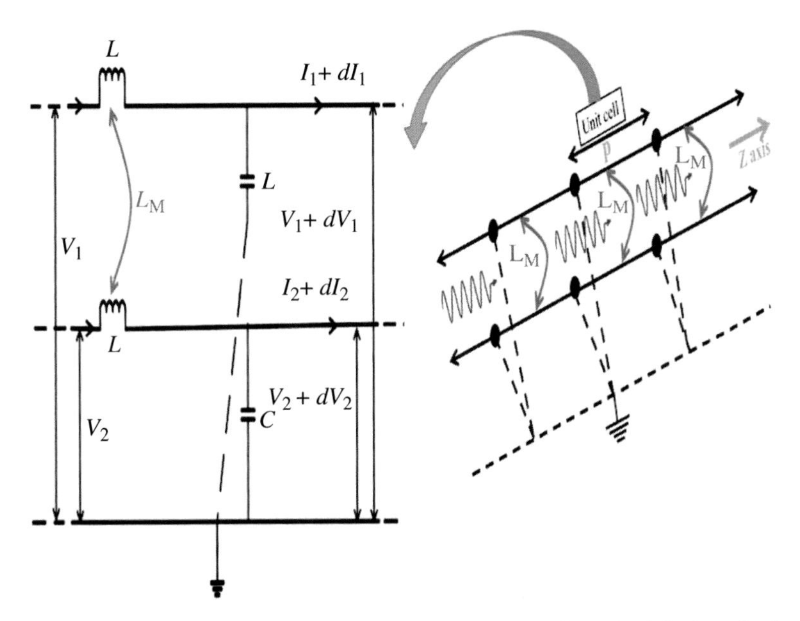

Figure 3.4 Coupled transmission lines with its unit cell to the left described using identical lumped circuit elements (*L*,*C*). A common ground is used forming a typical three-phase power transmission system.

3.2.3.1 Derivation

Let us assume that the two identical transmission lines (*i.e.*, straight wires) have the same per unit inductance of *L*, and per unit capacitance of *C*. Thus, individually, each transmission line supports a wave that has the same phase velocity, $1/\sqrt{LC}$, ideally the speed of light for free-space propagation. When the pair of transmission lines is close to each other, they exhibit coupling. This coupling can be described by an inductor placed between the two transmission lines when forming their equivalent circuit. Using this unit cell model, the telegrapher's equation takes this form:

$$\frac{\partial V_1^2}{\partial z^2} = LC\frac{\partial V_1^2}{\partial t^2} + L_M C\frac{\partial V_2^2}{\partial t^2} \tag{3.6a}$$

$$\frac{\partial V_2^2}{\partial z^2} = L_M C\frac{\partial V_1^2}{\partial t^2} + LC\frac{\partial V_2^2}{\partial t^2} \tag{3.6b}$$

$$\frac{\partial I_1^2}{\partial z^2} = LC\frac{\partial I_1^2}{\partial t^2} + L_M C\frac{\partial I_2^2}{\partial t^2} \quad (3.6c)$$

$$\frac{\partial I_2^2}{\partial z^2} = L_M C\frac{\partial I_1^2}{\partial t^2} + LC\frac{\partial I_2^2}{\partial t^2} \quad (3.6d)$$

In the above, the voltages $V_{1,2}$ and currents $I_{1,2}$ refer to the excitation values as measured between each transmission line and the ground lines. We also note the presence of additional terms due to L_M. These terms do not exist in the telegrapher's equations and will be responsible for the more complex dispersion curve.

To solve Eq. (3.6), we shall assume a time dependence of $e^{j(\omega t-\beta z)}$, where ω is angular frequency and β is wavenumber along the direction (z-axis) of propagation. Applying the time dependence on a pair of equations (3.6a and 3.6b), the resulting dispersion relation takes the form

$$(\omega^2 LC - \beta^2)^2 - \omega^4 L_M^2 C^2 = 0 \quad (3.7)$$

The solution to this fourth-order equation leads to four possible constants:

$$\beta_1 = \pm\omega\sqrt{(L - L_M)C} \quad (3.8a)$$

$$\beta_2 = \pm\omega\sqrt{(L + L_M)C} \quad (3.8b)$$

The corresponding phase velocities are:

$$v_1 = \frac{\omega}{\beta_1} = \sqrt{\frac{1}{(L - L_M)C}} > c \quad (3.9a)$$

$$v_2 = \frac{\omega}{\beta_2} = \sqrt{\frac{1}{(L + L_M)C}} < c \quad (3.9b)$$

Clearly, the presence of L_M leads to slow wave formation as depicted in Fig. 3.5. But more importantly, this β_2 implies a much slower wave and, therefore, more control to phase velocity and more control in designing smaller antennas, couplers, RF signal dividers, smaller TWTs, and even low- and high-power BWOs. Clearly, the added mutual inductance has led to the realization of slow waves using circuit parameters rather than actual dielectrics. Therefore, this analysis demonstrates that the dielectric behavior can be emulated by CTLs.

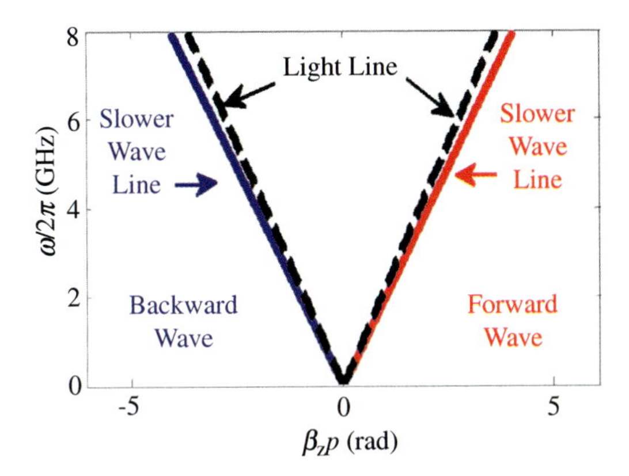

Figure 3.5 Coupled transmission lines can support slow waves. The plotted dispersion lines (red and blue lines) provide an approach for artificial dielectrics. The CTL parameters are $(L,C) = (1.61722\ \mu\text{H}, 6.8798\ \text{pF})$ and $L_M = 0.37308\ \mu\text{H}$.

3.2.3.2 Coupling of modes

In reality, two oppositely propagating waves ($n = 0$ or forward mode and $n = -1$ or backward mode) couple together to realize slow wave propagation. This leads to passbands and stopbands [16, 19], and coupling of the transmission lines is important in controlling the dispersion relation.

To better understand the coupling between the $n = 0$ and $n = -1$ modes, we consider a simple periodic transmission line system supporting forward wave and backward wave modes. The forward wave is denoted by P_n and has a propagation constant β_p. Similarly, the backward wave is denoted by Q_n and has a propagation constant β_q. The latter (Q_n) is coupled to the forward wave (P_n) with a coupling constant k. This coupling is assumed to be linear, and the resulting combined wave is assumed to have a propagation constant β. We note that the uncoupled forward waves differ from one cell to other via a phase delay $e^{-j\theta p}$. Similarly, the uncoupled backward wave differs from cell to cell via the delay $e^{-j\theta q}$. The pair of modes supported by the CTL system can be expressed as

$$P_{n+1} = P_n e^{-j\theta_p} + \kappa Q_n e^{-j\theta_q} \tag{3.10a}$$

$$Q_{n+1} = \kappa P_n e^{-j\theta_p} + Q_n e^{-j\theta_q} \tag{3.10b}$$

Rewriting these equations as a function of the differential distance dz, we have

$$P_n \rightarrow P(z)$$

$$Q_n \rightarrow Q(z)$$

$$\theta_p \rightarrow \beta_p dz$$

$$\theta_q \rightarrow \beta_q dz$$

$$P_{n+1} \rightarrow P(z) + dP(z)$$

$$Q_{n+1} \rightarrow Q(z) + dQ(z)$$

$$\kappa \rightarrow K dz$$

With these replacements, Eqs. (3.10a) and (3.10b) give the differential equations

$$\frac{dP}{dz} = -j\beta_p P + KQ \tag{3.11a}$$

$$\frac{dQ}{dz} = KP - j\beta_q Q \tag{3.11b}$$

To solve Eq. (3.11), we assume that the solutions are of the form: $P = P_0 e^{-j\beta}$ and $Q = Q_0 e^{-j\beta}$. As a result, Eq. (3.11) can be rewritten as

$$\begin{bmatrix} j(\beta-\beta_p) & K \\ K & j(\beta-\beta_q) \end{bmatrix} \begin{bmatrix} P_0 \\ Q_0 \end{bmatrix} = \begin{bmatrix} 0 \\ 0 \end{bmatrix} \tag{3.12}$$

The determinant of this 2×2 matrix leads to the quadratic

$$\beta^2 - (\beta_p + \beta_q)\beta + (\beta_p\beta_q + K^2) = 0 \tag{3.13}$$

For computing the propagation constants of the supported CTL waves, the solutions/roots of Eq. (3.13) are

$$\beta_{c1} = \frac{\beta_p + \beta_q}{2} + \sqrt{\left(\frac{\beta_p - \beta_q}{2}\right)^2 - K^2} \tag{3.16}$$

$$\beta_{c2} = \frac{\beta_p + \beta_q}{2} - \sqrt{\left(\frac{\beta_p - \beta_q}{2}\right)^2 - K^2} \tag{3.17}$$

Here, β_{c1} and β_{c2} are the propagation constants of the coupled wave and are a result of the coupling between the P (forward) and

Q (backward) modes. We note that P and Q have the individual propagation constants $\beta_p = \omega\sqrt{(L+L_{\mathrm{M}})C}$ and $\beta_q = -\omega\sqrt{(L+L_{\mathrm{M}})C}$. The final dispersion diagram obtained from the above relations is given in Fig. 3.6. The dispersion curve derived is of second order, and the band edge is called regular band edge (RBE) near the βp = π point.

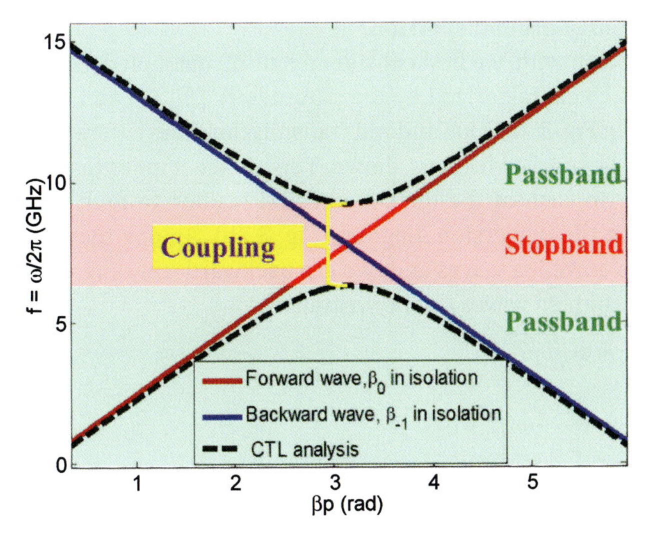

Figure 3.6 CTL modes due to coupling enabled by L_{M} of two identical transmission lines. Coupling is demonstrated via the formation of passband and stopband, a unique feature of periodic structures. The coupling coefficient for calculation was K = 20.85.

3.2.4 Higher-Order Dispersion Engineering

3.2.4.1 Graphical analysis

As noted earlier, second-order dispersion is a consequence of simple periodic stacks of isotropic dielectrics. Further, these stacks can be modeled as CTLs. A very special property of the CTLs is that each of the lines is identical to each other and, therefore, associated with the same phase and group velocity. However, if the transmission lines forming the CTLs are not electrically identical, we would then expect different behavior in the dispersion curve. In this section, we consider the analysis of such coupled non-identical transmission lines. The goal is to generate higher-order dispersion curves. As a first step, we proceed to formulate the fourth-order dispersion (DBE mode) since it is symmetric in nature, *i.e.*, $\omega(\beta) = \omega(-\beta)$ and comparatively easy to derive. Unlike the second-order dispersion, the calculation of fourth-order dispersion is complex and cumbersome. However, the second-order dispersion provides a building block for the derivation of the fourth-order dispersion.

To begin with, we first consider a pair of uncoupled transmission lines, each characterized by the parameters (L_1, C_1) and (L_2, C_2), as shown in Fig. 3.7a (blue and red transmission lines). Here we choose $L_2 = 0.4L_1$ and $C_2 = 0.4C_1$ as shown. This choice implies that the wave velocity in each of the lines is unequal leading to four uncoupled waves in the dispersion diagram (Fig. 3.7b). Among these, two are forward traveling waves and two are backward traveling waves. The four uncoupled waves can be written as:

$$\beta_a = \omega\sqrt{L_1C_1} \tag{3.18a}$$

$$\beta_b = \frac{2\pi}{p} - \omega\sqrt{L_1C_1} \tag{3.18b}$$

$$\beta_c = \omega\sqrt{L_2C_2} \tag{3.18c}$$

$$\beta_d = \frac{2\pi}{p} - \omega\sqrt{L_2C_2} \tag{3.18d}$$

Now let us proceed to assume that the transmission lines are coupled. The zones where possible coupling can occur are marked by green circles and denoted by '1' and '2'. As shown in Fig. 3.7b, a coupling is considered between a forward wave mode $\beta_a = \omega\sqrt{L_1C_1}$ and a backward wave mode $\beta_d = \dfrac{2\pi}{p} - \omega\sqrt{L_2C_2}$ in a similar manner as derived in Eqs. (3.16) and (3.17). The coupling coefficient chosen for zone '1' is K_1. Similarly, another coupling is considered between a forward wave mode $\beta_c = \omega\sqrt{L_2C_2}$ and a backward wave mode $\beta_b = \dfrac{2\pi}{p} - \omega\sqrt{L_1C_1}$. The coupling coefficient chosen for zone '2' is K_2. The coupling zones exhibit that uncoupled waves are not expected to couple at the $\beta p = \pi$ point (marked as pink vertical line in Fig. 3.7). Rather, they couple at two separate locations equally distanced from the π point. This is the consequence of unequal phase velocities of two non-identical lines. Unlike the case of second-order dispersion, a forward wave and a backward wave with different phase velocities are expected to couple. The coupling coefficients (K_1, K_2) can be attributed to the natural coupling between a forward and a backward wave mode associated with any periodic structures, and (K_1, K_2) can be modeled by coupled inductance L_M or capacitance C_M between two transmission lines similar to the circuit model shown in Fig. 3.8a. However, if the coupled modes experience anisotropic medium while propagating along the transmission lines, they proceed to couple further and generate a new fourth-order dispersion characteristic. To model and enumerate the coupling due to anisotropy, we employ a third kind of coupling coefficient, K_3, that couples two second-order modes together. The final dispersion relations of the resultant coupled modes are given in Eq. (3.19). It turns out that setting all K's equal to each other, *i.e.*, $K_1 = K_2 = K_3$, leads to a special type of fourth-order dispersion curve associated with the double band edge (DbBE) mode (Fig. 3.8b). The name double band edge is attributed to the two different band edges observed at the same frequency.

$$\beta_1 = \frac{\pi}{p} + \sqrt{\left[\frac{\omega}{2}\left(\sqrt{L_1C_1} - \sqrt{L_2C_2}\right) - \sqrt{\left\{\frac{\pi}{p} - \frac{\omega}{2}\left(\sqrt{L_1C_1} + \sqrt{L_2C_2}\right)\right\}^2 - K_1^2}\right]^2 - K_3^2} \tag{3.19a}$$

$$\beta_2 = \frac{\pi}{p} - \sqrt{\left[\frac{\omega}{2}\left(\sqrt{L_1C_1} - \sqrt{L_2C_2}\right) - \sqrt{\left\{\frac{\pi}{p} - \frac{\omega}{2}\left(\sqrt{L_1C_1} + \sqrt{L_2C_2}\right)\right\}^2 - K_1^2}\right]^2 - K_3^2} \tag{3.19b}$$

$$\beta_3 = \frac{\pi}{p} + \sqrt{\left[\frac{\omega}{2}\left(\sqrt{L_1C_1} - \sqrt{L_2C_2}\right) + \sqrt{\left\{\frac{\pi}{p} - \frac{\omega}{2}\left(\sqrt{L_1C_1} + \sqrt{L_2C_2}\right)\right\}^2 - K_2^2}\right]^2 - K_3^2} \tag{3.19c}$$

$$\beta_4 = \frac{\pi}{p} - \sqrt{\left[\frac{\omega}{2}\left(\sqrt{L_1C_1} - \sqrt{L_2C_2}\right) + \sqrt{\left\{\frac{\pi}{p} - \frac{\omega}{2}\left(\sqrt{L_1C_1} + \sqrt{L_2C_2}\right)\right\}^2 - K_2^2}\right]^2 - K_3^2} \tag{3.19d}$$

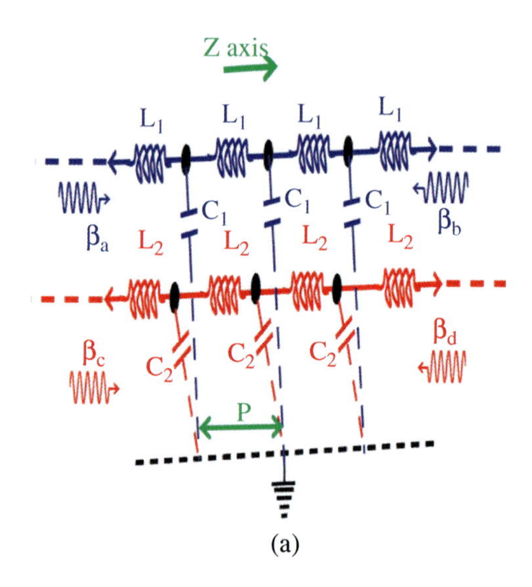

(a)

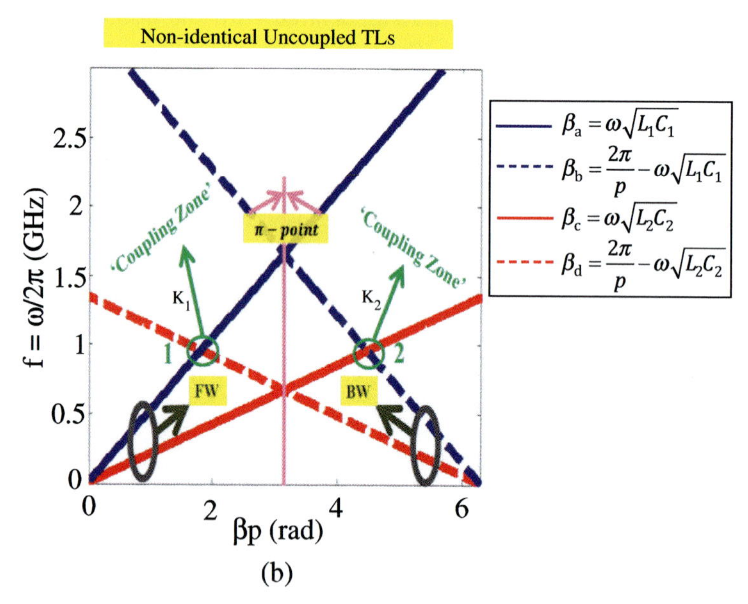

(b)

Figure 3.7 Two non-identical uncoupled transmission lines with four solutions of the propagation constants. (a) Two non-identical uncoupled transmission lines are marked by blue and red colors with circuit elements per unit length characterized by (L_1, C_1) and (L_2, C_2). (b) Uncoupled forward and backward waves corresponding to blue and red transmission lines. Possible coupling is marked by green circles as '1' and '2' with coupling coefficients K_1 and K_2, respectively.

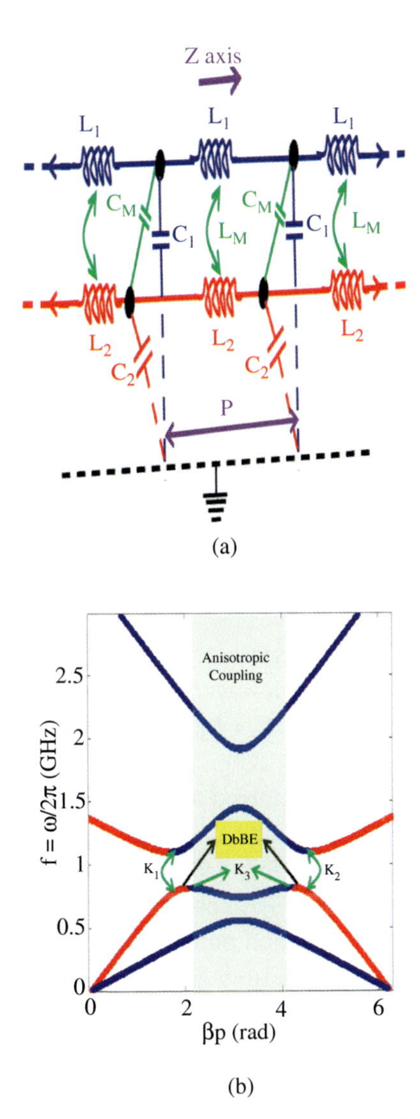

Figure 3.8 Two non-identical coupled transmission lines with coupling introduced by anisotropy and the formation of DbBE modes. (a) Two non-identical coupled transmission lines are marked by blue and red colors with circuit elements per unit length characterized by (L_1, C_1) and (L_2, C_2). The coupling inductance and capacitance are (L_M, C_M) and can be attributed to the coupling coefficients (K_1, K_2). (b) Formation of DbBE mode is shown with coupling due to anisotropy depicted by the shadowed box. The coupling coefficient is marked as K_3, which relates to the coupling between two second-order modes. For DbBE formation, all three *K*'s are chosen as equal, *i.e.*, $K_1 = K_2 = K_3$.

Although the coupling parameters (K_1, K_2) are found from the coupled elements (L_M, C_M), the third coupling coefficient K_3 is attributed only to the existence of anisotropy, as mentioned by Figotin et al. [7]. In fact, it is related to the polarization of two oppositely directed waves propagating in an appropriate anisotropic medium. Usually, anisotropy is realized when using magnetic materials to form the propagating medium, e.g., ferrites with DC bias. Actually, the dispersion relations written above provide for three of the known dispersion behaviors, depending on the choice of (K_1, K_2, K_3), but more possibilities exist. Specifically, when all there coupling coefficients, *i.e.*, $K_1 = K_2 = K_3$, are the same, a DbBE is observed (Fig. 3.8b). But if we choose the case of $K_1 = K_2$, $K_1 \neq K_3$, a fourth-order maximally flat dispersion profile results. The associated mode is referred to as the DBE mode with the corresponding dispersion curve shown in Fig. 3.9a. This flat profile of the DBE modes is useful for antenna miniaturization and increased directivity, but leads to narrow bandwidths.

It is noted that spectral symmetry, *i.e.*, $\omega(\beta) = \omega(-\beta)$ is inherent in the DbBE and DBE modes. However, when the coupling parameters are chosen such that $K_1 \neq K_2$, $K_1 = K_3$, an unusual dispersion behavior is found, as shown in Fig. 3.9b. The resulting modes from the solution to Eqs. (3.19) are referred to as MPC modes. This is because in volumetric media, magnetic materials are required to realize these modes. As shown in Fig. 3.9b, the MPC mode dispersion is asymmetric, *i.e.*, $\omega(\beta) \neq \omega(-\beta)$, in nature and of third order with an inflection point. This inflection point is evocative of the stationary inflection point (SIP) in a typical steepest decent path and contributes to bandwidth improvements for antennas. Table 3.1 relates DbBE, DBE, and MPC modes to the three aforementioned parameters.

Table 3.1 Classification of dispersion engineering based on K parameters

Case 1	Double band edge (DbBE) mode (Fig. 3.8a)	$(K_1 = K_2 = K_3)$
Case 2	Degenerate band edge (DBE) mode (Fig. 3.8b)	$(K_1 = K_2, K_1 \neq K_3)$
Case 3	Magnetic photonic crystals (MPC) mode (Fig. 3.9)	$(K_1 \neq K_2, K_1 = K_3)$

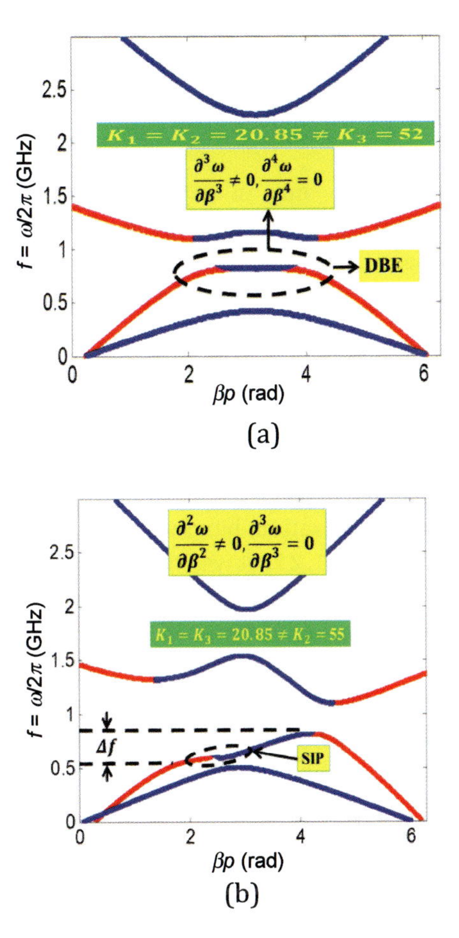

Figure 3.9 Higher-order dispersion attributed to the coupled, but non-identical, transmission lines. (a) Degenerate band edge (DBE) mode dispersion diagram, leading to a fourth-order curve. (b) Third-order dispersion diagram of the MPC mode showing the presence of the stationary inflection point.

3.2.4.2 Realizations of higher-order dispersion

Figotin et al. [7] noted that dielectric stacks of anisotropic materials can generate third- or even fourth-order dispersion. In this case, anisotropy relates to the use of ferrite layers within the dielectric stacks. It has been demonstrated that spectral asymmetry, *i.e.*, $\omega(\beta) \neq \omega(-\beta)$, actually leads to third-order dispersion. As such, it can include an inflection point within its curve, referred to as the SIP in the ω–β diagram. The presence of SIP within the dispersion curve implies a number of properties:

- Wave group velocity slows down and even vanishes at the SIP [15].
- Waves propagate only in a single direction due to the asymmetry of the dispersion curve. Specifically, one mode can propagate, but the other in the opposite direction is "frozen" at the SIP. That is, unidirectional propagation occurs in asymmetric media [10].
- Dispersion curves are of third order. Also, at the inflection point, $\frac{\partial \omega}{\partial \beta} \neq 0, \frac{\partial^2 \omega}{\partial \beta^2} \neq 0, \frac{\partial^3 \omega}{\partial \beta^3} = 0$ (see Figs. 3.9b and 3.10a).

The MPC dispersion and related wave slowdown have been used for RF antenna miniaturization, bandwidth enhancements, and for photonics applications [11, 21]. It was shown in Ref. [20] that at least two misaligned anisotropic layers (misalignment angle other then 0 or $\pi/2$) inside one period of the stack can lead to maximally flat dispersion behavior at the band edge. This is depicted in Figs. 3.9a and 3.11b. Four propagating waves are involved to create the subject ω–β diagram, implying a fourth-order dispersion. This special band edge is referred to as DBE. Its maximally flat dispersion curve is associated with the conditions: $\frac{\partial \omega}{\partial \beta} \neq 0, \frac{\partial^2 \omega}{\partial \beta^2} \neq 0, \frac{\partial^3 \omega}{\partial \beta^3} \neq 0, \frac{\partial^4 \omega}{\partial \beta^4} = 0,$ as depicted in Fig. 3.9a. The implied field at amplification at the band edge was applied to improve the directivity of antennas.

In 2008, Yarga et al. [22] demonstrated an experimental realization of the DBE modes using volumetric stacks of dielectrics. To introduce anisotropy in the stack layers, thin metal stripes with different orientations were employed, as depicted in Fig. 3.2b. MPC "frozen modes" were also realized by Stephenson et al. [15] using CTLs on ferrite substrates. Mumcu et al. [23] were the first to realize DBE modes using coupled microstrip lines and used the concept in antenna miniaturization. Further, to achieve emulation of anisotropy, the dispersion relation of the MPC modes in Ref. [15] was calculated using the same approach as in Figotin's work [20]. However, the first experimental demonstration of the "frozen mode" using transmission lines was done by Apaydin et al. [5] and is depicted in Fig. 3.12. Two CTLs consisting of eight or nine periods were printed

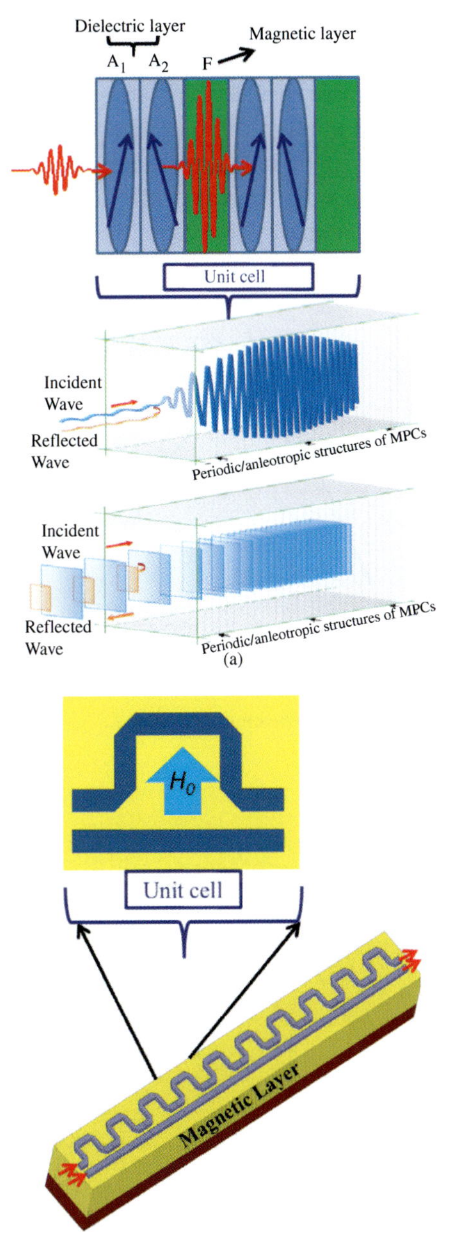

Figure 3.10 (a) MPC with anisotropic layers (A_1 and A_2) and a third ferrite layer. Field enhancement and unidirectionality is also illustrated. (b) Realization of the MPC mode in printed form. Magnetic biasing is shown as circles and labeled as H_0.

on calcium vanadium garnet (CVG) and an external magnet was used for biasing. This successful experiment demonstrated the existence of the MPC modes and also verified the theoretical predictions of the supported modes. The measured group velocity at the SIP was 286 times slower than the speed of light. Further, the contrast between forward and backward wave transmission was 75%, verifying the theoretical prediction of unidirectionality [5].

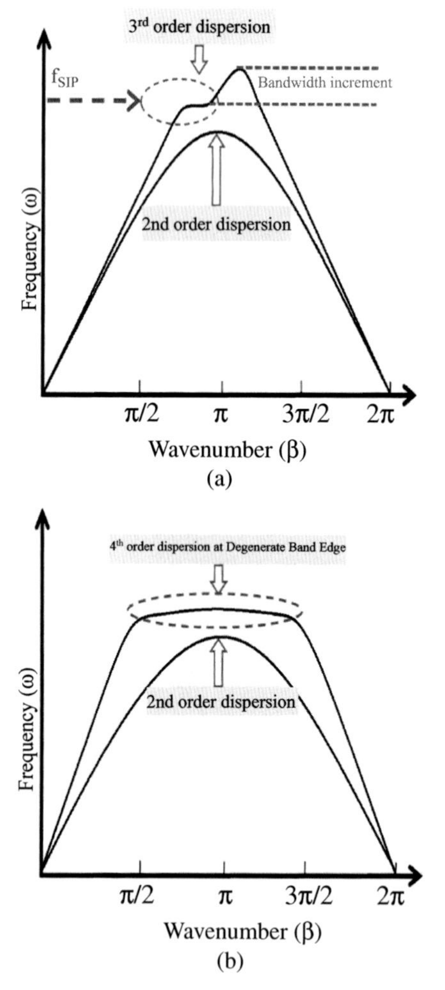

Figure 3.11 (a) Non-reciprocal dispersion diagram of the MPC, displaying the f_{SIP}. (b) Fourth-order dispersion diagram showing the DBE resonance at the band edge [15].

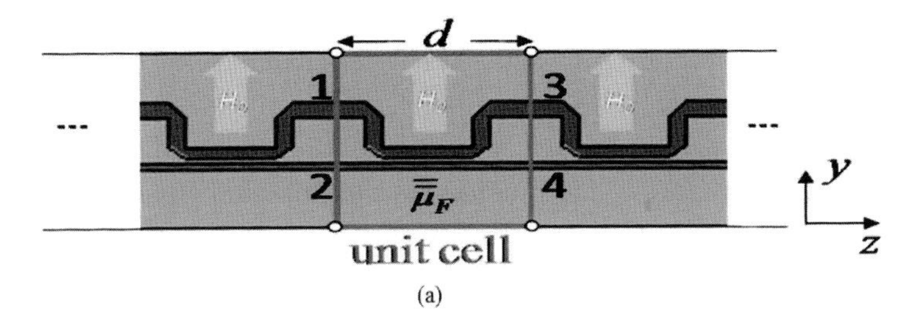

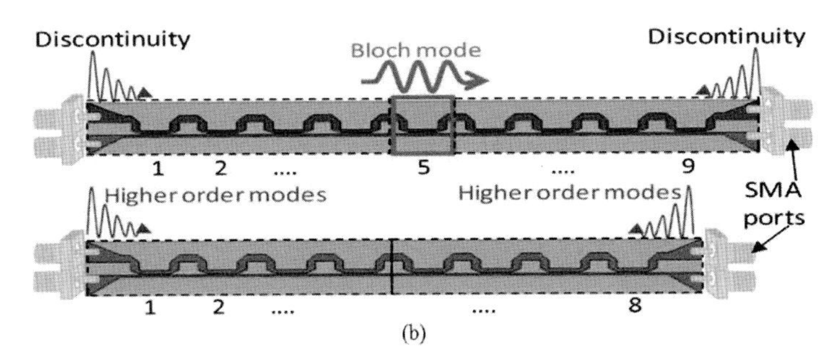

Figure 3.12 Coupled transmission lines printed on a magnetic substrate used to generate MPC modes. (a) Unit cell of the periodic transmission line structure. (b) Finite printed MPC prototypes comprising nine and eight unit cells for calculating the dispersion behavior of the coupled printed transmission lines. Reprinted with permission from Ref. 5, Copyright 2012, IEEE.

3.3 Applications of Slow Waves

Slow wave applications exploit dispersion engineering and the associated ω–β diagram. Antennas are often operated near resonance frequency and near the band edge of the dispersion diagram to achieve field enhancement [3, 20, 21]. However, microwave amplifiers such as TWTs may require broader bandwidth [24]. In that case, second-order dispersion may be appropriate. In the following, we discuss some microwave/RF applications.

3.3.1 Traveling Wave Tubes

The operation of TWTs is based on the Cherenkov radiation and involves the coupling of an electron beam to an RF wave [25].

These devices are basically amplifiers that amplify the incoming RF wave by modulating an injected electron beam from the cathode ray tube. To achieve strong coupling between the electron beam and the electromagnetic wave or mode of the waveguide, the TWT must support a slow TM_{01} mode [26] that matches the velocity of the electron beam. Also, the strength of the beam–wave coupling is dependent on the strength of the axial field component supported by the TM_{01} mode. Therefore, a dispersion-free TM_{01} mode with a strong axial electric field component facilitates efficient coupling of the electron beam. Such a dispersion-free TM_{01} mode can be introduced by designing slow wave structures (SWS). In this regard, SWSs play an important role in the process of electron beam to waveguide mode coupling.

Since the CTLs are made of metallic conductors and can emulate dielectric behavior in their passband, they have the potential for high-power microwave applications. Interestingly, several SWSs, e.g., helix, double-helix, ring-bar structures [24], have been considered and termed as slow wave circuits in the literature. To improve coupling to microwave power delivery, Fig. 3.13 shows the curved ring-bar (CRB) structure [27]. CRB is an upgrade of the typical helical structures and can be modeled as a pair of CTLs as discussed earlier. Specifically, the top and bottom elliptically bent lines refer to the pair of transmission lines. These transmission lines are coupled through the inner rings. More bending of the elliptic feature (*i.e.*, $m > 1$) of the transmission lines provides more coupling, implying a way to control the slow wave properties and interaction impedance of the TWT model (Figs. 3.1c and 3.13). In effect, the elliptic bent of the transmission line controls the effective permittivity of the propagating wave by reducing the phase velocity below c (Fig. 3.13). Indeed, the dispersion diagram in Fig. 3.13 (right) and Fig. 3.14 validates the second-order dispersion curve. The latter is, of course, the behavior caused by the inductively CTLs. The designed TWT, in this manner, can generate up to 1 MW of output power with a gain of around 30 dB, whereas typical pulsed helix TWTs can generate power up to several kilowatts [28]. This high-power enhancement is due to the slow wave phenomena and associated propagation constants caused by the CTLs.

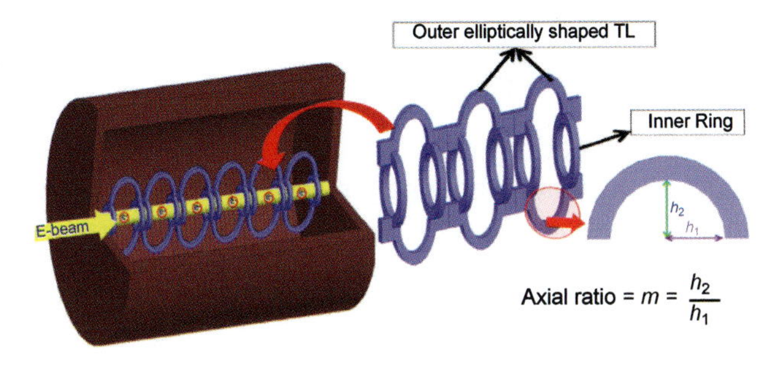

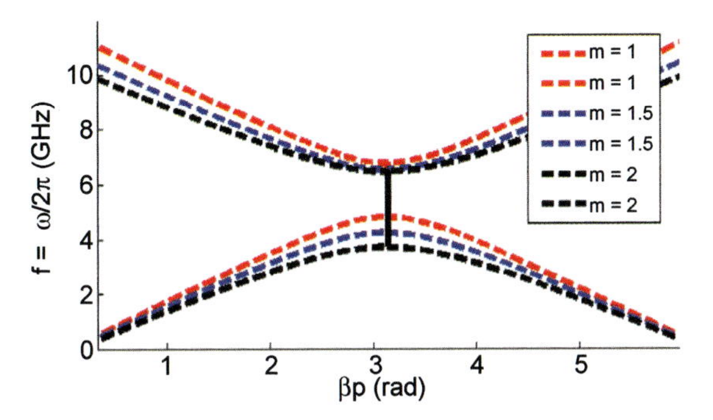

Figure 3.13 Curved ring-bar structure placed inside the waveguide to improve electron beam to waveguide mode coupling. The dispersion diagrams to the right show how the propagation constant and wave velocity can be controlled by the geometric property, *i.e.*, axial ratio m.

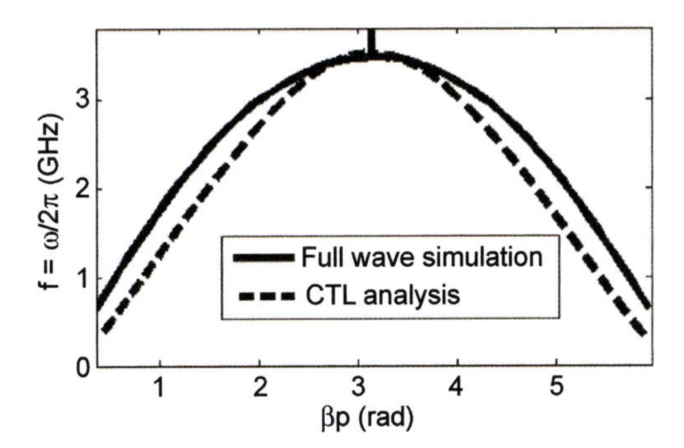

Figure 3.14 Dispersion diagram due to the curved ring-bar structure within the waveguide. Provided computations of the dispersion diagram were carried out via full wave simulations and the CTL model using appropriate coupling parameters.

3.3.2 Antenna Miniaturization, Directivity, and Bandwidth Improvement

Antenna miniaturization is a desirable property in antenna engineering. Reduction in antenna dimensions helps accommodate small mobile devices, such as smartphones and portable transceivers. A way to miniaturize the antenna is to lower its resonance frequency [29]. Typically, the resonance frequency occurs when βp = $K = \pm\pi$ in the dispersion diagram, and where the group velocity vanishes. For the RBE that occurs in second-order dispersion diagrams, reasonable bandwidth and miniaturization can be achieved. Lower resonances, viz. greater miniaturization, can be achieved using fourth-order dispersion engineering, but the bandwidth will be smaller. That is, DBE modes achieve maximally flat fourth-order dispersion [20] but lead to a more narrowband behavior.

Although the DBE, DbBE, and MPC modes can be exploited to slow down wave, in practice, there is a need to have several periods of the CTL. This requirement implies large structures (at least five periods next to each other) to create the mode, defeating the goal of miniaturization. A way to realize periodicity and overcome the need to stack unit cells next to each other is to adapt circular periodicity. Such an arrangement of unit cells is depicted in Fig. 3.15a and is key to exploiting dispersion engineering for antenna miniaturization. The dispersion diagrams to the left also depict how the dispersion diagrams can be modified to achieve resonance reduction and, therefore, miniaturization.

As already noted, DBE mode formation can be attributed to anisotropy in the dielectric layers, and a challenge is how to realize such an anisotropy in practice. Losses must also be considered since slowly propagating DBE modes are known to experience greater losses per unit cell. To realize DBE modes in practice, a naturally available material, referred to as 'rutile' was chosen due to its very low loss and large permittivity (ε_{xx} = 165, $\varepsilon_{yy} = \varepsilon_{zz}$ = 85, tanδ = 1 × 10^{-4} at X-band) [11], both important for small antennas. To create the DBE crystals, a periodic arrangement of printed metal strips was printed on the dielectric layers, as depicted in Fig. 3.16. These strips were printed on two of the three dielectric layers forming the unit cell. But the strip directions on each printing surface are rotated. This leads to polarization rotation of the propagating modes through the unit cell, implying an effective anisotropy [3,

30–32]. Also the dimensions of the strips can be varied to change the effective permittivity tensor of the overall medium, polarization rotation, and propagation properties. The arrangement and unit cell profile are depicted in Fig. 3.16. Similar arrangement can be done on different substrates to realize the DBE mode. Specifically, instead of "rutile," a thick Rogers RO4350 substrate was used to form the DBE crystal. A dipole antenna was then placed on the maximum field position of the crystal to enhance radiation. This enhancement was experimentally demonstrated, and the DBE resonance was verified, leading to a directivity as high as 18 dB [3].

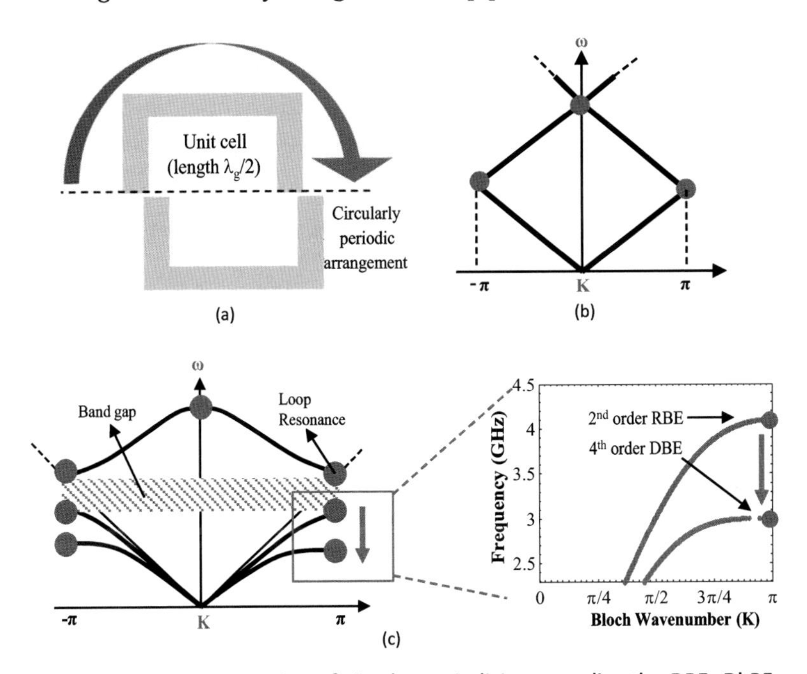

Figure 3.15 Demonstration of circular periodicity to realize the DBE, DbBE, and MPC modes for antenna miniaturization. (a) Unit cell arrangement, viz. flipping of the unit cell, to achieve circular periodicity. (b) Dispersion diagram associated with the unit cell forming the loop; resonances of this circularly periodic structure (two unit cells) are marked by dots, and occur at $\beta p = K = \pm\pi$. (c) Bending of the K–ω diagram to shift resonances to lower frequencies with the magnified view of the dispersion diagram around the band edge to the rightmost. Reprinted with permission from Ref. 23, Copyright 2009, IEEE.

Using the concept of rotated layer strips or rotated dielectric bars, a DBE resonator antenna was designed and experimentally demonstrated by Yarga et al. [33]. The antenna was a dielectric resonator antenna (DRA) type using a multilayered cavity. The layers were formed of $BaTiO_3$ and Al_2O_3 bars in an effort to create anisotropy in the unit cell. It was observed that the dielectric resonator modes TE_{101} and TM_{011} led to the formation of DBE modes. It was observed that DRA resonance was shifted to lower frequencies due to the maximally flat DBE mode (see Fig. 3.17). We remark that the misalignment angle between the bars forming the layers added more degree of freedom, leading to the creation of the DBE mode.

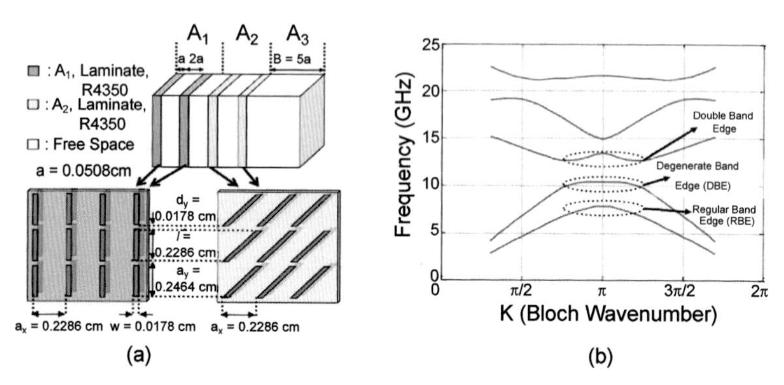

Figure 3.16 (a) DBE unit cell configurations. (b) Dispersion diagram of unit cell. Reprinted with permission from Ref. 3, Copyright 2008, IEEE.

Having an understanding of the DBE mode and its practical realization, as in Figs. 3.16 and 3.17, Mumcu et al. [23] proceeded to use the coupled dual transmission line to realize the DBE mode. Specifically, Fig. 3.18 shows the correspondence between the actual three-layer unit cell and the CTL coupling to realize anisotropy. Explicitly, the shown CTL unit cell includes three sections. Two of these are uncoupled transmission lines, and the middle sections refer to a coupled CTL having a coupling capacitance C_M. The propagation mechanism in the three-section unit cell is rather straightforward. Each of the transmission lines can be thought of as carrying one of the polarization components of the electric field propagating within the DBE crystal. In fact, the non-identical transmission lines of Fig. 3.18a are associated with the propagation constants in Eqs. (3.19) and Fig. 3.8b. Concurrently, the different line lengths provide a phase delay between the two polarizations to emulate diagonal anisotropy. Similarly, even–odd mode impedances and propagation constants on the coupled lines can be used to emulate a general anisotropic medium (*i.e.*, non-diagonal anisotropy tensor). Further, by cascading the uncoupled and coupled transmission line sections, as in Fig. 3.18a, an equivalent printed circuit is realized, which emulates the volumetric DBE crystal [34].

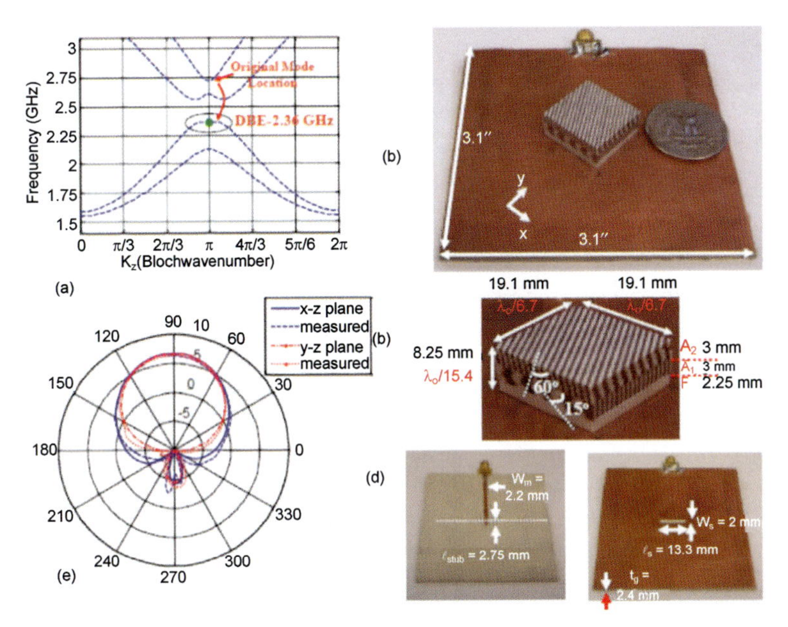

Figure 3.17 (a) DBE–DRA dispersion diagram and the realized DBE–DRA antenna: (b) top view; (c) side view; (d) bottom and top view showing the slot coupled microstrip line feed; (e) simulated and measured gain pattern in principle cuts. Reprinted with permission from Ref. 33, Copyright 2009, IEEE.

As can be realized, the tunable capacitance between the CTLs can play a key role in controlling the dispersion diagram, as in Fig. 3.18c. Specifically by referring to our earlier analysis, the coupling capacitance controls the value of the coupling coefficients K_1, K_2, K_3 in Eqs. (3.19). This approach was actually used to design the double loop antenna supporting the DBE mode in Ref. [37]. To realize C_M, the transmission lines were bent toward the center to keep the footprint as small as possible while concurrently modifying the property of the coupling coefficients K_1, K_2, K_3 (see Fig. 3.19). For additional control in creating the dispersion diagram, lumped loads were used to further modify the individual transmission line parameters. By doing so, the DBE mode was achieved with ease.

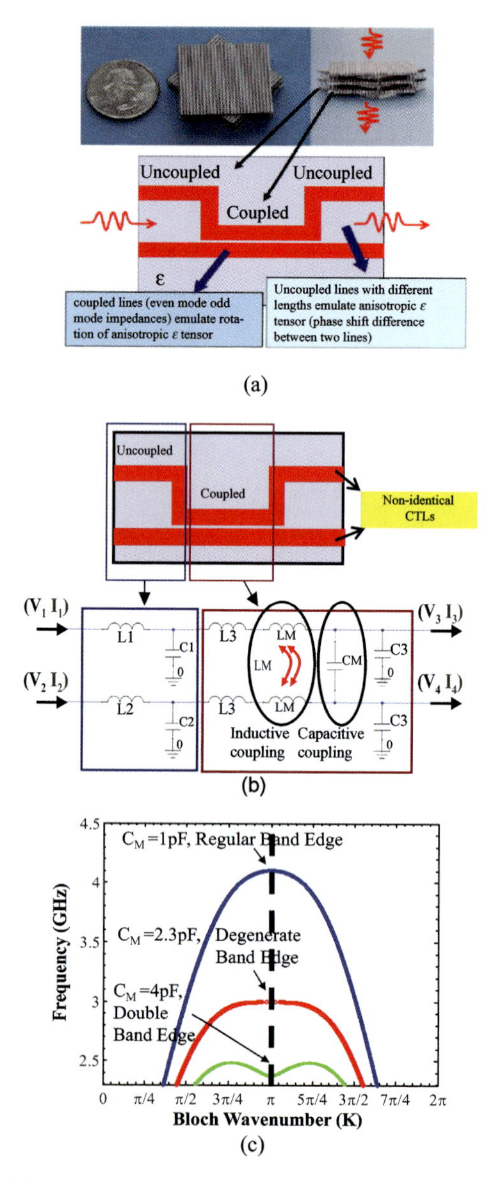

Figure 3.18 (a) Concept of emulating an anisotropic medium using cascaded coupled and uncoupled transmission lines, (b) lumped circuit model of the partially coupled lines, (c) dispersion diagrams obtained by changing the coupling capacitance C_M for transmission lines with $L_1 = L_2 = L_3 = 1\text{n}\ H$, $C_1 = 10$ pF, $C_2 = C_3$. Reprinted with permission from Ref. 23, 34, and 36, Copyright 2009, 2010, and 2009, IEEE, respectively.

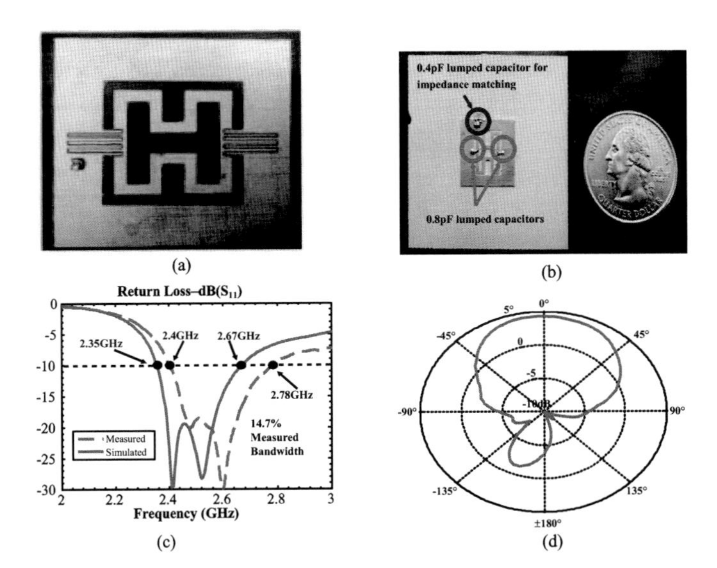

Figure 3.19 (a) Capacitively loaded double loop (CDL) antenna layout printed on a 2″ × 2″, 125 mil thick Duroid substrate (ε_r = 2.2, tan δ = 0.0009). This antenna is 1″ × 1″ in size and exhibits improved gain of 3.9 dB with 51% efficiency at 2.26 GHz. (b) Fabricated double loop antenna on a 250 mil thick 1.5″ × 1.5″ Rogers TMM 10i (ε_r = 9.8, tan δ = 0.002) substrate. An additional 0.4 pF capacitor is connected between the coaxial probe and the outer microstrip line to improve $S_{11} < -10$ dB. (c) Comparison of simulated and measured return loss. (d) 4.34 dB *x*-pol gain is measured at 2.65 GHz on the *y*–*z* plane. Antenna footprint is $\lambda_0/9.8 \times \lambda_0/9.8 \times \lambda_0/19.7$ at 2.4 GHz. Reprinted with permission from Ref. 37, Copyright 2011, IEEE.

But the DBE antennas resulting from the concept in Fig. 3.18 do not have significant bandwidth since the DBE resonances are maximally flat near the band edge. Thus, a modification on the dispersion diagram was considered by using magnetic substrates to create an MPC mode. In this case, the ferrite substrate was only placed below the coupled middle section of the three-section transmission line unit cell. This led to unequal coupling coefficients ($K_1 \neq K_2$) as discussed in the previous section (Fig. 3.9). Consequently, it provided the means for creating anisotropy. The ferrite blocks/substrates were made of CVG (from TCI Ceramics, $4\pi M_s$ = 1000 G, ΔH = 6 Oe, ε_r = 15, tan δ = 0.00014) [12] and were inserted as in Fig. 3.20. The ferrite biasing was also controlled to improve bandwidth by introducing an SIP in the dispersion diagram, viz. a third-order dispersion curve. A bias field of 1000 G was applied normal to the ground plane to saturate the magnetic inserts. This antenna achieved 3.1 dB realized gain with 8.1% bandwidth at 1.51 GHz. The radiation efficiency of the antenna was 73% due to losses associated with non-uniformly biased ferrite sections. Nevertheless, the antenna had a remarkably small footprint of $\lambda_0/9.8 \times \lambda_0/10.4$ on a $\lambda_0/16$ thick substrate, making it near optimal in terms of gain-bandwidth product with respect to the Chu–Harrington limit.

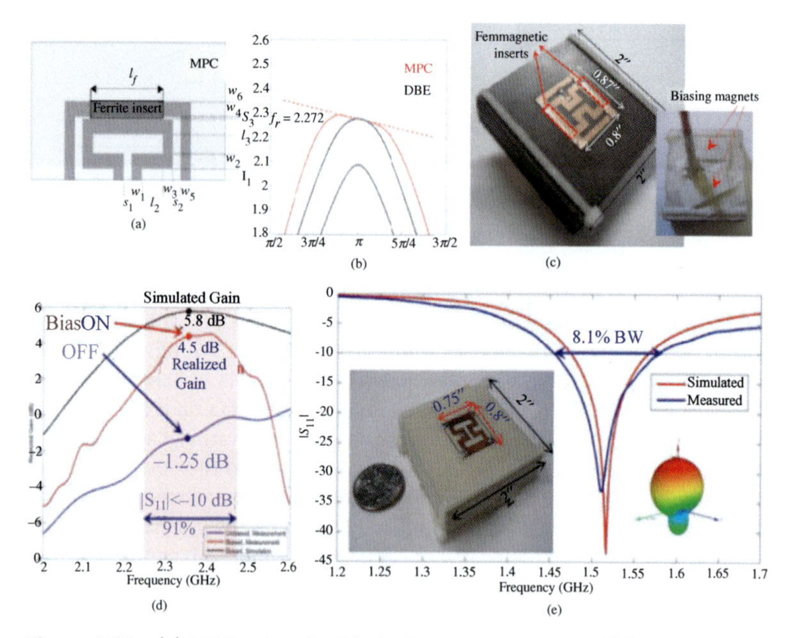

Figure 3.20 (a) MPC unit cell with design parameters (in mils): $w_1 = w_2 = w_3 = 100$, $w_4 = 20$, $w_5 = 80$, $w_6 = 120$, $s_1 = 50$, $s_2 = 70$, $s_3 = 10$, $l_1 = 50$, $l_2 = 60$, $l_3 = 100$, $l_f = 500$. (b) Corresponding third-order dispersion diagram of the unit cell. (c) Fabricated MPC antenna on composite substrate. Calcium vanadium garnet (CVG, $4\pi M_s = 1000$ G, $\Delta H = 6$ Oe, $\varepsilon_r = 15$, $\tan\delta = 0.00014$) sections are inserted into the low contrast Duroid substrate ($\varepsilon_r = 2.2$, $\tan\delta = 0.0009$). The bottom view of the antenna with magnets is shown on the bottom right inset. (d) Comparison of simulated and measured gains. (e) Miniature MPC antenna performance. The substrate is formed by inserting CVG sections into the high-contrast Rogers RT/Duroid 6010 laminate ($\varepsilon_r = 10.2$, $\tan\delta = 0.0023$). Reprinted with permission from Ref. 39, Copyright 2011, IEEE.

We can conclude that the DBE modes can lead to miniaturization and high directivity, but lack wideband characteristics. On the contrary, MPC modes can have moderate miniaturization with significant bandwidth enhancement. A comparison of the DBE and MPC mode antennas along with typical microstrip patch antennas is provided in Fig. 3.21.

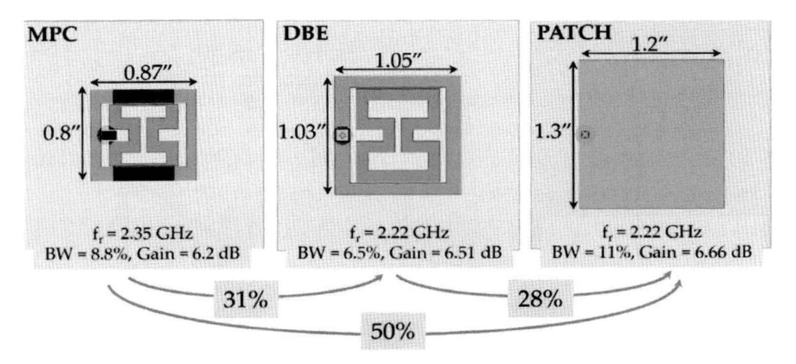

Figure 3.21 Comparison of the MPC, DBE, and patch antenna performance on a 2″×2″ in size and 500 mil thick grounded substrate. Reprinted with permission from Ref. 38, Copyright 2010, Elsevier.

3.3.3 Leaky-Wave Antenna

Leaky-wave antennas (LWAs) are guiding structures [40] that radiate due to leakage caused by a small slit or cut on the mode guiding structure [41]. As the wave travels on the guiding structure (see Fig. 3.22), it experiences exponential decay due to the aforementioned leakage. In general, the rate of energy leakage is very small along the guiding path, implying the need for long LWA antennas, viz. several periodic cells long. Early versions of LWAs were standard rectangular waveguides with a periodic set of slits on one of their faces [41]. Each slit radiates a small portion of the waveguide mode's energy as the mode travels, and its size, periodic arrangement, and orientation provide for bandwidth, beam, and radiation efficiency control. A key parameter is the propagation constant of the guided wave, β, and as noted in Fig. 3.22, β must be kept smaller than k_0, the free-space wave number. Typically, we write the LWA's propagation constant in its complex form, $k = \beta - ja$ [41]. The attenuation constant represents the loss of energy due to radiation. For uniform waveguides, the leakage is usually exponential and α has a small value. But for other guiding structures, α can attain a larger value, implying stronger radiated fields. Typically, the beam direction of the LWA array changes with frequency, and this is because the phase delay or period length also changes with frequency. Consequently, LWAs can be scanned by changing the frequency of operation.

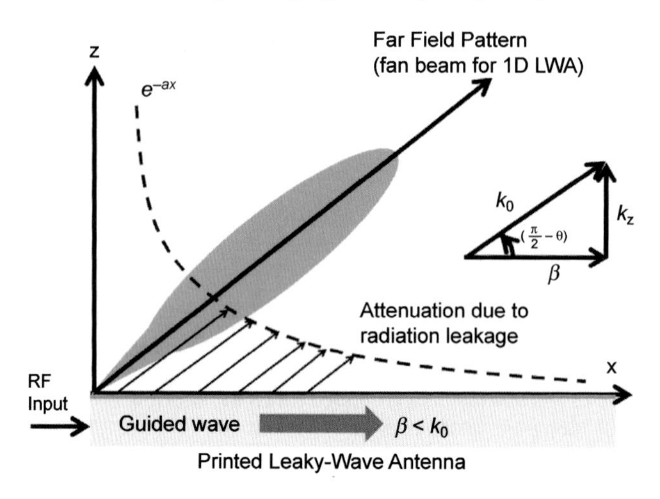

Figure 3.22 Operational principle of LWAs. Their radiation can be attributed to a complex propagation constant with radiation occurring when the LWA propagation constant is less than the free-space propagation wavenumber. Reprinted with permission from Ref. 41, Copyright 2007, McGraw Hill.

We can distinguish two types of LWAs. One class refers to uniform guided structures and another to LWA constructed of several periodic unit cells. A uniform LWA may be a waveguide that contains a longitudinal slot for radiation and supports a fast wave only. These antennas radiate as soon as the wave reaches the opening in the waveguide. Periodic LWAs can support forward and backward waves. Of course, the period and unit cell structure (see Fig. 3.23) can be adjusted to control radiation and frequency of operation. It is important to note that radiation from these LWAs is not due to the dominant DBE or MPC slow wave mode. Instead, it is necessary to also consider all harmonics of the dominant slow wave. These harmonics have the propagation constants

$$\frac{\beta_n}{k_0} = \frac{\beta_0}{k_0} + \frac{2n\pi}{k_0 p} = \frac{\beta_0}{k_0} + \frac{n\lambda_0}{p} \tag{3.20}$$

where p is the periodicity of the structure.

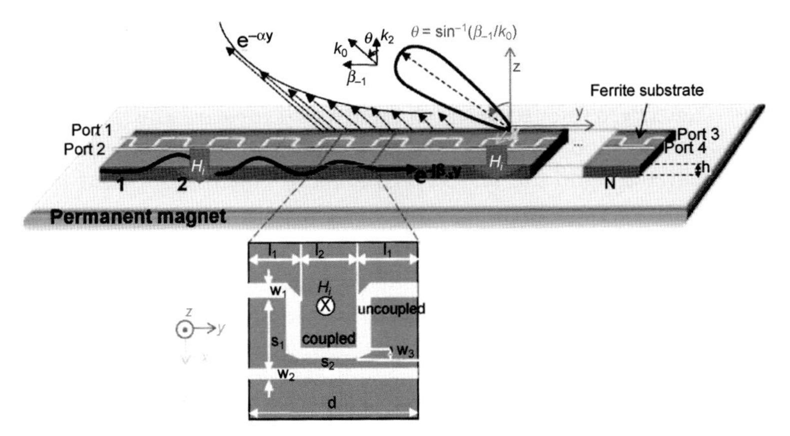

Figure 3.23 Coupled lines printed on a ferrite substrate having the material properties $4\pi M_s$ = 1000 G, loss linewidth ΔH = 10 Oe, relative permittivity ε_r = 14, and dielectric loss tangent tan δ_ε = 0.0002. The internal DC magnetic field H_i = 1450 Oe is assumed to be in −z-direction, normal to the ground plane. Unit cell dimensions are l_1 = 120, l_2 = 200, w_1 = 60, w_2 = 20, w_3 = 30, s_1 = 105, s_2 = 10 (mils). Reprinted with permission from Ref. 43, Copyright 2013, IEEE.

As noted above, LWAs support both forward and backward waves. But this leads to multiple radiation beams, one for the forward and another for the backward mode. A way to suppress the backward mode is to employ a periodic cell that is anisotropic. More specifically, the MPC unit cell has this property. With this in mind, a non-reciprocal magnetic-biased LWA was proposed in Ref. [42] and demonstrated by Apaydin et al. [43]. Non-reciprocity was achieved using the approach described in Fig. 3.9. That is, two non-identical transmission lines with magnetic biasing led to unequal coupling coefficients $K_1 \neq K_2$, which eventually provides non-reciprocal MPC modes. Figure 3.23 shows the unit cell design comprising a pair of CTLs on a biased ferrite substrate. The transmission lines are printed on a 100 mil thick commercially available CVG substrate. The material properties of the CVG, as specified by the manufacturer are as follows: saturation magnetization $4\pi M_s$ = 1000G, loss linewidth ΔH = 10 Oe, relative permittivity ε_r = 14, and dielectric loss tangent tan δ_ε = 0.0002. The narrow linewidth of the magnetic material was specifically chosen to minimize losses.

The unique feature of the MPC mode LWA antenna in Fig. 3.23 is its scanning capability via an external biasing voltage on the ferrite substrate. This external bias serves to change the effective constitutive parameters of the scanning angle versus frequency

[43]. The dispersion diagram of the propagating modes by the subject LWA is given in Fig. 3.24 and includes several harmonics. One of them represents the fast/leaky-wave mode. Also, the black and blue curves refer to the $n = \pm 1$ and $n = -1$ forward and backward propagating modes, respectively. We also observe the spectral asymmetry around the 3.7–3.78 GHz band. This is associated with the $n = -1$ fast wave.

To use the LWA as a transmitting antenna, as in Fig. 3.24, it can be fed at Ports 1 or 2 (Fig. 3.23) and terminated on Ports 3 or 4 with a matched load $Z_L = 50\ \Omega$. As already noted, spectral asymmetry suppresses backward radiation due to reflected waves $(\beta_{+1}^{-} > k_0)$. But in the receive mode, only forward waves are supported $(\beta_{-1}^{+} < k_0)$. These forward waves will, in turn, be guided with the leftover signal dissipated at the matched load termination Z_L.

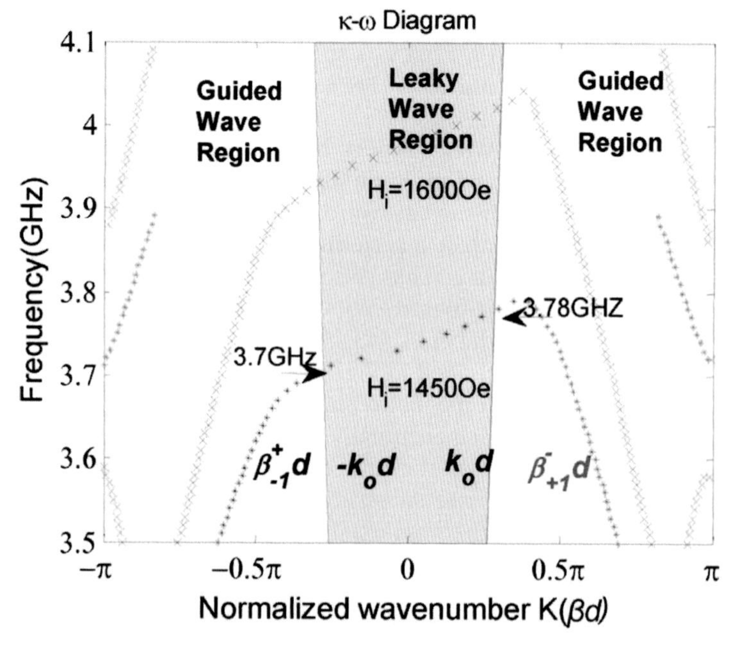

Figure 3.24 Dispersion diagram of the coupled microstrip lines unit cell in Fig. 3.23. Reprinted with permission from Ref. 43, Copyright 2013, IEEE.

Another important feature of the antenna is its capability of scanning at a fixed frequency by changing the external biasing magnetic field. The magnetic biasing can control the magnetic property of the ferrites and can shift the ω–β diagram to higher or lower. As shown in Fig. 3.25, the antenna gain remained reasonably constant with respect to frequency change. Thus, frequency-independent operation of scanning with magnetic biasing is expected to be observed in these types of MPC-based LWAs.

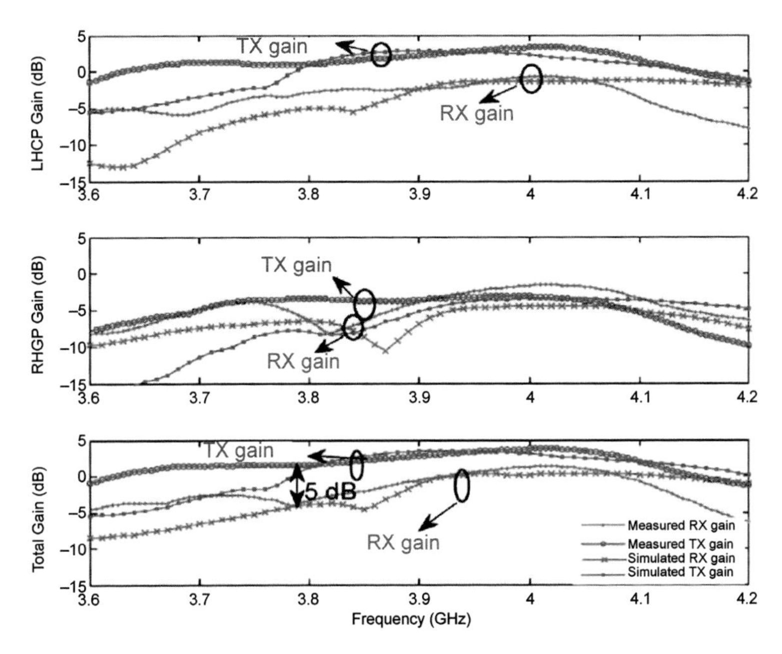

Figure 3.25 Simulated versus measured TX/RX antenna gain as a function of frequency. Reprinted with permission from Ref. 43, Copyright 2013, IEEE.

References

1. A. Figotin and I. Vitebsky, "Slow wave phenomena in photonic crystals," *Laser Photon. Rev.*, vol. 5, no. 2, pp. 201–213, 2006.
2. T. Baba, "Slow light in photonic crystals," *Nat. Photon.*, vol. 2, pp. 465–473, 2008.
3. S. Yarga, K. Sertel, and J. L. Volakis, "Degenerate band edge crystals for directive antennas," *IEEE Trans. Antennas Propag.*, vol. 56, no. 1, pp. 119–126, 2008.
4. B. Epsztein, "Slow-wave structures in microwave tubes," *IEDM Tech. Dig.*, pp. 486–489, 1984.
5. N. Apaydin, L. Zhang, K. Sertel, and J. L. Volakis, "Experimental validation of frozen modes guided on printed coupled transmission

lines," *IEEE Trans. Microw. Theory Tech.*, vol. 60, no. 6, pp.1513–1518, 2012.

6. Z. Zhang and S. Satpathy, "Electromagnetic wave propagation in periodic structures: Bloch wave solution of Maxwell's equations," *Phys. Rev. Lett.*, vol. 65, pp. 2650–2653, 1990.
7. A. Figotin and I. Vitebsky, "Nonreciprocal magnetic photonic crystals," *Phys. Rev. E*, vol. 63, no. 066609, pp. 1–20, 2001.
8. M. Caiazzo, S. Maci, and N. Engheta, "A metamaterial surface for compact cavity resonators," *IEEE Antennas Wireless Propag. Lett.*, vol. 3, pp. 261–264, 2004.
9. S. Yarga, K. Sertel, and J. L. Volakis, "Finite degenerate band edge crystals using barium titanate-alumina layers emulating uniaxial media for directive planar antennas," *2007 IEEE Antennas and Propagation Society International Symposium,* Honolulu, HI, pp. 1317–1320, 2007.
10. A. Figotin and I. Vitebsky, "Electromagnetic unidirectionality in magnetic photonic crystals," *Phys. Rev. B*, vol. 67, no. 165210, pp. 1–20, 2003.
11. J. L. Volakis, G. Mumcu, K. Sertel, C.-C. Chen, M. Lee, B. Kramer, D. Psychoudakis, and G. Kiziltas, "Antenna miniaturization using magnetic photonic and degenerate band-edge crystals," *IEEE Trans. Antenn. Propag.*, vol. 48, pp. 12–28, 2006.
12. E. Irci, K. Sertel, and J. L. Volakis, "Antenna miniaturization for vehicular platforms using printed coupled lines emulating magnetic photonic crystals," *Metamaterials*, vol. 4, no. 2–3, pp. 127–138, 2010.
13. G. Mumcu, K. Sertel, J. L. Volakis, I. Vitebskiy, and A. Figotin, "RF propagation in finite thickness unidirectional magnetic photonic crystals," *IEEE Trans. Antennas Propag.*, vol. 53, no. 12, pp. 4026–4034, 2005.
14. C. Locker, K. Sertel, and J. L. Volakis, "Emulation of propagation in layered anisotropic media with equivalent coupled microstrip lines," *IEEE Microw. Wireless Compon. Lett.*, vol. 16, no. 12, pp. 642–644, 2006.
15. M. B. Stephanson, K. Sertel, and J. L. Volakis, "Frozen modes in coupled microstrip lines printed on ferromagnetic substrates," *IEEE Microw. Wireless Compon. Lett.*, vol. 18, no. 5, pp. 305–307, 2008.
16. K. Zhang and D. Li, *Electromagnetic Theory for Microwave and Optoelectronics*, Springer, 2nd ed., 2007.
17. Robert E. Collin, *Field Theory of Guided Waves*, IEEE Press, 1991.
18. I. A. Sukhoivanov and I. V. Guryev, *Photonic Crystals: Physics and Practical Modeling*, Springer-Verlag, Berlin, 2009.

19. D. A. Watkins, *Topics in Electromagnetic Theory*, John Wiley & Sons, New York, 1958.
20. A. Figotin and I. Vitebsky, "Gigantic transmission band-edge resonance in periodic stacks of anisotropic layers," *Phys. Rev. E*, vol. 72-036619, pp. 1–12, 2005.
21. R. Chilton, K.-Y. Jung, R. Lee, and F. L. Teixeira, "Frozen modes in parallel-plate waveguides loaded with magnetic photonic crystals," *IEEE Trans. Microw. Theory Tech.*, vol. 55, no. 12, pp. 2631–2641, 2007.
22. L. Zhang, G. Mumcu, S. Yarga, K. Sertel, J. L. Volakis, and H. Verweij, "Fabrication and characterization of anisotropic dielectrics for low-loss microwave applications," *J. Mater. Sci.*, vol. 43, no. 5, pp. 1505–1509, 2008.
23. G. Mumcu, K. Sertel, and J. L. Volakis, "Miniature antenna using printed coupled lines emulating degenerate band edge crystals," *IEEE Trans. Antennas Propag.*, vol. 57, no. 6, pp. 1618–1624, 2009.
24. W. X. Wang, Y. Y. Wei, G. F. Yu, Y. B. Gong, M. Z. Huang, and G. Q. Zhao, "Review of the novel slow-wave structures for high-power traveling-wave tube," *Int. J. Infrared Millimeter Waves*, vol. 24, no. 9, pp. 1469–1484, 2003.
25. A. S. Gilmour Jr., "Traveling wave tubes," in *Klystrons, Traveling Wave Tubes, Magnetrons, Crossed-Field Amplifiers, and Gyrotrons*, Artech House, 2011.
26. J. R. Pierce, *Traveling-Wave Tubes*, New York: Van Nostrand, 1950.
27. S. K. Datta, V. B. Naidu, P. R. R. Rao, L. Kumar, B. N. Basu, "Equivalent circuit analysis of a ring–bar slow-wave structure for high-power traveling-wave tubes," *IEEE Trans. Electron Devices*, vol. 56, no. 12, pp. 3184–3190, 2009.
28. T. Munehiro, M. Yoshida, K. Tomikawa, A. Kajiwara, and K. Tsutaki, "Development of an S-band 1 KW pulsed mini-TWT for MPMs," *Vacuum Electronics Conference, 2007 IEEE International*, pp. 1–2, 2007.
29. J. L. Volakis and K. Sertel, "Narrowband and wideband metamaterial antennas based on degenerate band edge and magnetic photonic crystals," *Proc. IEEE*, vol. 99, no. 10, pp. 1732–1745, 2011.
30. W. E. Kock, "Metallic delay lenses," *Bell Syst. Tech. J.*, vol. 27, pp. 58–82, 1948.
31. A. Munir, N. Hamanaga, H. Kubo, and I. Awai, "Artificial dielectric rectangular resonator with novel anisotropic permittivity and its TE mode waveguide filter application," *IEICE Trans. Electron.*, vol. E88-C, no. 1, pp. 40–46, 2005.

32. I. Awai, H. Kubo, T. Iribe, D. Wakamiya, and A. Sanada, "An artificial dielectric material of huge permittivity with novel anisotropy and its application to a microwave BPF," *IEICE Trans. Electron.*, vol. E88-C, no. 7, pp. 1412–1419, 2005.

33. S. Yarga, K. Sertel, and J. L. Volakis, "Multilayer dielectric resonator antenna operating at degenerate band edge modes," *IEEE Antennas Wireless Propag. Lett.*, vol. 8, pp. 287–290, 2009.

34. G. Mumcu, K. Sertel, and J. L. Volakis, "Lumped circuit models for degenerate band edge and magnetic photonic crystals," *IEEE Microwave Wireless Components Lett.*, vol. 20, no. 1, pp. 4–6, 2010.

35. G. Mumcu, K. Sertel, and J. L. Volakis, "Printed coupled lines with lumped loads for realizing degenerate band edge and magnetic photonic crystal modes," *2008 IEEE Antennas and Propagation Society Symposium*, San Diego, CA, 2008.

36. G. Mumcu, K. Sertel, and J. L. Volakis, "Partially coupled microstrip lines for antenna miniaturization," *IEEE International Workshop on Antenna Technology: Small Antennas and Novel Metamaterials* (*IWAT*), Santa Monica, CA, 2009.

37. G. Mumcu, S. Gupta, K. Sertel, and J. L. Volakis, "Small wideband double-loop antennas using lumped inductors and coupling capacitors," *IEEE Antennas Wireless Propag. Lett.*, vol. 10, pp. 107–110, 2011.

38. E. Irci, K. Sertel, and J. L. Volakis, "Antenna miniaturization for vehicular platforms using printed coupled lines emulating magnetic photonic crystals", *Metamaterials,* vol. 4, no. 2–3, pp. 127–138, 2010.

39. N. Apaydin, G. Mumcu, E. Irci, K. Sertel, and J. L. Volakis, "Miniature antennas based on printed coupled lines emulating anisotropy," *IET Microwaves, Antennas and Propagation,* vol. 61, no. 7, 2013.

40. D. R. Jackson, C. Caloz, and T. Itoh, "Leaky-wave antennas," *Proc. IEEE,* vol. 100, no. 7, pp. 2194–2206, 2012.

41. A. A. Oliner and D. R. Jackson, "Leaky-wave antennas," in *Antenna Engineering Handbook,* J. L. Volakis, Ed. McGraw Hill, 2007.

42. N. Apaydin, L. Zhang, K. Sertel, and J. L. Volakis, "Nonreciprocal and magnetically scanned leaky-wave antenna using coupled microstrip lines," in *Proceedings of Antennas and Propagation Society International Symposium,* Chicago, IL, USA, 2012.

43. N. Apaydin, K. Serlel, and J. L. Volakis, "Nonreciprocal leaky-wave antenna based on coupled microstrip lines on a non-uniformly biased ferrite substrate," *IEEE Trans. Antennas Propag.*, vol. 61, no. 7, pp. 3458–3465, 2013.

Chapter 4

Design Synthesis of Multiband and Broadband Gap Electromagnetic Metasurfaces

Idellyse Martinez, Anastasios H. Panaretos, and Douglas H. Werner
Department of Electrical Engineering, Penn State University,
University Park, PA 16802, USA
dhw@psu.edu

This chapter presents a design methodology for the synthesis of multiband and broadband electromagnetic bandgap devices based on the mushroom or Sievenpiper structure. The methodology described herein is based on the non-uniform capacitive loading of mushroom structures using lumped capacitors. The documented analysis begins with the observation that the bandgap of a mushroom-type metasurface shifts to lower frequencies once a load capacitance is applied across the gaps defined between the patches of neighboring unit cells. Our analysis further reveals that if, instead of the uniform capacitive loading, a non-uniform scheme is applied, then the resulting structure is characterized by a significantly wider stopband. In the first part of this chapter, the theoretical basis behind

Broadband Metamaterials in Electromagnetics: Technology and Applications
Edited by Douglas H. Werner
Copyright © 2017 Pan Stanford Publishing Pte. Ltd.
ISBN 978-981-4745-68-0 (Hardcover), 978-1-315-36443-8 (eBook)
www.panstanford.com

the non-uniform capacitive loading of electromagnetic bandgap (EBG) surfaces is investigated and additionally the theoretical analysis of the proposed methodology is presented. Afterward its applicability is demonstrated through representative examples that involve the investigation of the isolation properties of EBGs embedded into transverse electromagnetic (TEM) waveguides. From a practical standpoint the key aspect of the proposed methodology is the multiport network representation of the EBG. This allows the structure to be analyzed using simple circuit-type calculations rather than full-wave simulations. In the second part, this design methodology is extended to the case of open structures and again its applicability is demonstrated through indicative examples. Finally, in the third part, the aforementioned circuit-based EBG synthesis approach is adapted for the design of tunable absorbers that utilize the mushroom unit cell loaded with tuning capacitive and resistive lumped elements.

4.1 Introduction

One of the devices that have revolutionized microwave engineering in the past two decades is the so called "mushroom" or "Sievenpiper structure." In his seminal paper [1], Dan Sievenpiper and colleagues introduced a simple yet powerful methodology to design a high impedance surface (HIS), which additionally exhibits surface wave suppression properties and, therefore, functions as an EBG device operating at microwave frequencies. This mushroom-type HIS can be classified as a metamaterial and, in particular, as a metasurface since, as will be shown later in this section, it exhibits artificial properties equivalent to that of a material characterized by negative dielectric permittivity. The layout of a mushroom-type metasurface is shown in Fig. 4.1. It comprises a 2D periodic arrangement of square metallic patches that extend along the x–y plane, etched on a grounded dielectric substrate. The patch of each unit cell is connected to the ground through shorting metallic vias, as depicted in Fig. 4.1b.

Undoubtedly, the Sievenpiper metasurface has been one of the most useful and most successful metamaterial configurations. Since its conception, there has been a tremendous research effort devoted to the analysis and improvement of the original design, while numerous examples have been documented that demonstrate its applicability and ease of implementation [2–12]. One of the most popular applications of the mushroom-type EBG is the radiation enhancement of low-profile antennas [13–19]. It is well known that when, for instance, a horizontal-wire-type antenna is placed in close proximity to a perfect electric conductor (PEC) ground plane (at a distance shorter than $\lambda/4$), its radiation properties are degraded due to destructive interference created by the out-of-phase radiation from the antenna's image. Now, as mentioned earlier, a mushroom-type metasurface behaves like a narrowband artificial magnetic conductor (AMC). As a consequence, when the same wire-type antenna is placed in close proximity to it, the resulting scattered field is in phase with the electromagnetic field radiated by the antenna, and thus its radiation properties are enhanced. Note, however, that this is an ideal scenario and in reality, when an actual antenna configuration, as the one described previously, is designed, there are several other parameters that need to be taken into account in order to achieve the desired radiation enhancement effects. These parameters are related to the coupling/interference between the metasurface and the antenna, and they manifest themselves as detuning of the antenna and of the AMC.

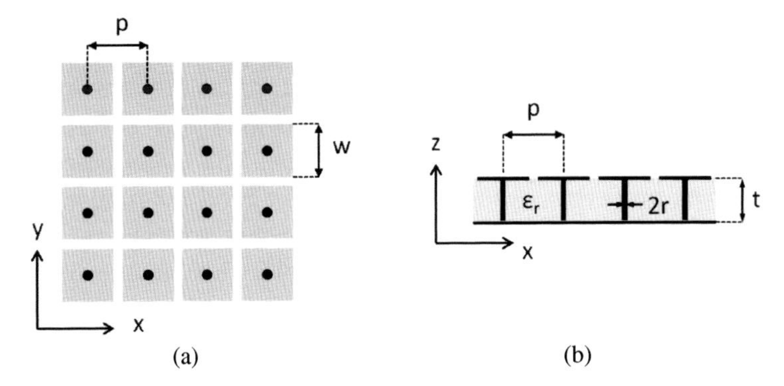

Figure 4.1 Two-dimensional periodic mushroom-type metasurface: (a) top view, (b) side view.

An equally popular and useful application of the mushroom-type metasurface, which is the main focus of this chapter, is surface wave suppression. Surface wave suppression is a topic of paramount importance in microwave and antenna engineering since the unwanted excitation of surface waves can be a major source of electromagnetic interference with detrimental effects on the operation of radio frequency (RF) devices. In the context of antenna design, a notable example of the EBG's applicability is for the minimization of the mutual coupling between neighboring patch antennas [20–27]. In particular, consider two patch antennas in close proximity to each other that form a two-element antenna array. Usually, the two radiating elements are etched on a grounded dielectric substrate. As a consequence, surface waves are excited along the air/dielectric interface. These surface waves propagate from one antenna to the other and result in unwanted interference that detunes and deteriorates their radiation performance. However, when a finite in length and, most importantly, judiciously tuned mushroom-type metasurface is placed in between these patch antennas, the excited surface waves can be significantly suppressed and thus the undisrupted radiation properties of the two-element antenna array are restored.

Although the principles of operation of a mushroom structure are well known, in what follows is a brief overview, which will help to put into perspective the design methodology presented later in this chapter. As mentioned previously, the Sievenpiper structure is essentially an HIS that additionally exhibits EBG properties. This chapter is focused on the latter functionality. A common misconception is that these two functionalities are related to each other; however, the underlying physics that govern the two phenomena are fundamentally different. The HIS is based on scattering properties, while the EBG surface behavior is governed by controlling wave propagation. Now let us first examine the case where the metasurface functions as an AMC. This functionality is enabled when the structure is illuminated by an incident wave; in other words, it is a typical electromagnetic scattering problem. In particular, if we consider a plane wave impinging normally on an AMC, then the impedance at the surface of the device can be represented by the impedance of the equivalent circuit shown in Fig. 4.2. The capacitance represents the field concentration between the metallic patches, while the shorted transmission line section represents the propagating wave in the grounded dielectric substrate. In the limit of an electrically thin substrate, the input admittance of this structure reduces to that of a parallel LC circuit (shown in Fig. 4.2b), which is given by:

$$Y_{\mathrm{HIS}} = j\omega C + \frac{1}{jZ_0 \tan(\beta t)} \overset{|\beta t|<<1}{\approx} j\omega C + \frac{1}{j\omega L} \tag{4.1}$$

Evidently, the preceding expression indicates that at the resonant frequency $\omega = (LC)^{-1/2}$, the metasurface reflects the impinging wave similar to a perfect magnetic conductor (PMC). However, there is no clear evidence how this resonant response justifies the surface wave suppression mechanism that characterizes an EBG device. Figure 4.3 shows the reflection coefficient phase variation of an actual mushroom-based AMC when it is illuminated by a normally impinging plane wave. It is commonly assumed that the frequencies defined within the ±90° phase variation range, shown in Fig. 4.3, correspond to the stopband of the EBG. However, it has been demonstrated by several researchers through very detailed analyses that this is not the case, while studies have been documented that have attempted to identify and quantify the relation between the two phenomena [28–32]. In conclusion, intuitively, one would expect that since a periodic arrangement of the circuit shown in Fig. 4.2 cannot possibly describe any propagation mechanism, then this circuit configuration would not suffice to represent the inherently propagating surface waves.

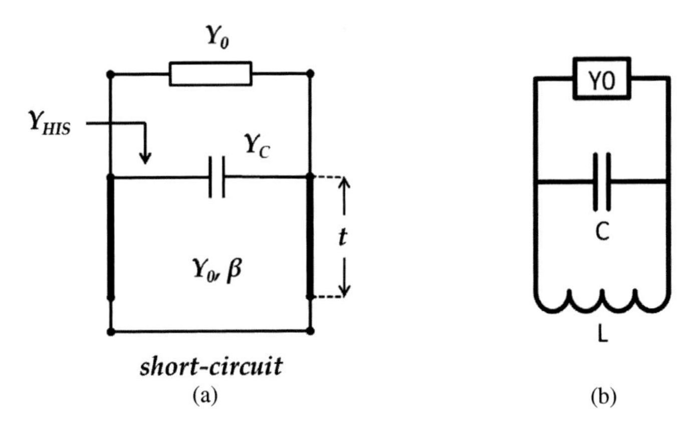

Figure 4.2 Equivalent circuit representation of an AMC's input impedance when the metasurface is illuminated by a normally incident plane wave.

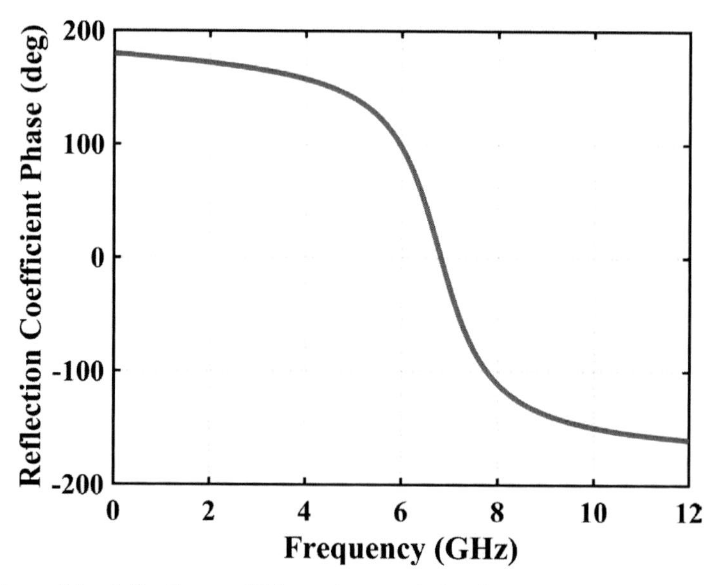

Figure 4.3 Reflection coefficient phase variation from an AMC with t = 1.6 mm, p = 6 mm, w = 5.7 mm, r = 0.2 mm, and ε_r = 4.4.

The reality is that the stopband exhibited by the mushroom-type metasurface is the resulting effect of the interaction between the oppositely directed TM_0 surface wave and the left-handed or backward propagating wave supported by the structure [33, 34]. While the excitation of surface waves across an impedance surface is a well-known phenomenon, the excitation of left-handed waves is a feature that uniquely characterizes the structure under consideration. As a matter of fact, because of this particular property, the mushroom structure is categorized as a metamaterial, or as a metasurface.

The mushroom unit cell is one realization of what is usually referred to as a composite right-/left-handed (CRLH) transmission line [35, 36]. In particular, negative refraction in microwave frequencies has been demonstrated by devising 2D transmission lines through a periodic arrangement of mushroom-type unit cells, also referred to as flat lenses. The circuit topology of a CRLH unit cell is shown in Fig. 4.4. The indicated inductance L_R is attributed to the currents induced on the surface of the metallic patch, while the capacitance C_R is due to the electric field confinement between the metallic patches and the ground plane. Moreover, the capacitance C_L is due to the field confinement between the metallic patches of adjacent unit cells, while the inductance L_L is attributed to the effective negative permittivity of the wire/Brown medium that the vias constitute. Note that subscript "R" indicates that the corresponding lumped element contributes to forward or right-handed wave propagation. In contrast, subscript "L" indicates that the corresponding lumped element contributes to backward or left-handed wave propagation. It should be emphasized that among the four lumped elements, only L_L is an artificial/engineered quantity. It is important to note that the circuit schematic in Fig. 4.4 does not represent the TM_0 surface wave mode that the structure supports. It only captures the backward and forward guided modes of the mushroom-type EBG.

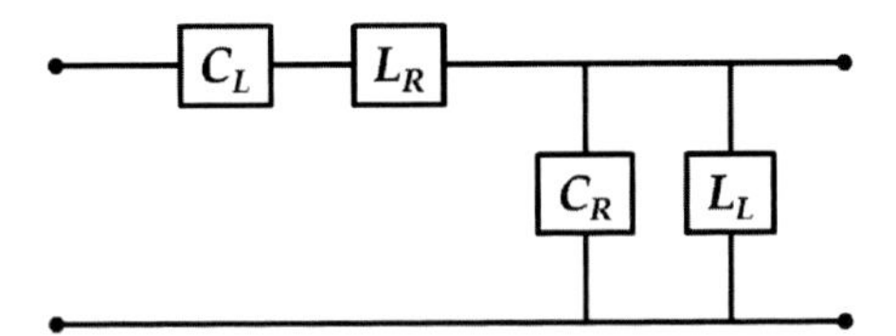

Figure 4.4 CRLH transmission line unit cell.

This chapter is concerned with techniques that allow the manipulation of the surface wave isolation performance of mushroom-type EBG structures. This task is equivalent to the design of a stopband microwave filter, where it is well known that an important figure of merit for such a structure is the manipulation of its bandwidth. Two design specifications that usually are of great practical importance are (a) the realization of a wide stopband and (b) the realization of multiple isolation bands at specific design frequencies. The former design ensures robust and undisrupted isolation performance, while the latter allows a designer to accommodate different functionalities with the same isolation device.

A cursory examination of standard microwave filter architectures, such as the binomial or the Chebyshev, immediately reveals that their performance is based on introducing spatial inhomogeneity along the direction of propagation. This approach has also been adopted in the development of wideband EBG devices [37–44]. In the context of mushroom-type EBGs, the introduction of longitudinal inhomogeneity requires the design of a structure where along the direction of propagation, at least one of the four geometric factors that determine the electromagnetic properties of a unit cell varies as a function of space. Namely, these factors include the following: the patch size or shape (L_R), the diameter size of the via (L_L), the gap size between adjacent patches (C_L), and finally, the height and constitution of the substrate (C_R).

Although in principle a mushroom-based EBG that exhibits inhomogeneity in any of the above parameters can be a candidate for the realization of wideband isolation, from a practical standpoint, this is not a viable solution. Ideally, it is desired to have a baseline structure and introduce spatial inhomogeneity with no geometric modifications. Preferably any modification should be introduced using electrical/electronic based architectures because this could potentially allow the creation of tunable mushroom-type EBGs. For these reasons, the proposed design methodology employs a standard geometrically uniform mushroom EBG and the required inhomogeneity is introduced by capacitive loading using lumped capacitors placed in the gaps defined between the patches of neighboring unit cells. This approach does not require any geometrical modification of the baseline structure, and it can be extended to a tunable EBG by substituting the capacitors with varactors. With respect to the schematic shown in Fig. 4.4, this loading scheme corresponds to adding a capacitor in parallel to C_L. It turns out that such loading offers great flexibility for the manipulation of the EBG's isolation performance. In the next sections, the theoretical justification for this approach is presented along with several indicative examples that clearly demonstrate its utility.

4.2 Capacitively Loaded Mushroom-Type EBG

This section presents a compact methodology that allows wideband or multiband surface wave suppression through capacitive loading of a mushroom-type EBG. The capacitive loading is applied across the gaps defined between the patches of neighboring unit cells. The documented analysis begins by giving first the theoretical justification behind the choice of this particular loading scheme. Once the necessary theoretical basis has been established, the practical and computational aspects of the problem are discussed.

As mentioned previously, the objective in the proposed design methodology is to determine the loading capacitor combination that yields some desired isolation performance. Obviously, if one attempts to determine the necessary capacitor values through repeated 3D full-wave simulations, the problem becomes computationally intractable even for an EBG structure with a moderate number of unit cells. The beauty of the proposed methodology is that it exploits the fact that the EBG configuration under study can be conveniently represented as a multiport microwave device and thus it can be effectively described in terms of an *S*-matrix. As a result, its analysis does not require computationally intensive 3D full-wave simulations but rather simple circuit-type calculations that drastically reduce the simulation/design time. In addition, the design time is further accelerated by integrating the aforementioned circuit-based analysis of the EBG into an optimization procedure, which can be utilized to determine the appropriate values of the loaded lumped elements based on some targeted design criteria.

4.2.1 Theory

We begin our analysis by considering the EBG configuration shown in Fig. 4.5. The characteristics of the unit cell are the following: square patch length 6.5 mm, via radius 12 mil, periodicity 7 mm, substrate height 1.52 mm, and substrate constitution Rogers RO3203 with ε_r = 3.02. For the numerical experiments documented in this section, the metasurface is embedded into a TEM waveguide with height equal to 3.04 mm. This is realized by defining PEC boundary conditions on the top and bottom faces of the computational model's bounding box and PMC boundary conditions on its side faces. Note that this numerical configuration models a metasurface that extends infinitely only along the transverse direction, in this case being the *x*-axis. Obviously, the structure is finite along the direction of propagation and, in this case, it comprises 12 unit cells. The TEM waveguide is excited at its two ends (using wave ports), and the isolation performance of the EBG is gauged by the frequency response of the transmission coefficient S_{12} defined between the two excitation ports. It needs to be emphasized that although the aforementioned propagation scenario is rather simplified, it serves to provide important insight into the underlying physics that governs the electromagnetic response of such EBG structures. Furthermore, through this specific example, it becomes very straightforward to justify and illustrate the theoretical basis behind the proposed capacitive loading scheme.

As mentioned previously, it is proposed to improve the isolation performance of the EBG configuration shown in Fig. 4.5 by loading, along the direction of propagation, the gaps defined between the mushroom patches with lumped elements consisting of capacitors. The most instructive way to introduce the proposed loading scheme is by first determining how a uniform capacitive loading affects the surface wave performance of an EBG metasurface.

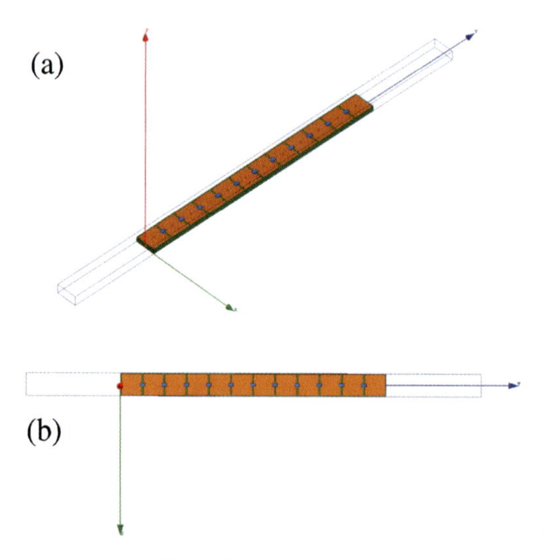

Figure 4.5 TEM waveguide with a capacitively loaded 12-unit-cell long mushroom-type EBG embedded into it: (a) perspective view, (b) top view.

The most informative way to perform this study is by computing the Brillouin diagram of the corresponding structure. Such a diagram provides all the necessary information regarding the dispersion properties of the various modes that the structure supports. Figure 4.6 shows the Brillouin diagrams corresponding to the cases where the EBG is unloaded and loaded, respectively, between its patches along the direction of propagation with 1 pF capacitors. Note here that since the metasurface is embedded within a TEM waveguide, the propagation of electromagnetic waves occurs only along the longitudinal direction (y-axis). For this reason, the phase sweep in the dispersion diagrams shown in Fig. 4.6 corresponds only to phase variation from lattice point Γ to lattice point X ($\beta_y p = 0 \rightarrow \pi$, $\beta_x p = 0$).

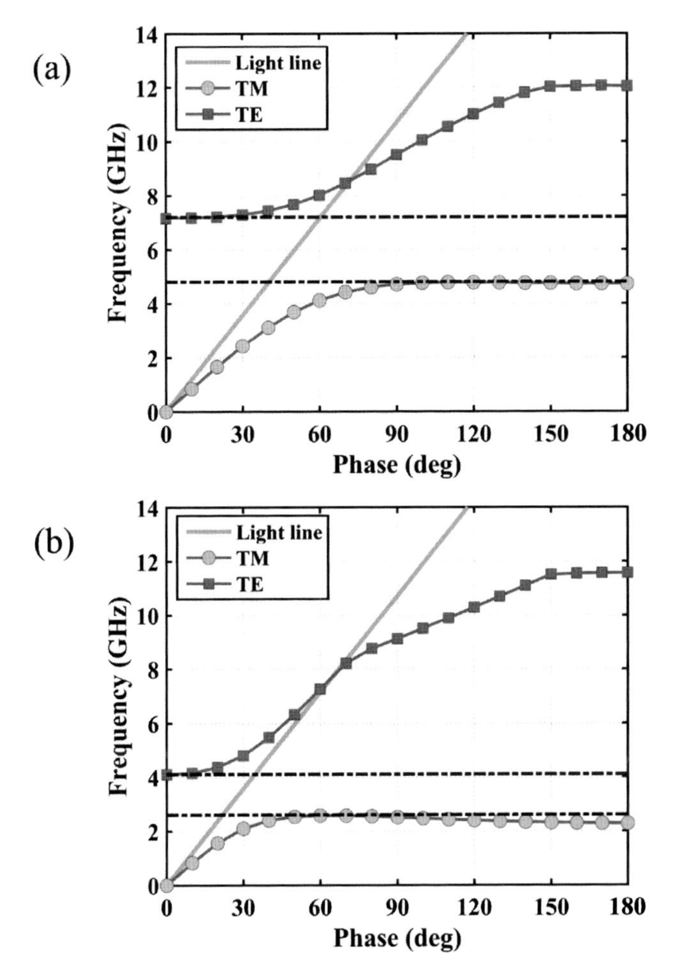

Figure 4.6 Brillouin diagrams of a 2D periodic mushroom-type EBG: (a) unloaded structure, (b) capacitively loaded EBG with 1 pF capacitors. The loads are applied along only those gaps that lie perpendicular to the x-axis.

The two Brillouin diagrams include the first TM and TE modes that, for this particular unit cell configuration, determine the first stopband of the EBG metasurface, denoted by the horizontal dash-dotted lines. In particular, for the unloaded EBG, its first stopband occurs in the frequency range from 4.8 GHz to 7.2 GHz, while for the 1 pF loaded EBG, the stopband shifts to the lower frequency range from 2.6 GHz to 4.1 GHz. From this numerical experiment, it can be concluded that the EBG's first stopband can be shifted to lower frequencies and thus controlled by simply increasing the value of the loading capacitance. However, the disadvantage of such uniform capacitor loading is that it does not further allow control of the bandwidth of the stopband. As a matter of fact, from the dispersion diagrams it can be observed that the lower in frequency the stopband shifts, the narrower it becomes. So the question becomes what pattern should the capacitive loading follow in order to control the suppression bandwidth of the EBG?

At this point, we recall that according to small reflection theory, a plane wave can undergo wideband reflection when it impinges upon a judiciously designed multilayered dielectric structure. This suggests that some sort of longitudinal inhomogeneity should be introduced along the EBG structure under consideration. Given that the geometrical characteristics of each unit cell as well as the substrate material constitution cannot be modified, the only viable solution to introduce the desired longitudinal inhomogenity is by loading the gap defined between neighboring unit cells with different capacitor values. The aforementioned statement is not as straightforward to prove by directly applying small reflection theory. However, in what follows we provide an alternative approximate but very insightful proof that clearly demonstrates the effects of non-uniform capacitive loading to the isolation performance of the EBG metasurface.

Let us consider the 12-unit-cell long EBG structure, shown in Fig. 4.6, uniformly loaded by a capacitance C_1. Let us now assume that the isolation between the two excitation wave ports can be written as

$$S_{12}^{(1)}(f) = \exp(-\alpha_y^{(1)} N_p)\exp(-j\beta_y^{(1)} N_p) \tag{4.2}$$

where N = 12 is the number of unit cells, and p is their periodicity so that N_p represents the total longitudinal length of the structure. The preceding expression tacitly assumes that the EBG-loaded TEM waveguide along the direction of propagation is characterized by an effective propagation constant $\beta_y^{(1)}$ and an effective attenuation constant $\alpha_y^{(1)}$. As a result, when electromagnetic waves travel though this loaded TEM waveguide, they undergo a cumulative attenuation equal to $\exp(-\alpha_y^{(1)} N_p)$. Now, it is well known that the qualitative characteristics of $\alpha_y^{(1)}$ are determined by the stopband of the EBG. Put differently, $\alpha_y^{(1)}$ attains non-zero values for those frequencies that lie within the stopband. This means there is a direct mapping between the frequency response of the effective attenuation constant $\alpha_y^{(1)}$ and the frequency range where the stopband occurs. Therefore, if $\alpha_y^{(1)}$ is known, then the frequency range where the stopband occurs can be determined.

Now, given the expression in Eq. (4.2), an estimate, in the mean average sense, of the effective attenuation constant associated with a single unit cell loaded with a capacitance C_1 is given by

$$\alpha_y^{(1)} p = -\frac{1}{N} \ln \left| S_{12}^{(1)}(f) \right| \tag{4.3}$$

In the same manner, we can derive the attenuation constant that corresponds to any uniform capacitive loading scenario. The importance of the previous outcome is that if we cascade (in this case) 12 unit cells, each one loaded by a different capacitor, and provided that there is no pronounced inter-unit-cell coupling, then an estimate of the resulting 12-unit-cell-long structure's isolation performance is given by

$$\left| S_{12}(f) \right| = \exp\left(-\sum_{i=1}^{12} \alpha_y^{(i)} p \right) \tag{4.4}$$

The beauty of the preceding result stems from the fact that it predicts that the total isolation of the non-uniformly loaded mushroom-type EBG is determined by the superposition of the effective attenuation constants of the individual unit cells. Moreover, since the attenuation constant is a positive real number that attains non-zero values within the frequency range that the stopband occurs, Eq. (4.4) shows that the stopbands of each unit cell that comprises the EBG superimpose. Note also that since the expression in Eq. (4.4) is an exponential function, the more unit cells there are along the direction of propagation, the higher the level of isolation that can be achieved.

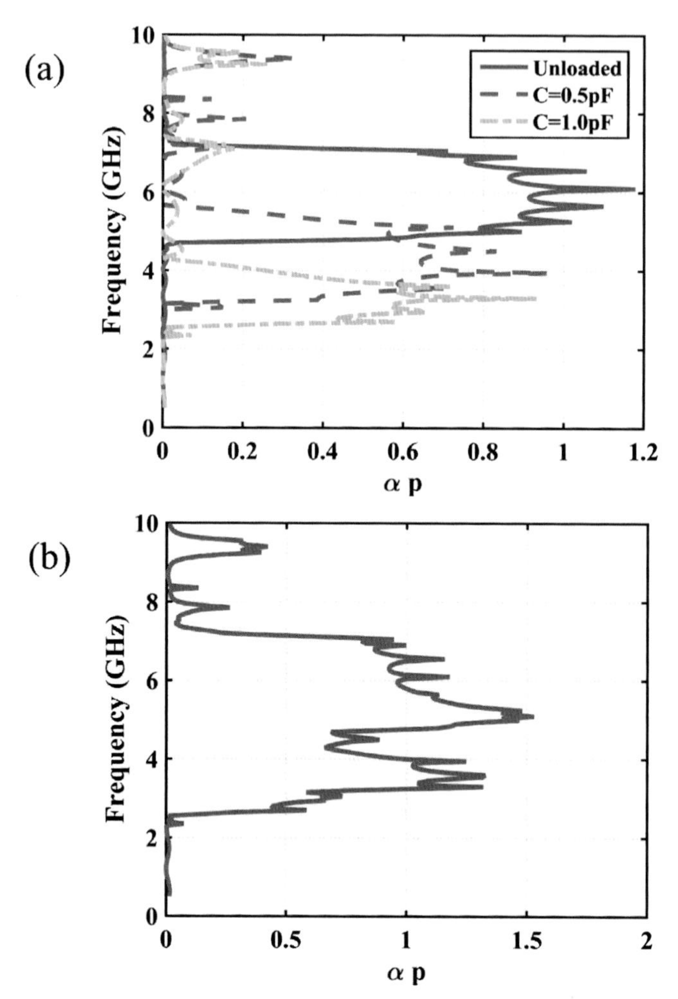

Figure 4.7 (a) Attenuation constants of mushroom-type EBG unit cells for three different capacitive loadings. (b) Attenuation factor of an EBG created by cascading the unit cells from (a).

In order to demonstrate the applicability of the previously described scenario, we performed three simulations for three different uniform capacitive loadings: (a) unloaded EBG, (b) 0.5 pF loads, and (c) 1.0 pF loads. From the computed S_{12} responses, the corresponding attenuation factors were derived, according to Eq. (4.4), and they are displayed in Fig. 4.7a. Note that, as expected, the non-zero attenuation constant values that indicate the existence of a stopband shift to lower frequencies as the loading capacitance increases. Most importantly, due to the additive effect in the structure's attenuation constant predicted by Eq. (4.4), the three responses shown in Fig. 4.7a add constructively, resulting in the total attenuation response shown in Fig. 4.7b. Evidently, the cumulative attenuation response attains non-zero values over a wider frequency range. In other words, by cascading four EBG unit cells and applying the aforementioned capacitive loading (no load, 0.5 pF, 1 pF) between them, we can in principle increase the isolation performance bandwidth from approximately 5.0–7.5 GHz to approximately 3.0–7.5 GHz.

Therefore, the question becomes, what is required so that an isolation performance as wide as possible can be achieved? First, intuitively one expects that the larger the number of unit cells along the direction of propagation, the greater the flexibility one has to manipulate the structure's isolation performance. Second and most important, given that an exhaustive trial of all possible capacitor combinations is not practically feasible, a fast and efficient selection algorithm is required in order to determine the judicious capacitor combination that yields the desired isolation performance. The latter requirement is analyzed in the following subsection.

4.2.2 Circuit Representation of Capacitively Loaded Mushroom-Type EBG

The methodology presented herein exploits the fact that the loaded EBG is a two-port microwave device whose performance is completely described by its S_{12} scattering parameter. A more general description of the loaded EBG is obtained if we substitute the capacitor loads with lumped ports. This substitution renders the EBG an (N+2)-port network where N is the number of the available loading ports.

The latter observation is of paramount importance because (a) it allows us to conveniently describe the electromagnetic response of the EBG in terms of an $(N + 2) \times (N + 2)$ S-matrix (which is obviously independent of any loading), and (b) the circuit representation of an electromagnetic device reduces the computational complexity of its modeling considerably. This is especially important for the problem under investigation where only a small part of the geometry is modified, and therefore re-simulating the entire structure becomes computationally redundant. In particular, without a circuit representation of the EBG, the effects of a certain capacitor loading (where we recall that the capacitors in this case are lumped elements occupying a minute fraction of the total geometry) could only be examined through computationally demanding 3D full-wave simulations. It should be emphasized here that the circuit/network representation of the EBG can fully describe its response only because the device under study is a guiding structure and not an antenna. In the case of an antenna, its complete electromagnetic characterization requires knowledge of both its *S*-matrix and its radiation efficiency.

It is helpful to begin our analysis by considering first a two-port network where the elements of its 2 × 2 *S*-matrix are denoted by S_{ij}. If we terminate its second port at some impedance load Z_L, then the reflection coefficient at port-1 is given by [45]:

$$\Gamma_{\text{in}} = S_{11} + \frac{S_{12}S_{21}\Gamma_L}{1 - S_{22}\Gamma_L} \tag{4.5}$$

where Γ_L is the reflection coefficient created when a 50 Ω transmission line is terminated at the load Z_L. From a practical standpoint, the importance of the expression in Eq. (4.5) can be summarized as follows: once the *S*-parameters of the two-port network are known (a task that requires a computationally intensive 3D full-wave simulation), then the reflection coefficient at port-1, for any load applied at port-2, can be trivially calculated by simply substituting in Eq. (4.5) the corresponding reflection coefficient Γ_L. Second, Eq. (4.5) reflects a more general property of multiport networks usually referred to as "port reduction." This property simply states that if the second port of a two-port network is loaded, then this reduces to a single-port network, where its *S*-parameter matrix has a single element, $S_{11} \equiv \Gamma_{\text{in}}$.

From a computational standpoint, the aforementioned observations become more appreciated when the number of ports increases. As a matter of fact, it can be shown that the previously described port reduction method can be generalized in the case of an N-port network. Specifically, an N-port network can be reduced to any number of ports $M<N$, by simply loading the remaining $(N-M)$ ports. Most importantly, the *S*-parameter matrix of the reduced

M-port network can be obtained recursively according to the formula [46, 47] (which is the generalization of Eq. (4.5)):

$$S_{ij}^{(k)} = S_{ij} + \frac{S_{ik}S_{kj}\Gamma_k}{1 - S_{kk}\Gamma_k} \tag{4.6}$$

In the preceding formula, it is assumed that the kth port is terminated by a load Z_L, while as before Γ_k is the reflection coefficient created when this load terminates a 50 Ω transmission line.

Now let us stitch all the pieces of the previous analysis together and show how this methodology can be applied to the problem under investigation here. First, we recall that the EBG is characterized in terms of its S_{12} parameter, and therefore it eventually needs to be treated as a two-port network. Second, as mentioned previously, a more general description of the EBG is in terms of an $(N+2)\times(N+2)$ S-matrix where N is the number of the available ports to load (in this case $N = 11$). Subsequently, given a combination of loading capacitors and after applying the recursive formula in Eq. (4.6), the $(N+2)$-port network is reduced to a two-port system. The S_{12} parameter of the reduced two-port network describes the transmission between the two excitation ports of the EBG-loaded TEM waveguide and thus effectively characterizes the isolation performance of the metasurface.

Finally, in addition to the aforementioned convenient circuit-based description of the EBG, it is also required to utilize a methodology that would allow fast determination of the values of the loading capacitors so as to meet a specific isolation goal. A global optimization strategy would be well suited to perform this task. In particular, for the numerical examples reported herein, the covariance matrix adaptation evolutionary strategy (CMA-ES) [48, 49] has been employed.

4.2.3 Numerical Examples

In this subsection, we present two indicative numerical examples that demonstrate the capabilities of the proposed loading scheme. Both numerical experiments are performed with respect to the computational model shown in Fig. 4.5. The design methodology consists of the following steps:

1. Modify the computational model and define N lumped ports, terminated to 50 Ω loads, across the N gaps defined between the patches of neighboring unit cells.
2. Simulate the EBG structure inside the TEM waveguide (using some numerical full-wave solver such as Ansys HFSS) and extract its $(N + 2) \times (N + 2)$ S-matrix.
3. Define the desired design goals as well as the range of the available capacitance values.
4. Using some post-processing software, such as MATLAB, implement the port reduction method and derive the reduced 2×2 S-matrix, for a combination of N loading capacitors.
5. Repeat the previous step using an optimization driven search for the combination of capacitances that meets the desired design criteria.
6. Substitute the lumped ports in the computational model with the actual lumped capacitors and verify the predictions of the optimizer.

In the first numerical experiment, the optimization objective was to achieve an isolation performance as wide as possible. For the problem under study, the size of the S-matrix from Step 2 is equal to 13×13. In this case, the capacitor values were set to discretely vary from 0.1 pF to 1.0 pF in increments of 0.1 pF. The scenario that a certain gap need not be capacitively loaded was designated by letting the search algorithm choose a capacitance value equal to 0.001 pF.

The search algorithm converged to the following capacitance values for the 11 available loading ports: $[10^{-3}, 1.0, 10^{-3}, 0.2, 1.0, 0.3, 1.0, 1.0, 1.0, 1.0, 1.0] \times 1$ pF. Figure 4.8a shows the isolation performance of the unloaded 12-unit-cell EBG, along with the isolation performance of the loaded structure optimized to achieve a stopband as wide as possible. The unloaded structure exhibits a stopband from approximately 5 GHz to 7 GHz. However, when the same structure is capacitively loaded according to the aforementioned optimal capacitor combination, its stopband becomes wider and it extends from approximately 2.5 GHz to 7 GHz. In the same figure, we have also included for validation purposes the full-wave result of the structure's isolation performance when it is loaded with the actual lumped capacitors. The full-wave result is in excellent agreement with the isolation response that the optimization methodology predicts.

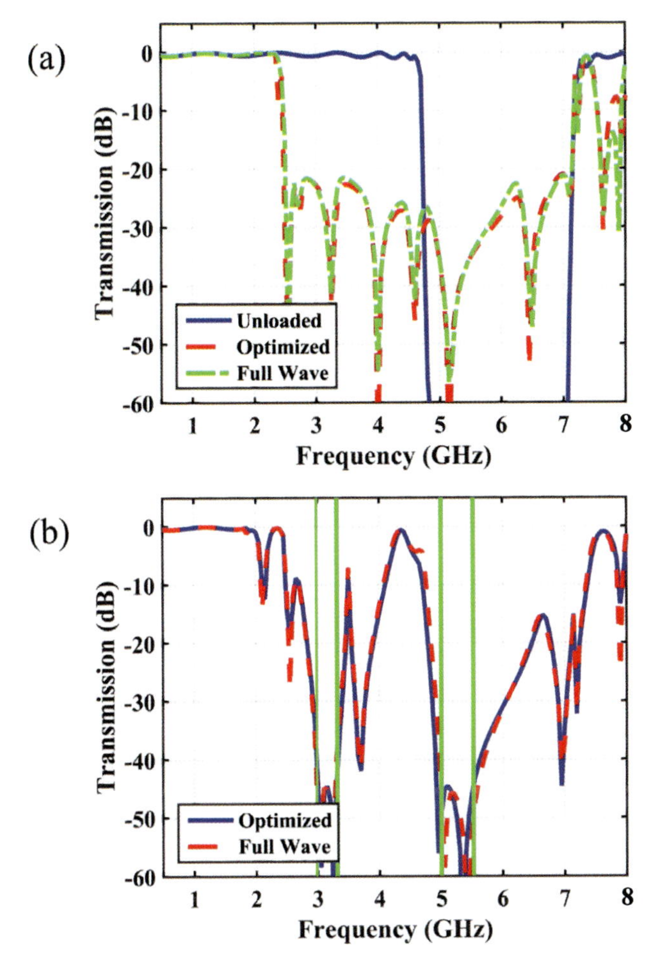

Figure 4.8 (a) Optimized, with respect to bandwidth, isolation performance for the 12-unit-cell-long EBG shown in Fig. 4.5. (b) Isolation performance of the same EBG structure, optimized for dual-band operation.

It should be noted that although the optimally loaded EBG exhibits superior wideband isolation performance, the level of isolation is significantly smaller than that of the unloaded EBG. This is to be expected since as described previously for the unloaded case, the non-zero values of the attenuation factor of all unit cells occur at the same frequency range. These non-zero values add up and create a cumulative effect that results in the observed very high isolation. In contrast, when the EBG is non-uniformly loaded, the non-zero values of the attenuation factor are distributed across different frequency ranges. Therefore, the cumulative effect is not as pronounced and this is manifested by the lower isolation level depicted in Fig. 4.8a. In the non-uniform loading case, if higher isolation levels are desired, one needs to employ an EBG with more unit cells along the direction of propagation.

As a second example, we examined whether a dual-band isolation performance can be achieved with the same EBG structure. For this numerical experiment, the optimization goals were to determine the capacitor combination that would yield the maximum isolation performance in the frequency range 3.0–3.3 GHz as well as in the frequency range 5.0–5.5 GHz. In this case, we let the capacitance values vary discretely from 0.1 pF to 2.0 pF in increments of 0.1 pF, in addition to 0.001 pF representing the unloaded case. The optimal capacitor values were determined by the optimization algorithm as $[10^{-3}, 0.6, 0.4, 10^{-3}, 10^{-3}, 0.7, 1.0, 1.7, 1.9, 0.9, 0.9] \times 1$ pF. The corresponding results in Fig. 4.8b show very high isolation within the two desired frequency bands. As before, we have also included the full-wave result obtained after numerically simulating the EBG with the actual capacitor loads. Again, the full-wave result is in excellent agreement with the optimization prediction, further validating the accuracy of the proposed methodology.

4.2.4 Experimental Verification

It should be stressed that one of the main advantages of this optimization methodology is that any type of circuit model can be added into each loading port, allowing for a more realistic response of the metasurface. In this way, we can account for the non-ideal behavior of the capacitors, in particular their series parasitic inductance. This series inductance is incorporated into the optimization process using the self-resonance frequency provided by the manufacturer's data sheet. In order to verify this design process, a prototype of the optimized broadband structure was built and tested. The capacitor values were determined by the optimization algorithm as [0.9, 1.2, 0.8, 0.5, 0.8, 1.0, 0.2, 0.2, 10^{-3}, 0.8, 10^{-3}] × 1 pF. The fabricated metasurface is illustrated in Fig. 4.9. To mimic the parallel plate waveguide setup used in the full-wave simulation, a grounded dielectric was placed on top of the metasurface. In addition, an absorbing material was put along the edges of the structure in order to alleviate the effects caused by its finite size. A comparison of the simulated and measured results for the optimized metasurface is presented in Fig. 4.10. In general, the measurements agree well with the optimization results validating the previously discussed methodology.

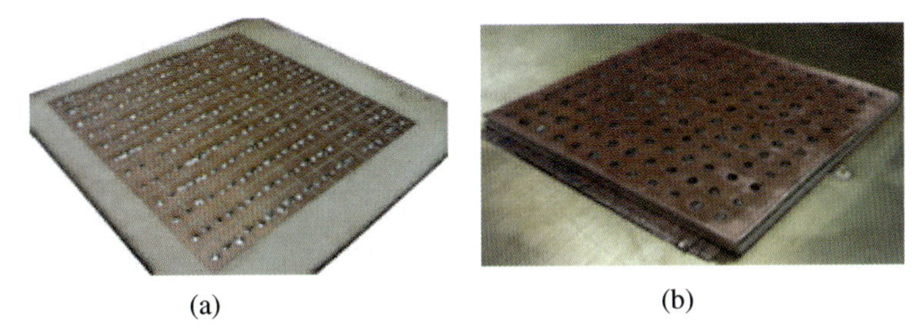

Figure 4.9 Photograph of (a) the manufactured capacitively loaded metasurface and (b) the waveguide setup used for testing. Reprinted, with permission, from Ref. 3, Copyright 2014, IEEE.

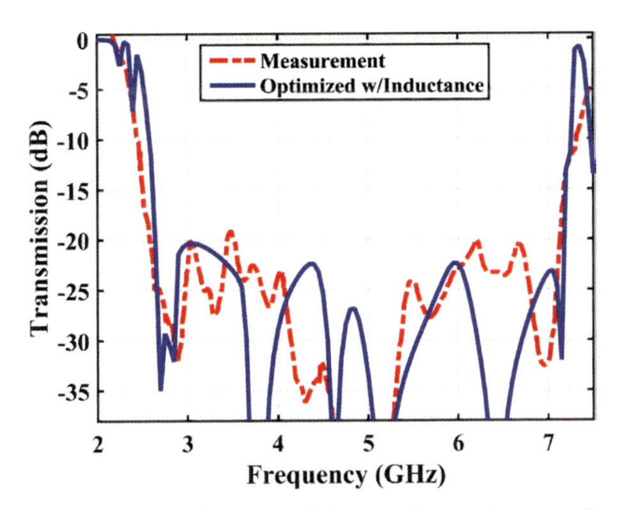

Figure 4.10 Isolation performance of the manufactured metasurface taking into consideration the parasitic inductance.

4.2.5 Free-Space Setup

Now instead of the waveguide scenario studied previously, a metasurface in free space will be considered. Similar to the TEM waveguide, the transmission is considered across the metasurface, but the difference lies in trying to mimic the free-space scenario, which is a more practical testing environment. For example, in many situations, it is required to improve the isolation between two antennas, which cannot be enclosed in a waveguide. For this type of setup, the top parallel waveguide layer is removed and replaced by an open boundary. Wave port excitations are placed at the edges of a single row of the EBG with PMC boundary condition in the transverse direction, as shown in Fig. 4.11. In order to validate this scenario, measured results for the transmission of the unloaded EBG were compared to that of a single row formed by 12 unit cells placed in this test environment. The geometry of the unit cell remains the same as in the previous section. Figure 4.12 shows the transmission results for the full-wave simulation and measurements of the unloaded base structure for the free-space setup. In order to reduce the measurement variations, both the simulation and the measurement were normalized to the results obtained when no structure was present in the waveguide setup.

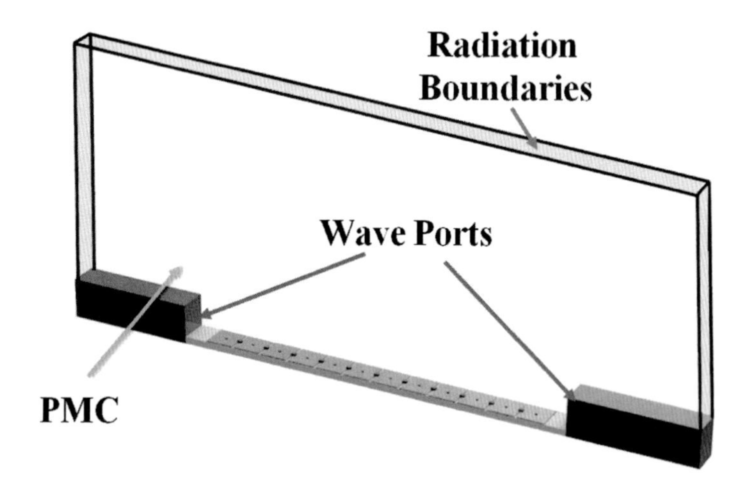

Figure 4.11 Free-space setup.

As can be observed in Fig. 4.12, the level of isolation in this setup is reduced when compared to the waveguide setup. This is due to the fact that the surface wave is not as closely bounded to the surface as was the case with the waveguide setup. As a consequence, the probes tend to be more difficult to isolate. As discussed earlier, this can be alleviated by simply increasing the number of unit cells and loading ports used for the underlying structure. For this reason, in the remainder of the examples corresponding to the free-space setup, the underlying structure will consist of 25 unit cells, which allows for 24 capacitive loading values to be optimized. A comparison between the design methodology discussed previously and the full-wave simulation for both the unloaded and loaded metasurface is presented in Fig. 4.13.

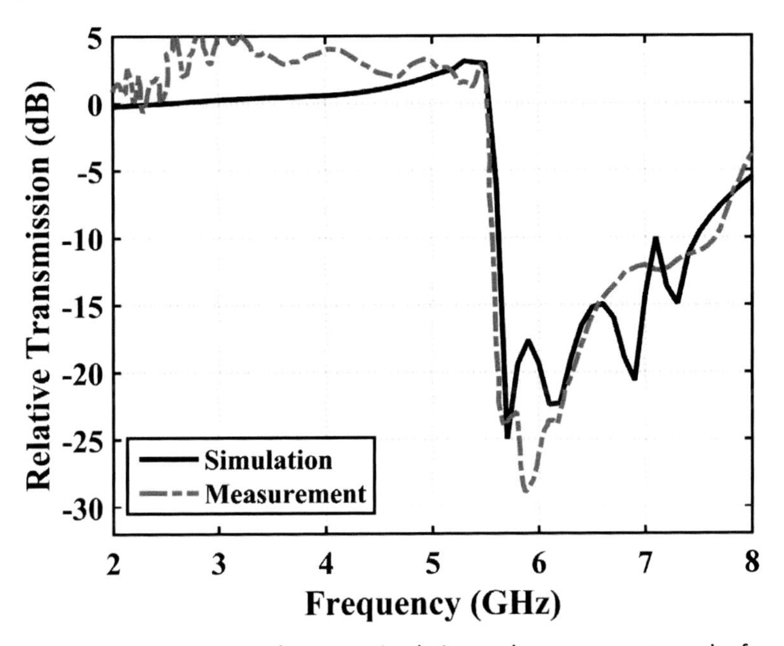

Figure 4.12 Comparison between simulation and measurement results for the free-space setup design.

As can be observed, both the test setup simulation and the unloaded structure, as well as the port reduction and the full-wave results, match very closely. As stated above, the energy in this type of test setup is less tightly bound to the surface, resulting in a reduction in the level of isolation that can be accomplished. Therefore, the bandwidth for optimization purposes is defined as a continuous frequency range with less than −10 dB of transmission. Also in order to achieve a more realistic design performance, the non-ideal capacitor model, which includes the parasitic series inductance, will be incorporated into the optimization process. Two optimization scenarios were examined for the realization of a broadband response. In the first case, the capacitance values were discretely varied from no capacitor (0.001 pF) to 0.8 pF every 0.1 pF. In the second case, the capacitance values were continuously varied in the same range. As can be observed in Fig. 4.14, the unloaded metasurface has a bandgap of 2.15 GHz (from 5.63 to 7.78 GHz), while the continuous capacitively loaded optimized metasurface has a bandgap of 3.29 GHz (from 3.96 to 7.25 GHz). The discrete capacitance case has a 0.11 GHz bandwidth reduction when compared to the continuous capacitor case. Since the improvement in bandwidth is minimal, it can be concluded that the implementation of commercially available capacitors will not limit the performance of the metasurface.

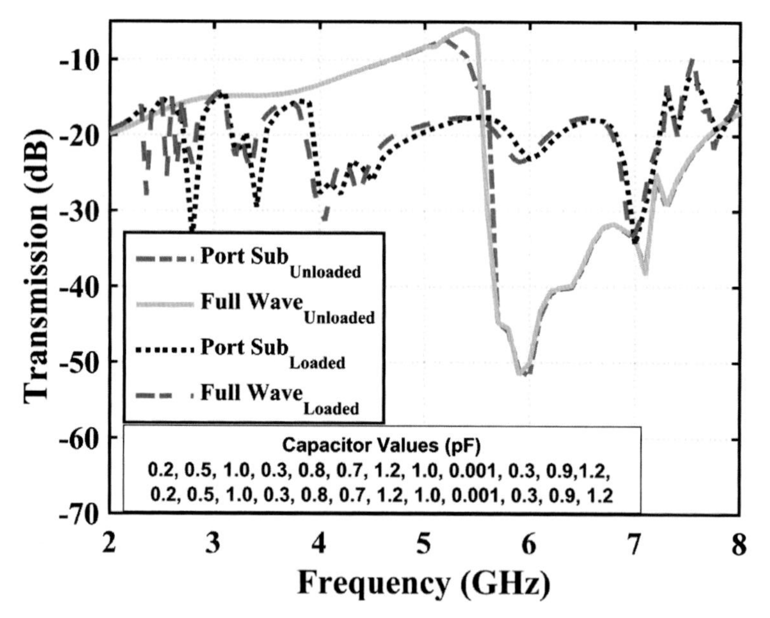

Figure 4.13 Comparison of the port-substitution and the full-wave simulation for the unloaded and arbitrary capacitively loaded metasurface.

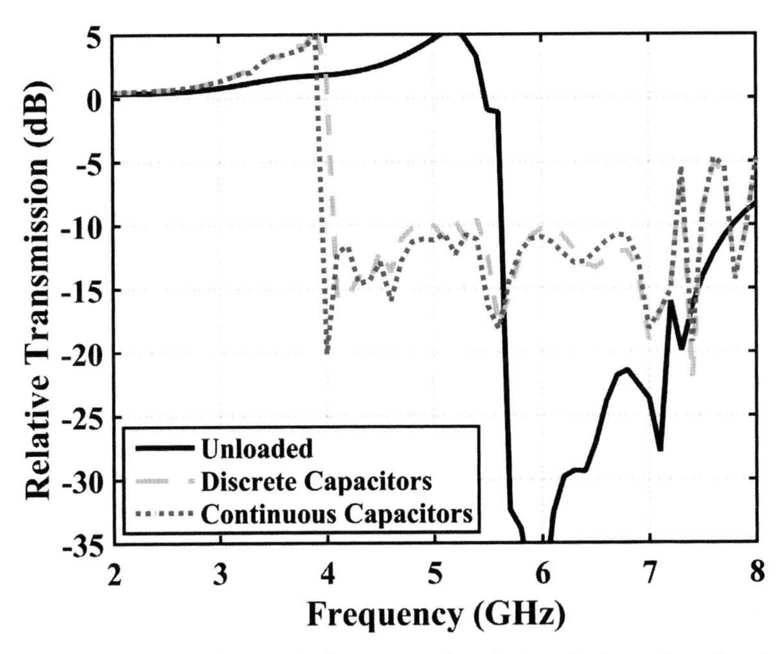

Figure 4.14 Optimized results for a metasurface designed to have a broadband response for both continuous and discrete capacitance values.

The next optimized designs attempt to reduce the relative transmission for the 3.6 and 5 GHz wireless local area network (WLAN) bands. Two scenarios are investigated; the first goal is to create a dual-band structure, while the second relates to a tunable type of metasurface for the same frequency bands of interest. For the dual-band response, the capacitor values were continuously varied from no capacitor (0.001 pF) to 1.2 pF. As can be observed in Fig. 4.15a, there is more than −14 dB of relative transmission (isolation) for both bands of interest. For the tunable metasurface, a varactor with a capacitance value ranging from 0.14 to 1.1 pF was implemented in the optimization. Given that the capacitor values for each case correspond to a particular band, a greater level of isolation can be realized when compared with the results of the dual-band design. For the tunable metasurface, less than −20 dB of relative transmission for both frequency bands of interest was achieved (see Fig. 4.15b).

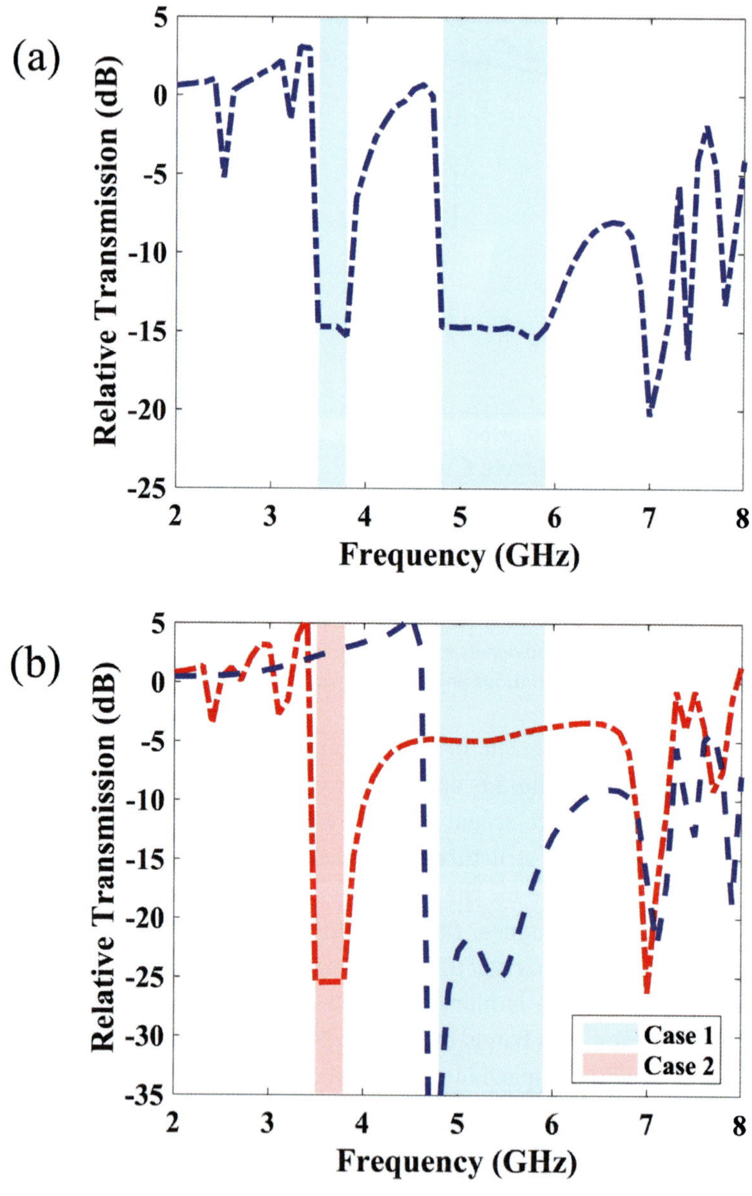

Figure 4.15 Simulation results of the capacitively loaded metasurface comprises 25 unit cells designed to target (a) a dual-band response and (b) a tunable response for the same frequency bands.

4.2.6 Omnidirectional EBG Metasurface

The previous examples deal with trying to increase the level of isolation for a single direction, but sometimes isolation in all directions is required. A design that has a square unit cell will need to employ capacitive loading on all its sides, which consequently increases both the size and the complexity of the underlying structure. To reduce these effects, the underlying unit cell was chosen to have a hexagonal geometry. The underlying base structure design is depicted in Fig. 4.16. The hexagonal unit cells have a periodicity of 9 mm, a separation between patches of 1.5 mm, a length for each of the sides of the patches of 4.1858 mm, a via radius of 12 mil, and a 1.52 mm thick Rogers RO3203 substrate. To reduce the amount of capacitors being optimized, a configuration with eight distinct capacitor values was chosen, as represented by the color scheme present in Fig. 4.16. This arrangement of the capacitors was selected because the energy traveling across the surface (0°) and at 60° will detect the same set of capacitors values. To optimize for all directions, only two critical angles are needed for this setup. First is the 0° angle, representing the worst case scenario with propagation parallel to the capacitors, and the second is the 30° angle, representing the angle away from the parallel that encounters the first set of the capacitors. A comparison between the port reduction optimization design and a full-wave simulation for the unloaded and an arbitrary capacitively loaded metasurface is shown in Fig. 4.17. There are some minor discrepancies between the full-wave simulation results and the optimization methodology; however, these are not significant enough as to have an impact on the optimization process.

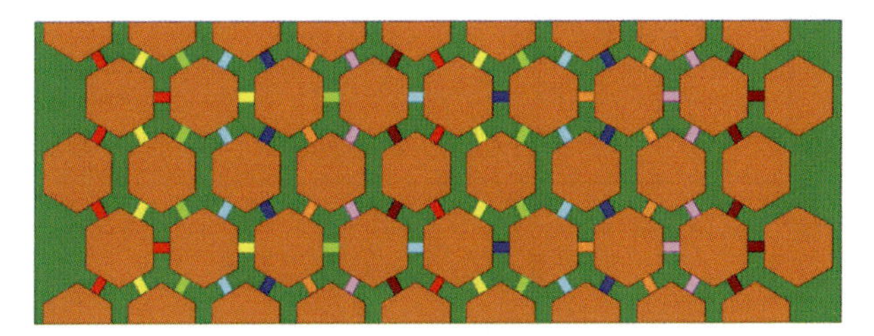

Figure 4.16 Ominidirectional base structure design. Each color represents a distinct capacitance value used in the optimization scheme.

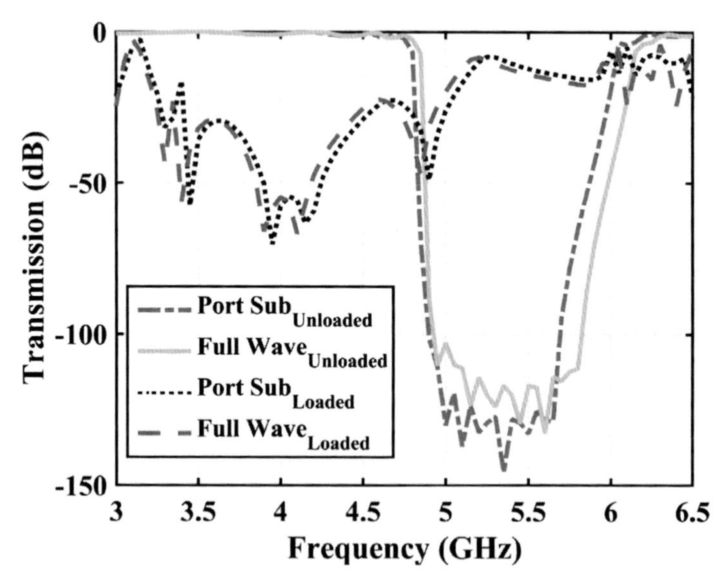

Figure 4.17 Comparison between the port-substitution method and the full-wave simulation for the omnidirectional design.

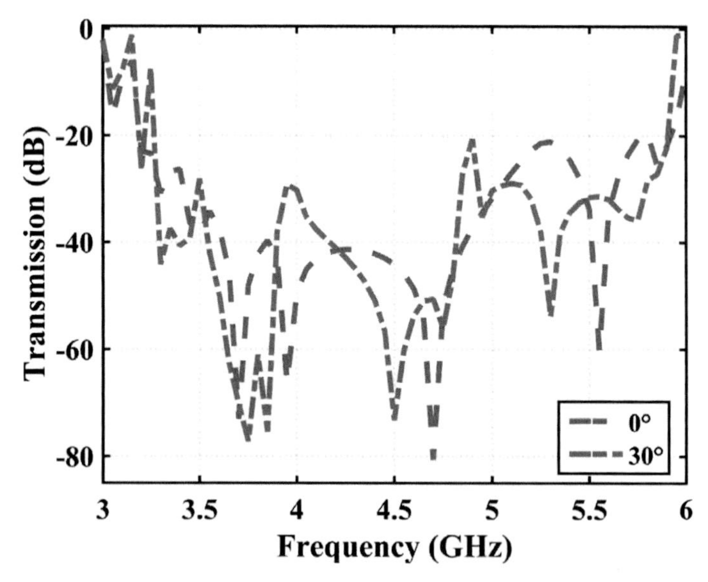

Figure 4.18 Optimization results for the omnidirectional metasurface design to maximize the broadband response for both the 0° and 30° angle cases.

This underlying structure was used in the design of a broadband metasurface. For the optimization, the values of capacitance were continuously varied from 0.1 to 1 pF and the cost function used attempted to minimize isolation bandwidth for both the 0° and the 30° cases. Figure 4.18 shows the results of this optimization. The unloaded metasurface has less than −20 dB of transmission from 4.8 to 6.1 GHz (bandgap of 1.3 GHz), while the continuously capacitively loaded optimized metasurface has a stopband from 3.25 to 5.95 GHz (bandgap of 2.7 GHz).

4.3 Tunable Absorbers Based on Mushroom-Type Metasurfaces

So far our analysis has focused on demonstrating how to manipulate the bandgap of mushroom-type metasurfaces through non-uniform capacitive loadings. As clarified earlier in this chapter, periodic structures based on the mushroom metasurface, apart from EBGs, can also be employed to realize AMCs. However, it is important to emphasize once again that the underlying physics that govern the two functionalities are fundamentally different. This is simply because EBGs are responsible for controlling the surface wave propagating along the metasurface. In contrast, AMCs are responsible for controlling the properties of the electromagnetic field scattered off of them when they are illuminated by an impinging electromagnetic wave. In this section, we borrow elements from the EBG design methodology presented in the previous sections, and in particular the port reduction technique as summarized by Eq. (4.6), and we apply them toward the development of tunable AMCs and elementary tunable microwave absorbers. Note here that AMCs and absorbers are inherently related since once a loss mechanism is introduced in an AMC, it becomes an absorbing device.

Although research on microwave absorbers dates back to World War II, it still remains a very active research area due to its applicability in the development of stealth technology, and due to the specifications of modern telecommunication systems that require to isolate multifunctional and highly sensitive antenna platforms from either intentional or unintentional electromagnetic radiation/interference. The design of an electromagnetic absorber is essentially a matching problem where it is desired to match the input impedance of the absorbing device to that of free space, so that an incident wave can be completely or partially absorbed by a lossy medium [50]. An absorbing device usually comprises a multilayer structure where frequency-selective surfaces (FSS) are sandwiched in between its layers. The number, material constitution, and height of the layers as well as the filtering response of the FSSs are factors that determine the absorption performance of the device. In principle, it is desirable to devise an as electrically thin as possible absorber, which also exhibits wide absorption bandwidth. However, in reality, these two design goals contradict each other. For this reason, usually the designer needs to favor one of the two features and then attempt to improve the other by custom engineering the material constitution and the geometrical characteristics of the device.

A very popular absorber design is based on the periodic arrangement of the mushroom-type unit cell [51, 52]. It should be mentioned here again that the via of the mushroom unit cell is responsible for the propagation of backward guided waves across the metasurface, but this is irrelevant to the absorption mechanism. The absorption mechanism, as will be demonstrated later in this section, is solely dependent on the loss mechanism of the device and on the AMC properties of the shorted metasurface, where in this case the metasurface is simply a 2D periodic arrangement of metallic patches. These properties are essentially determined by the zero crossings of the reflection coefficient phase of the AMC. In principle, the only effect that the vias have on the performance of mushroom-type unit cell absorbers is to perturb the Q factor of the shorted spacer, and therefore they can be omitted.

The popularity of this type of absorber stems from their structural simplicity as well as from their ultra-thin profile. This type of absorber also has disadvantages. First, when a loss mechanism is introduced by either loading some resistors across their metallic patches or applying a uniform Ohmic sheet on their FSS, reasonable absorption can be achieved only for a very narrow frequency bandwidth, centered around the frequency where the phase of the corresponding AMC's reflection coefficient phase goes to zero. Second, mushroom-based absorbers lack the ability to reconfigure their response since none of their structural components can be electrically tuned. Consequently, it becomes evident that if wider or tunable/reconfigurable absorption performance is required, then the absorber needs to be appropriately modified.

One way to introduce the aforementioned capabilities in a mushroom structure is by judiciously introducing lumped elements across the metasurface [53–55]. Note here that similar to the EBG configurations examined previously, a lumped element loaded absorber can also be described in terms of the *S*-matrix of a multiport network, and therefore the optimization procedure and design methodology presented in the previous sections can be directly applied for the synthesis of absorbers as well. Given this strategy, in the following subsections, we present a tunable narrowband, a multiband, and a broadband absorber where all three designs are based on different flavors of a lumped element loaded metallic square patch metasurface.

4.3.1 Narrowband Reconfigurable Absorber

For the realization of this type of absorber, we employ a mushroom unit cell with the following characteristics: periodicity 7.5 mm, top square patch width 7 mm, via radius 12 mil, substrate thickness 1.52 mm, and relative permittivity of the dielectric substrate ε_r = 3.02. The material constitution of the metallic patch as well as the via is defined as a perfect electric conductor. The corresponding HFSS schematic of the unit cell is illustrated in Fig. 4.19a. Along the transverse boundary faces, periodic boundary (master/slave) conditions are defined, while the unit cell is excited by a normally impinging linearly polarized plane wave (Floquet port). Note that, as also indicated in Fig. 4.19a, there are lumped elements across the gaps between the metallic patches of neighboring unit cells. Through the modification of these lumped elements, we can load the structure with an equivalent circuit that corresponds to the parallel connection of a resistor and a capacitor.

In particular, with respect to the equivalent circuit shown in Fig. 4.19b, the tuning role of the loading RC circuit can be justified as follows: the parallel lumped capacitance C_L increases the unit cell's intrinsic capacitance (defined across the gaps between neighboring patches) and thus shifts the resonance of the unit cell toward lower frequencies. The resonance of the capacitively loaded unit cell is defined as the frequency where the imaginary part of the unit cell's input admittance goes to zero, or

$$\text{Im}\{Y_{\text{in}} + j\omega C_{\text{L}}\} = 0 \Rightarrow \omega C - \frac{1}{\omega L} + \omega C_{\text{L}} = 0 \Rightarrow \omega = \frac{1}{\sqrt{L(C + C_{\text{L}})}} \quad (4.7)$$

Consequently, in order to achieve absorption at the same frequency, the real part of the loaded structure's input admittance needs to be as close as possible to the characteristic admittance of free space, or

$$\text{Re}\left\{Y_{\text{in}} + j\omega C_{\text{L}} + \frac{1}{R_{\text{L}}}\right\} = Y_0 \Rightarrow \frac{1}{R_{\text{L}}} = Y_0 \quad (4.8)$$

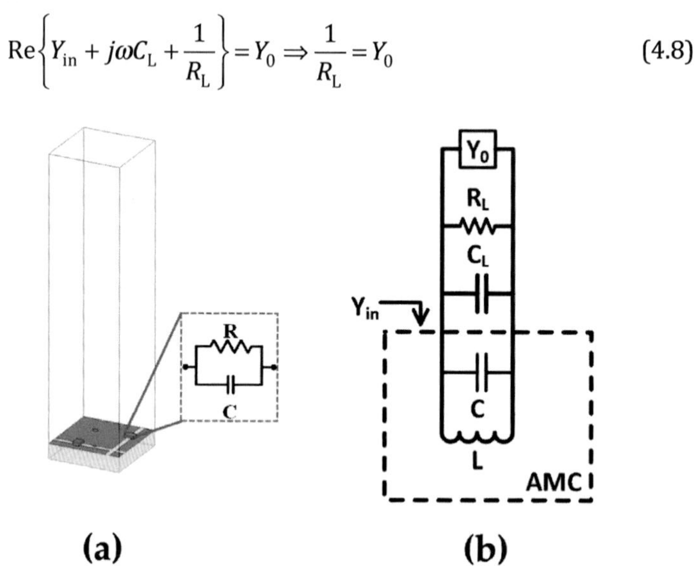

Figure 4.19 (a) CAD model of a mushroom-type unit cell absorber loaded with lumped capacitors and resistors connected in parallel. (b) Circuit equivalent model of the absorber configuration in (a).

Figure 4.20a shows the reflection coefficient phase from the capacitively loaded but lossless unit cell, *i.e.*, when $C_{\text{L}} \neq 0$ and $R_{\text{L}} = 0$. In particular, three different capacitor loads are examined: $C_{\text{L}} = 0.5$ pF, $C_{\text{L}} = 0.75$ pF, and $C_{\text{L}} = 1.5$ pF. The unloaded unit cell exhibits a zero reflection coefficient phase at around 7 GHz. As expected, when the value of the loading capacitance increases, the structure's resonance frequency shifts to frequencies lower than 7 GHz, as predicted by Eq. (4.7). When a lumped resistor is additionally applied to the parallel RC circuit, the response shown in Fig. 4.20b is obtained. Evidently, for the same lumped loading resistor, the absorption performance can be switched between different frequencies by simply changing the lumped capacitor value. Also, by direct comparison between Figs. 4.20a,b, it is evident that for each one of the three loading capacitances, absorption occurs at the same frequency where the phase of the reflection coefficient of the AMC exhibits its zero crossing, exactly as Eqs. (4.7) and (4.8) predict.

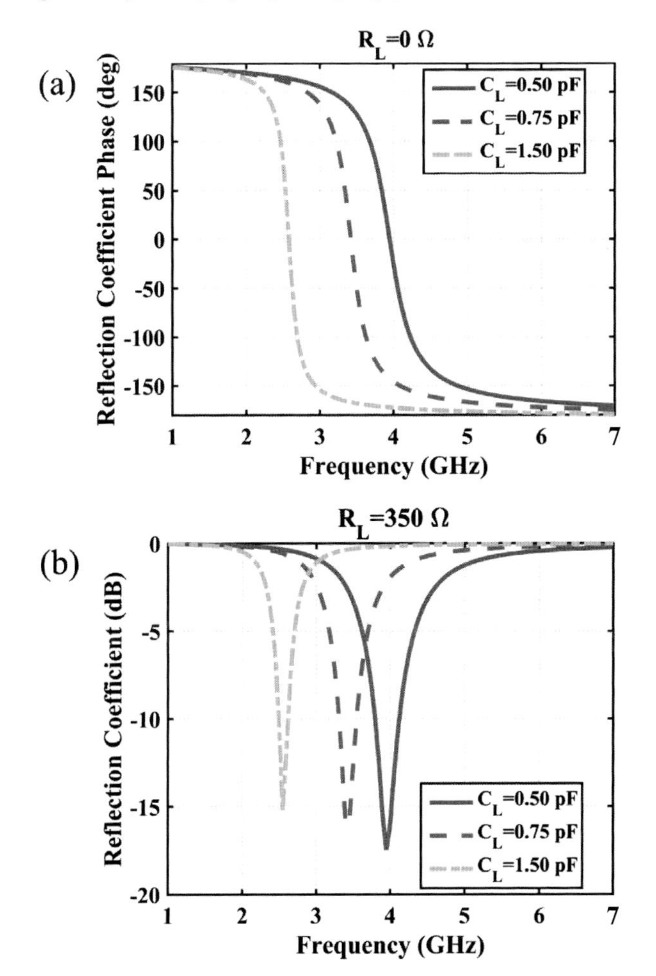

Figure 4.20 (a) Reflection coefficient phase of the unit cell in Fig. 4.19 for three different capacitor loads and zero resistance. (b) Reflection coefficient magnitude for the capacitively loaded unit cell from (a) and a resistor load equal to 350 Ω.

At this point, it should be emphasized that the above analysis can be trivially performed by simple circuit-type calculations instead of repetitive full-wave simulations. Similar to the procedure described in Section 4.2, one needs to perform a single full-wave simulation where the lumped port that corresponds to the loading parallel RC circuit is substituted by a lumped port terminated to a 50 Ω load. Subsequently, the 2×2 S-matrix of the multiport network representation of the excited unit cell is extracted. Port #1 corresponds to the excitation port, while Port #2 corresponds to the parallel RC loading circuit. Then using the port reduction approach, the reflection coefficient $\Gamma \equiv S_{11}$ at Port #1 can be trivially computed for any loaded RC circuit.

4.3.2 Multiband Absorber

In the preceding example, it was shown that if one wants to design a narrowband/single-frequency absorber using a mushroom-type unit cell, then the lossless AMC simply needs to be tuned so that it resonates at the desired frequency and then to introduce loss in the system using lumped resistors. Consequently, it becomes obvious that if absorption at multiple frequencies is desired, then the mushroom unit cell needs to be appropriately modified so that it exhibits a multi-resonant response or multiple phase zero crossings. In order to introduce additional resonances in a mushroom AMC, usually some structural modification is required, such as an FSS shaped differently than a square patch.

In the present study, an alternative approach is proposed to introduce the additional resonances. In particular, given the mushroom configuration defined in the previous subsection, we introduce a larger unit cell that consists of four metallic patches, as shown in Fig. 4.21a. The advantage that this offers is that more degrees of freedom can be added and thus more flexibility to the system by applying the transverse non-uniform lumped element loading scheme shown in Fig. 4.21b. Essentially, this unit cell configuration allows us to accommodate two different AMC resonances: one frequency is controlled by the green capacitors, shown in Fig. 4.21b, while the second is controlled by the set of dark blue capacitors. It needs to be emphasized here that the proposed approach is solely based on lumped capacitor loadings, and therefore the port reduction design methodology is directly applicable in this case.

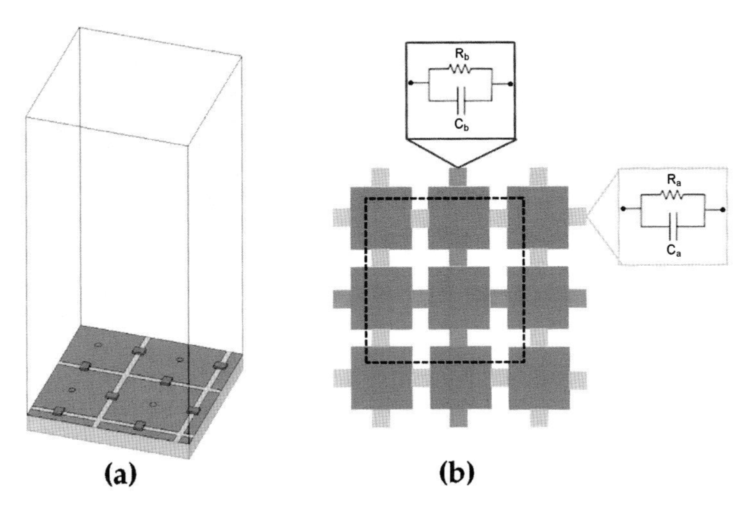

Figure 4.21 (a) Multi-frequency absorber unit cell configuration. (b) Transversely non-uniform lumped element loading scheme for the realization of the multi-frequency absorber.

Figure 4.22a shows the reflection coefficient phase of the multi-frequency AMC when C_a = 1.49 pF and C_b = 0.71 pF. As expected, the structure exhibits two distinct resonances around 2.9 GHz and 4.5 GHz, respectively. Subsequently, in order to create absorption at these frequencies, the structure was additionally loaded with lumped resistors equal to R_a = 350 Ω and R_b = 1.8 kΩ. In Fig. 4.22b, it can be seen that at the aforementioned resonant frequencies of the resistively loaded AMC, the desired dual frequency absorption has been achieved.

4.3.3 Broadband Tunable Absorber

In this subsection, the previous methodology is employed to devise a broadband tunable absorber. As clearly demonstrated in the two previous subsections, the absorption bandwidth that can be achieved using the mushroom-type unit cell is rather limited. This is primarily due to the reduced capacitance tunability that the patch-type FSS offers. For this reason, in order to introduce more degrees of freedom in the system, we modify the patch-type FSS of the unit cell, as illustrated in Fig. 4.23. In particular, we surround the center patch with two concentric loops. The patch and the two loops are

0.02 mm thick, while their material constitution is defined as PEC. This FSS is placed on top of an air-filled, PEC-backed substrate. The geometrical characteristics of the unit cell are summarized in Table 4.1. Note that the role of the two concentric loops is to introduce two AMC resonances as far apart as possible. Indeed, the structure under consideration exhibits two zero-phase crossings occurring at 19.5 GHz and 33.5 GHz, respectively.

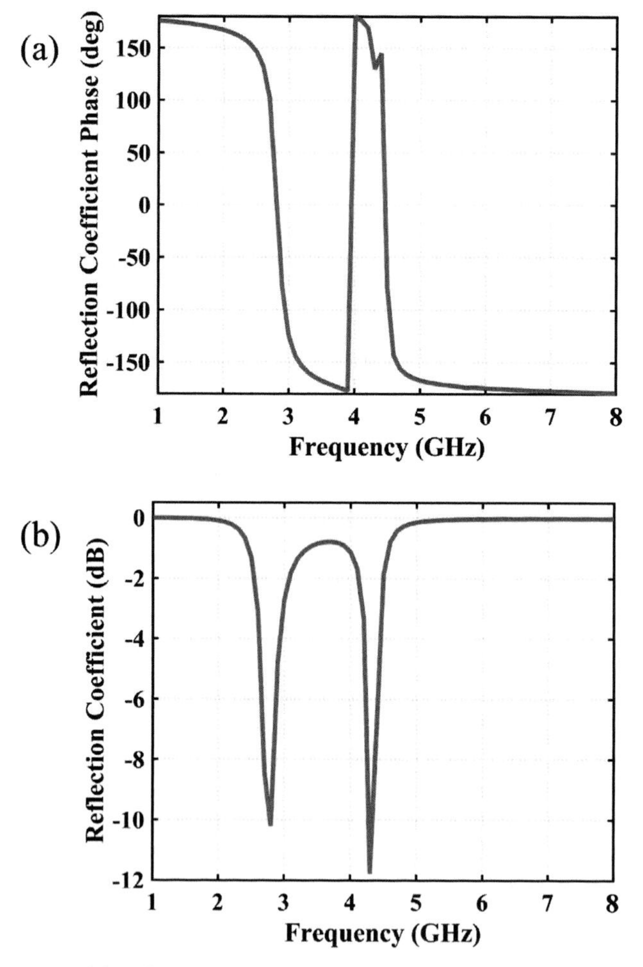

Figure 4.22 (a) Reflection coefficient phase for the unit cell in Fig. 4.19 assuming $R_a = R_b = 0\ \Omega$, $C_a = 1.49$ pF, and $C_b = 0.71$ pF. (b) Reflection coefficient magnitude for the multi-frequency AMC from (a) when $R_a = 350\ \Omega$ and $R_b = 1.8\ \text{k}\Omega$.

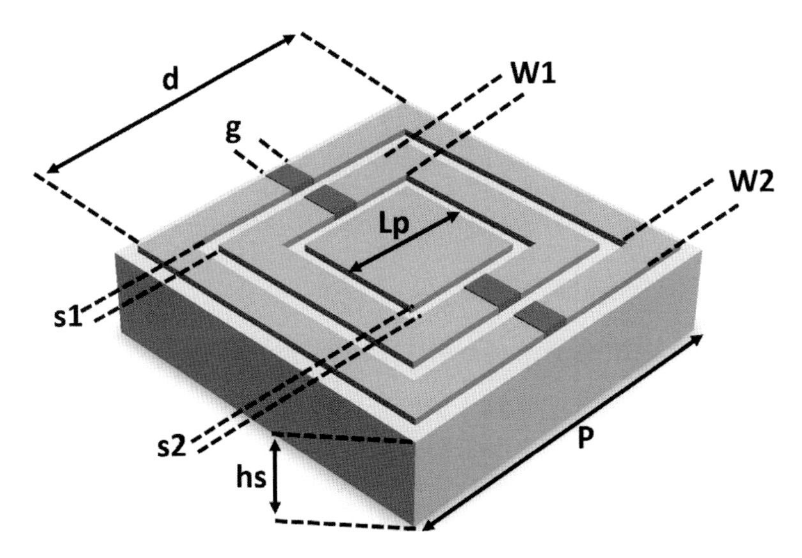

Figure 4.23 The broadband absorber unit cell comprises a center patch and two concentric loops. This FSS is placed on top of a shorted air-filled substrate.

Table 4.1 Characteristics of the unit cell illustrated in Fig. 4.21

	Length (mm)	Width (mm)	Spacing (mm)
Loop #1	6.00 (d)	0.60 (w_1)	0.30 (s_1)
Loop #2	4.20 (d−2×(w_1−s_1))	0.70 (w_2)	0.25 (s_2)
Patch	2.30 (L_p)	NA	NA
Gap	0.50 (g)	NA	NA
Unit cell periodicity	6.5 mm (p)	NA	NA
Unit cell thickness	1.52 (h_s)	NA	NA

The loss mechanism is introduced in the unit cell by loading it with four lumped resistors, as shown in Fig. 4.21. In order to identify the widest absorption that can be achieved, an optimization was performed with the four resistors varying independently and with the objective being the maximization of the absorption bandwidth. The corresponding lumped resistance values are listed in the first row of Table 4.2. For these values, the return loss of the absorber is depicted in Fig. 4.24a. Less than −10 dB absorption has been achieved from 14.6 to 37 GHz.

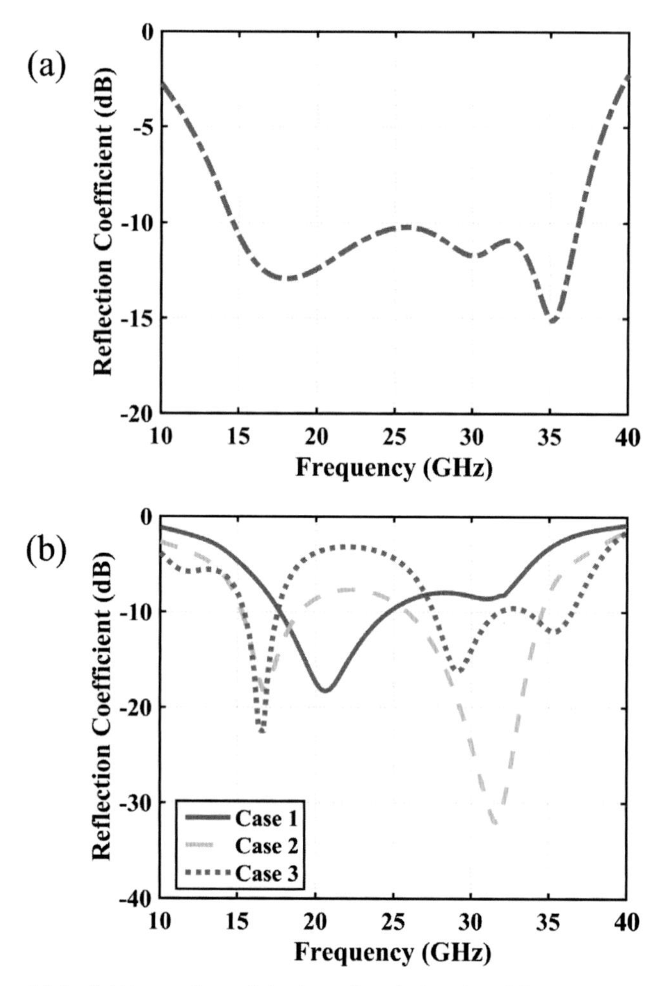

Figure 4.24 (a) Return loss of the broadband absorber. (b) Return loss of the tuned absorbers.

The performance of this absorber was further studied, and in particular, the feasibility of tuning its response was examined. Additional optimizations were performed, and three representative cases are shown in Fig. 4.24b. The lumped resistor values that result in these responses are summarized in Table 4.2. Figure 4.24b shows that by appropriately changing the FSS resistive loadings, the absorption bands can be tuned. In particular, for each of the combinations corresponding to Cases 1, 2, and 3, we can either introduce one, two, or three narrow absorption bands. The specific characteristics of these bands are summarized in Table 4.2.

Table 4.2 Absorber performance

	Resistor values (Ω)	Relative BW (%)	Frequency range (GHz) < −10 dB return loss
Optimized wideband	300, 43, 470, 620	87	14.6–37
Case 1	620, 1.5k, 620, 1.5k	33	17.7–24.7
Case 2	150, 1.5k, 150, 1.5k	22 (band 1)	15.3–19
		29 (band 2)	25.8–34.5
Case 3	51, 5.5k, 51, 1.5k	13 (band 1)	15.5–17.6
		14 (band 2)	27.6–31.8
		8 (band 3)	33.6–36.4

4.4 Conclusion

In summary, we have presented a very efficient methodology to synthesize broadband and multiband EBGs. The approach exploits the fact that non-uniform capacitively loaded EBGs exhibit, in principle, wider bandgaps compared to those of the same EBG with uniform capacitive loads. Additionally, the documented methodology exploits the fact that EBGs loaded with lumped capacitors can be conveniently represented as multiport networks whose response is fully characterized by their S-matrix. Consequently, their analysis can be performed via simple circuit-based calculations rather than through computationally expensive full-wave simulations. Finally, it was demonstrated that this circuit-based analysis can be extended for the design of mushroom-based absorbers loaded with lumped tuning capacitors and resistors.

References

1. Sievenpiper, D., Zhang, L., Broas, R., Alexopolous, N., and Yablonovitch, E. (1999). High–impedance electromagnetic surfaces with a forbidden

frequency band, *IEEE Trans. Microwave Theory Tech.*, **47**, pp. 2059–2074.

2. Kern, D. J., Werner, D. H., Monorchio, A., Lanuzza, L., and Wilhelm, M. J. (2005). The design synthesis of multiband artificial magnetic conductors using high impedance frequency selective surfaces, *IEEE Trans. Antennas Propag.*, **53**, pp. 8–17.
3. Martin, S. H., Martinez, I., Turpin, J. P., Werner, D. H., Lier, E., Bray, M. G. (2014). The synthesis of wide- and multi-bandgap electromagnetic surfaces with finite size and nonuniform capacitive loading, *IEEE Trans. Microw. Theory Tech.*, **62**, pp. 1962–1972.
4. Diaz, R. E., Sanchez, V., Caswell, E., and Miller, A. (2003). Magnetic loading of artificial magnetic conductors for bandwidth enhancement, *Proc. IEEE Antennas Propag. Soc. Int. Symp.*, **2**, pp. 431–434.
5. Kern, D. J., and Werner, D. H. (2006). Magnetic loading of EBG AMC ground planes and ultra-thin absorbers for improved bandwidth performance and reduced size, *Microw. Opt. Technol. Lett.*, **48**, pp. 2468–2471.
6. Gousettis, G., Feresidis, A. P., and Vardaxoglou, Y. C. (2005). Tailoring the AMC and EBG characteristics of periodic metallic arrays printed on grounded dielectric substrate, *IEEE Trans. Antennas Propag.*, **54**, pp. 82–89.
7. Simovski, C. R., De Maagt, P., and Melchakova, I. V. (2005). High-impedance surfaces having stable resonance with respect to polarization and incidence angle, *IEEE Trans. Antennas Propag.*, **53**, pp. 908–914.
8. Feresidis, A. P., Gousettis, G., Wang, S., and Vardaxoglou, C. J. (2005). Artificial magnetic conductor surfaces and their application to low-profile high-gain planar antennas, *IEEE Trans. Antennas Propag.*, **53**, pp. 209–215.
9. Caloz, C., and Itoh, T. (2005). *Electromagnetic Metamaterials: Transmission Line Theory and Microwave Applications* (Wiley-IEEE Press, USA).
10. Yazdi, M., and Komjani, N. (2011). Design of a band-notched UWB monopole antenna by means of an EBG structure, *IEEE Antennas Wireless Propag. Lett.*, **10**, pp. 170–173.
11. Bianconi, G., Costa, F., Genovesi, S., and Monorchio, A. (2011). Optimal design of dipole antennas backed by a finite high-impedance screen, *Prog. Electromagn. Res. C*, **18**, pp. 137–151.
12. Yang, F., and Rahmat-Samii, Y. (2009). *Electromagnetic Band Gap Structures in Antenna Engineering* (Cambridge University Press, UK).

13. Bell, J. M., and Iskander, M. F. (2004). A low-profile Archimedean spiral antenna using an EBG ground plane, *IEEE Antennas Wireless Propag. Lett.*, **3**, pp. 223–226.

14. Best, A., and Hanna, D. (2008). Design of a broadband dipole in close proximity to an EBG ground plane, *IEEE Antennas Propag. Mag.*, **50**, pp. 52–64.

15. Azad, M., and Ali, M. (2008). Novel wideband directional dipole antenna on a mushroom like EBG structure, *IEEE Trans. Antennas Propag.*, **56**, pp. 1242–1250.

16. Yang, F., and Rahmat-Samii, Y. (2003). Reflection phase characterizations of the EBG ground plane for low profile wire antenna applications, *IEEE Trans. Antennas Propag.*, **51**, pp. 2691–2703.

17. Yang, F., and Rahmat-Samii, Y. (2004). Bent monopole antennas on EBG ground plane with reconfigurable radiation patterns, *IEEE Proc. AP-S Int. Symp.*, **2**, pp. 1819–1822.

18. Baggen, R., Martinez-Vazquez, M., Leiss, J., Holzwarth, S., Drioli, L. S., and de Maagt, P. (2008). Low profile Galileo antenna using EBG technology, *IEEE Trans. Antennas Propag.*, **56**, pp. 667–674.

19. Yousefi, L., Mohajer-Iravani, B., and Ramahi, O. M. (2007). Enhanced bandwidth artificial magnetic ground plane for low-profile antennas, *IEEE Microw. Wireless Compon. Lett.*, **6**, pp. 289–292.

20. Rajo-Iglesias, E., Quevedo-Teruel, O., and Inclan-Sanchez, L. (2008). Mutual coupling reduction in patch antenna arrays by using a planar EBG structure and a multilayer dielectric substrate, *IEEE Trans. Antennas Propag.*, **56**, pp. 1648–1655.

21. Rajo-Iglesias, E., Quevedo-Teruel, O., Inclan-Sanchez, L., and Garcia-Munoz, L.-E. (2007). Design of a planar EBG structure to reduce mutual coupling in multilayer patch antennas, *Antennas Propag. Conf., LAPC 2007*, pp. 149–152.

22. Llombart, N., Neto, A., Gerini, G., and de Maagt, P. (2005). Planar circularly symmetric EBG structures for reducing surface waves in printed antennas, *IEEE Trans. Antennas Propag.*, **53**, pp. 3210–3218.

23. Abhari, R., and Eleftheriades, G. V. (2003). Metallo-dielectric electromagnetic bandgap structures for suppression and isolation of the parallel-plate noise in high-speed circuits, *IEEE Trans. Microw. Theory Tech.*, **51**, pp. 1629–1639.

24. Chen, X., Li, L., Liang, C. H., Su, Z. J., and Zhu, C. (2012). Dual-band high impedance surface with mushroom-type cells loaded by symmetric meandered slots, *IEEE Trans. Antennas Propag.*, **60**, pp. 4677–4687.

25. Exposito-Dominguez, G., Fernandez-Gonzalez, J.-M., Padilla, P., and Sierra-Castaner, M. (2012). Mutual coupling reduction using EBG in steering antennas, *IEEE Antennas Wireless Propag. Lett.*, **11**, pp. 1265–1268.
26. Yang, L., Fan, M., Chen, F., She, J., and Feng, Z. (2005). A novel compact electromagnetic-bandgap (EBG) structure and its applications for microwave circuits, *IEEE Trans. Microw. Theory Tech.*, **53**, pp. 183–190.
27. Yang, F., and Rahmat-Samii, Y. (2003). Microstrip antennas integrated with electromagnetic band-gap (EBG) structures: A low mutual coupling design for array applications, *IEEE Trans. Antennas Propag.*, **51**, pp. 2936–2946.
28. Clavijo, S., Diaz, R., and McKinzie, W. E. (2003). Design methodology for Sievenpiper high-impedance surfaces: An artificial magnetic conductor for positive gain electrically small antennas, *IEEE Trans. Antennas Propag.*, **51**, pp. 2678–2690.
29. Li, L., Chen, Q., Yuan, Q., Liang, C., and Sawaya, K. (2008). Surface wave suppression band gap and plane-wave reflection phase band of mushroom-like band gap structures, *J. Appl. Phys.*, **103**, pp. 023513/1–10.
30. Maci, S., Caiazzo, M., Cucini, A., and Casaletti, M. (2005). A pole-zero matching method for EBG surfaces composed of a dipole FSS printed on a grounded dielectric slab, *IEEE Trans. Antennas Propag.*, **53**, pp. 70–81.
31. Samani, M. F., Borji, A., and Safian, R. (2011). Relation between refection phase and surface-wave bandgap in artificial magnetic conductors, *IEEE Trans. Microwave Theory Tech.*, **59**, pp. 1901–1908.
32. Aminian, A., Yang, F., and Rahmat-Samii, Y. (2003). In-phase reflection and EM wave suppression characteristics of electromagnetic band gap ground planes, *IEEE Proc. AP-S Int. Symp.*, **4**, pp. 430–433.
33. Grbic, A., and Eleftheriades, G. V. (2003). Dispersion analysis of a microstrip based negative-refractive-index periodic structure, *IEEE Microwave Wireless Compon. Lett.*, **13**, pp. 155–157.
34. Elek, F., and Eleftheriades, G. V. (2004). Dispersion analysis of the shielded Sievenpiper structure using multiconductor transmission-line theory, *IEEE Microw. Wireless Compon. Lett.*, **14**, pp. 434–436.
35. Veselago, V. G. (1968). The electrodynamics of substances with simultaneously negative values of ε and μ, *Sov. Phys. Usp.*, **10**, pp. 509–514.

36. Eleftheriades, G. V., Iyer, A. K., and Kremer, P. C. (2002). Planar negative refractive index media using periodically L-C loaded transmission lines, *IEEE Trans. Microw. Theory Tech.*, **50**, pp. 2702–2712.
37. Abedin, M. F., Azad, M. Z., and Ali, M. (2008). Wideband smaller unit-cell planar EBG structures and their application, *IEEE Trans. Antennas Propag.*, **56**, pp. 903–908.
38. Kim, T., and Seo, C. (2000). A novel photonic bandgap structure for low-pass filter of wide stopband, *IEEE Microw. Wireless Compon. Lett.*, **10**, pp. 13–15.
39. Karmakar, N. C. (2002). Improved performance of photonic band-gap micro-stripline structures with the use of Chebyshev distributions, *Microwave Opt. Technol. Lett.*, **33**, pp. 1–5.
40. Mosallaei, H., and Sarabandi, K. (2005). A compact wide-band EBG structure utilizing embedded resonator circuits, *IEEE Antennas Wireless Propag. Lett.*, **4**, pp. 5–8.
41. Karim, M. F., Liu, A. Q., Alphones, A., and Zhang, X. J. (2004). Low-pass filter using a hybrid EBG structure, *Microwave Opt. Technol. Lett.*, **45**, pp. 95–98.
42. Chappell, W. J., Little, M. P., and Katehi, L. P. B. (2001). High isolation, planar filters using EBG substrate, *IEEE Microwave Wireless Compon. Lett.*, **11**, pp. 246–248.
43. Liang, L., Liang, C., Zhao, X., and Su, Z. (2008). A novel broadband EBG using multi-period mushroom-like structure, *Int. Conf. Microw. Millimeter Wave Technol., ICMMT 2008*, **4**, pp. 1609–1612.
44. Chen, L., Wang, C., Zhang, Q., and Yang, X. (2011). A novel wide-band cascaded EBG structure with chip capacitor loading, *IEEE Int. Symp. Microw., Antenna, Propag. EMC Technol. Wireless Commun., MAPE 2011*, pp. 289–291.
45. Pozar D. (2005). *Microwave Engineering*, 3rd ed. (Wiley).
46. Davidovitz, M. (1995). Reconstruction of the S-matrix for a 3-port using measurements at only two ports, *IEEE Microwave Guided Wave Lett.*, **5**, pp. 349–350.
47. Lu, H., and Chu, T. (2000). Port reduction methods for scattering matrix measurement of an n-port network, *IEEE Trans. Microwave Theory Tech.*, **48**, pp. 959–968.
48. Hansen, N., and Ostermeier, A. (2001). Completely derandomized self-adaptation in evolutionary strategies, *Evol. Comput.*, **9**, pp. 159–195.
49. Gregory, M. D., Bayraktar, Z., and Werner, D. H. (2011). Fast optimization of electromagnetic design problems using the covariance matrix

adaptation evolutionary strategy, *IEEE Trans. Antennas Propag.*, **59**, pp. 1275–1285.

50. Munk, B. (2000). Frequency selective surfaces: Theory and design (Wiley Press, USA).
51. Engheta, N. (2002). Thin absorbing screens using metamaterial surfaces, *Proc. IEEE Antennas Propag. Soc. Int. Symp.*, San Antonio, TX, **2**, pp. 392–395.
52. Kern, D. J., and Werner, D. H. (2003). A genetic algorithm approach to the design of ultra-thin electromagnetic bandgap absorbers, *Microw. Opt. Technol. Lett.*, **38**, pp. 61–64.
53. Gao, Q., Yin, Y., Yan, D.-B., and Yuan, N.-C. (2005). Application of metamaterials to ultra-thin radar-absorbing material design, *Electron. Lett.*, **41**, pp. 1311–1313.
54. Li, Y.-Q., Zhang, H., Fu, Y.-Q., and Yuan, N.-C. (2008). RCS reduction of ridged waveguide slot antenna array using EBG radar absorbing material, *IEEE Antennas Wireless Propag. Lett.*, **7**, pp. 473–476.
55. Simms, S., and Fusco, V. (2006). Tunable thin radar absorber using artificial magnetic ground plane with variable backplane, *Electron. Lett.*, **42**, pp. 1197–1198.

Chapter 5

Temporal and Spatial Dispersion Engineering Using Metamaterial Concepts and Structures

Shulabh Gupta,[a] Mohamed Ahmed Salem,[b] and Christophe Caloz[b,c]

[a]*Department of Electronics, Carleton University, Ottawa, Ontario, Canada*
[b]*Poly-Grames, Polytechnique de Montréal, Montréal, Canada*
[c]*Electrical and Computer Engineering Department, King Abdulaziz University, Jeddah, Saudi Arabia*
shulabh.gupta@cunet.carleton.ca

5.1 Introduction

Electromagnetic metamaterials (MTMs) are broadly defined as artificial effectively homogeneous electromagnetic structures with exotic properties not readily available in nature. They consist of an arrangement of subwavelength scattering particles emulating the atoms or molecules of real materials with enhanced properties. The scattering particles are typically arranged in a periodic lattice with the unit cell size $p \ll \lambda_g$, where λ_g is the guided wavelength inside the MTM. Under such operating conditions, the structure behaves as

Broadband Metamaterials in Electromagnetics: Technology and Applications
Edited by Douglas H. Werner
Copyright © 2017 Pan Stanford Publishing Pte. Ltd.
ISBN 978-981-4745-68-0 (Hardcover), 978-1-315-36443-8 (eBook)
www.panstanford.com

a real material so that the electromagnetic waves sense the average, or effective, macroscopic and well-defined constitutive parameters, which depend on the nature of the unit cell. Their 2D counterparts are known as metasurfaces, which are thin layers of subwavelength resonant scatters that interact strongly with electromagnetic waves for achieving unique wavefront processing functionalities.

One of the classical functionalities of MTMs is negative refraction, which requires simultaneous occurrence of negative ε and μ within the desired frequency range. However, it is physically impossible to achieve negative ε and μ in a non-dispersive medium to satisfy energy conditions, and thus metamaterials are inherently dispersive with frequency-dependent material parameters [$\varepsilon(\omega)$ and $\mu(\omega)$] to exhibit such exotic effects [Caloz and Itoh (2006)].

The first MTM implementations were dominantly volumetric arrangements of resonant particles consisting of split-ring resonators and wired mediums to realize effective $\varepsilon(\omega)$ and $\mu(\omega)$ for achieving negative refraction. However, the resonant nature of the particles gave them narrowband and lossy characteristics. Later, Caloz and Itoh (2006) developed the concept of composite right/left-handed (CRLH) transmission lines, which act as a left-handed transmissions line at low frequencies and right-handed transmissions line at high frequencies. They were based on tightly coupled particles, which resulted in broadband planar structures with low losses.

The rich electromagnetic characteristics of metamaterial CRLH lines combined with their broadband nature recently led to the novel paradigm of radio-analog signal processing (R-ASP) for processing broadband signals using exotic and unique dispersion functionalities of such MTMs, metasurfaces, and MTM-inspired structures [Gupta and Caloz (2009); Caloz (2009)]. This chapter focuses on this aspect of MTMs and develops a conceptual foundation for a unique class of R-ASP systems: real-time spectrum analyzers (RTSAs).

The chapter is organized as follows. Section 5.2 introduces R-ASP and its core signal processing component called phaser. These phasers are typically of two types, temporal and spatial, and the spatial types are expanded upon in Section 5.3 for real-time spectrum analysis applications. This section discusses two commonly used spatial phasers: a diffraction grating and leaky-wave antenna (LWA), with a detailed description of metamaterial LWAs as a practical device to be used in the rest of the chapter. Once the phaser components are presented and explained, Section 5.4 presents 1D/2D RTSAs using LWAs and discusses their typical system features and characteristics. Finally, a metasurface phaser-based spatial 2D RTSA is presented in Section 5.5.

5.2 Radio-Analog Signal Processing

The exploding demand for faster, more reliable and ubiquitous radio systems in communication, instrumentation, radar, and sensors poses unprecedented challenges in microwave and mm-wave engineering. Recently, the predominant trend has been to place an increasing emphasis on digital signal processing (DSP). However, while offering device compactness and processing flexibility, DSP suffers from fundamental drawbacks, such as high-cost analog-digital conversion, high power consumption, and poor performance at high frequencies. To overcome these drawbacks, and hence address the aforementioned challenges, one might possibly get inspiration from ultrafast optics [Saleh and Teich (2007)]. In this area, ultrashort and thus huge-bandwidth electromagnetic pulses are efficiently processed in real time using analog and dispersive materials and components. It has led to a wealth of new applications at microwaves such as analog RTSAs for the measurement and characterization of complex non-stationary signals [Gupta et al. (2009)], tunable delay lines [Abielmona et al. (2007)], compressive receivers [Abielmona et al. (2009)], real-time Fourier transformers [Muriel et al. (1999); Schwartz et al. (2006)], convolvers and convoluters [Campbell (1989)], frequency meters and discriminators [Nikfal et al. (2011)], spectrum sniffers [Nikfal et al. (2012)], direction-of-arrivals systems [Abielmona et al. (2011)], chipless RFID systems [Gupta et al. (2011)], pulse-position modulators [Nguyen and Caloz (2008)], and Dispersion Code Multiple Access system [Nikfal et al. (2013); Gupta et al. (2015b)].

5.2.1 R-ASP Paradigm

R-ASP might be defined as the manipulation of signals in their pristine analog form and in real time to realize specific operations enabling microwave or mm-wave and terahertz applications [Caloz et al. (2013)]. The essence of ASP might be best approached by considering the two basic effects described in Fig. 5.1, chirping with time spreading and frequency discrimination in the time domain. Both effects involve a linear element with transfer function $H(\omega) = e^{j\phi(\omega)}$, which is assumed to be of unity magnitude and whose phase, $\phi(\omega)$, is a nonlinear function of frequency, or whose group delay, $\tau(\omega) = \partial\phi(\omega)/\partial\omega$, is a function of frequency. Such an element, with frequency-dependent group delay, is called temporally dispersive. The bandwidth of $H(\omega)$ is assumed to cover the entire spectrum of the input signal.

In the first case, depicted in Fig. 5.1a, a pulse modulated at an angular frequency ω_0 is passed through the dispersive element, which is assumed here to exhibit a positive linear group delay slope over a frequency band centered at the frequency ω_0, corresponding to a group delay τ_0. Due to the dispersive nature of this element, the different spectral components of the pulse experience different phase-shifts and, therefore, emerge at different times. Here, the lower-frequency components are less delayed and, therefore, emerge earlier than the higher-frequency components, while the center-frequency component appears at the time $\tau_3 = \tau_0$. This results in an output pulse whose instantaneous frequency is progressively increasing, a phenomenon called "chirping," and which has experienced time spreading ($T_{out} > T_0$), accompanied with reduced amplitude due to energy conservation.

In the second case, depicted in Fig. 5.1b, the input pulse is modulated by a two-tone signal, with frequencies ω_{01} and ω_{02}, and passed through a dispersive element $H(\omega)$ exhibiting a positive stepped group delay, with two steps, centered at ω_{01} and ω_{02}, respectively. Based on this dispersive characteristic, the part of the pulse modulated at the lower frequency, ω_{01}, is delayed less than the part modulated at the higher frequency, ω_{02}, and hence emerges earlier in time. As a result, the two pulses are resolved (or separated) in the time domain, and their respective modulation frequencies may be deduced from their respective group delays from the dispersive transfer function, $H(\omega)$. Note that with the flat-step group delay considered here, the pulses are not time spread ($T_{out} = T_0$), assuming that the pulse bandwidth fits in the flat bands of the steps.

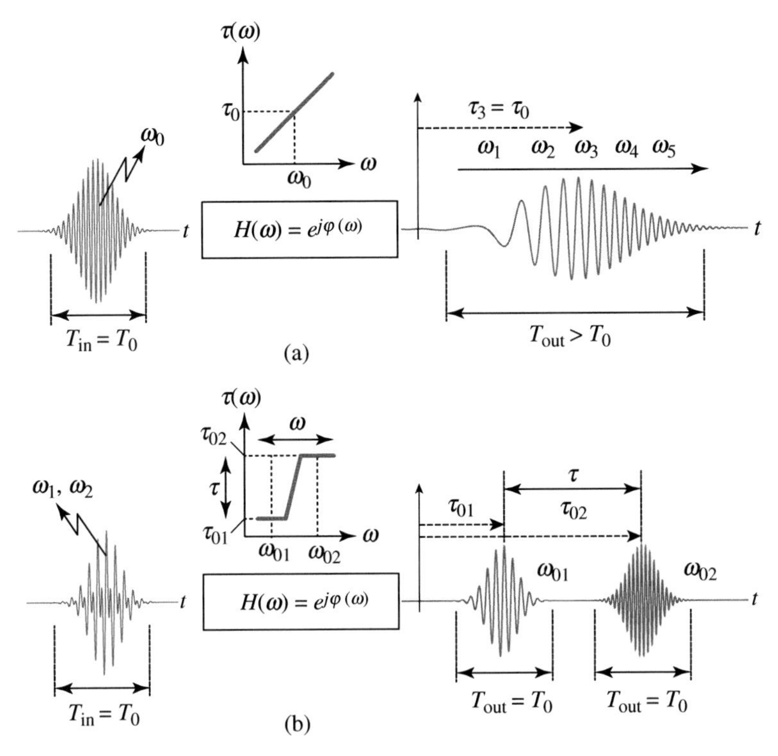

Figure 5.1 Basic effects in ASP. (a) Chirping with time spreading. (b) Frequency discrimination in the time domain. Reprinted with permission from Caloz et al., 2013, Copyright 2013, IEEE.

5.2.2 Phasers

The dispersive element with a transfer function $H(\omega)$ in Fig. 5.1, which manipulates the spectral components of the input signal in the time domain, is the core of an ASP system and is called a phaser [Gupta et al. (2015a); Caloz et al. (2013)]. A phaser "phases" as a filter: It modifies the phase of the input signal following phase—or more fundamentally group delay—specifications, with some magnitude considerations, just in the same way as a filter modifies the magnitude of the input signal following magnitude specifications, with some phase considerations. An ideal phaser exhibits an arbitrary group delay with flat and lossless magnitude over a given frequency band. The first phasers were inspired from metamaterial

CRLH lines exploiting their inherently dispersive and broadband characteristics of these structures [Gupta and Caloz (2009); Caloz (2009)]. Since then various types of phasers have been reported in the literature based on two-port network approaches, realized in transmission lines and waveguide technologies [Gupta et al. (2012); Zhang et al. (2012, 2013); Gupta et al. (2010, 2013, 2015a)] and good review is available in [Caloz et al. (2013)].

The essence of group delay engineering in phasers can be understood using a relatively simple structure, a microwave C-section. A microwave C-section is an end-connected coupled transmission line structure with an all-pass frequency response, as shown in Fig. 5.2(a) [Matthaei et al. (1980); Steenaart (1963)]. The transfer function of a C-section may be written as [Steenaart (1963)]

$$S_{21}(\theta) = \left[\frac{1 + j\rho\tan\theta}{1 - j\rho\tan\theta}\right], \tag{5.1}$$

with

$$\theta = \frac{\omega}{c}\sqrt{\varepsilon_{\text{eff}}\mu_{\text{eff}}}\,\ell, \tag{5.2a}$$

$$\rho = \left(\sqrt{\frac{1-k}{1+k}}\right). \tag{5.2b}$$

where ω is the angular frequency, c is the speed of light in vacuum, ε_{eff} is the effective permittivity, μ_{eff} is the effective permeability, ℓ is the physical length of the structure, and k is the voltage coupling coefficient. The C-section's transmission magnitude is unity ($|S_{21}(\theta)| = 1$), and its group delay response is obtained from (5.1) as

$$\tau(\omega) = \frac{d}{d\omega}\arg\{S_{21}(\omega)\} = \frac{2\rho}{\rho^2 + (1-\rho^2)\cos^2\theta}\frac{d\theta}{d\omega}. \tag{5.3}$$

Due to the commensurate nature of the C-section's structure, $\tau(\omega)$ is periodic in frequency. Since, according to (5.2a), $d\theta/d\omega = \sqrt{\varepsilon_{\text{eff}}\mu_{\text{eff}}}\,\ell/c$ is a constant, the maxima of $\tau(\omega)$ simply correspond to the zeros of the cosine function in (5.3), *i.e.*, $\theta_p = (2p+1)\pi/2$, where p is an integer. In terms of the length, these maxima correspond to odd multiples of the quarter wavelength, *i.e.*, $l_p = (2p + 1)\lambda/4$. The group delay swing, $\Delta\tau(\omega)$, of a C-section may be conveniently obtained from (5.3) with (5.2a) as

$$\Delta\tau(\omega) = (\tau_{\max} - \tau_{\min}) = 2\left(\frac{1}{\rho} - \rho\right)\sqrt{\varepsilon_{\text{eff}}\mu_{\text{eff}}}\,\ell. \tag{5.4}$$

Thus, the group delay swing can be controlled by varying ρ, through k according to (5.2b), and l. Alternatively, for a given $\Delta\tau$, the requirement on k can be relaxed by increasing l. However, increasing l shifts ω_0 downward, and therefore higher-order group delay peaks ($p > 0$), instead of the fundamental one ($p = 0$), have to be used in order to keep the frequency of maximum group delay unchanged, as illustrated in Fig. 5.2b.

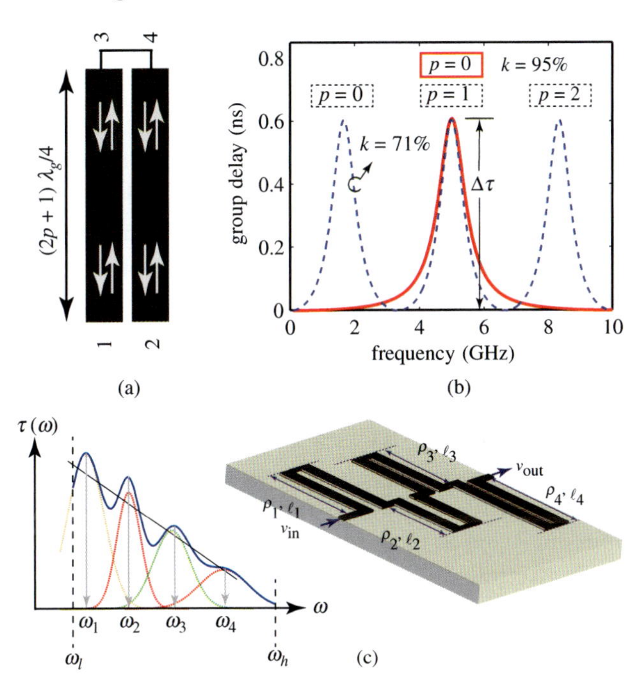

Figure 5.2 C-sections for dispersion engineering. (a) Single C-section layout. (b) Corresponding typical group delay response, computed by (5.3), for two different lengths l, corresponding to the first-maximum frequencies f_0 = 5 GHz (solid curve) and f_0 = 5/3 GHz (dashed curve), and two different coupling coefficients, k = 0.95 (solid curve) and k = 0.71 (dashed curve), yielding the same group delay swing $\Delta\tau$ = 0.6 ns, according to (5.4). (c) Principle of group delay synthesis by superposition of group delays with maxima at different frequencies realized by a cascaded C-section structure. Reprinted with permission from Gupta et al., 2013, Copyright 2013, John Wiley and Sons.

Finally, using C-section as a basic unit, a quasi-arbitrary group delay response with a specified bandwidth can be realized as series of individual all-pass C-sections with different parameters cascaded together. Specifically, as illustrated in Fig. 5.2c, a desired group delay response is achieved as a superposition of various group delay functions provided by coupled-line all-pass networks with peak group delays τ_n centered at ω_n. One application of this structure will be shown in Section 5.5.

The phasers are fundamentally of two types:

(1) A temporal phaser that discriminates the temporal frequency components of a signal in the time domain, and
(2) A spatial phaser that discriminates the temporal frequency components of a signal in the spatial domain.

While a microwave C-section described above belongs to temporal phasers, it is the second type of phasers that is the subject of this chapter, and forthcoming sections of this chapter will focus on a very specific class of applications of these phasers: ultrafast realtime spectrum analysis of broadband signals.

5.3 Spatial Phasers for Real-Time Spectrum Analysis

Real-time spectrum analysis is a ubiquitous signal processing operation in science and engineering. It involves real-time frequency discriminating devices that separate the various spectral components of a signal in either the space domain or the time domain. Typical applications include spectral analysis for instrumentation, electromagnetics and biomedical imaging [Metz et al. (2014); Lee and Wight (1986a); Gupta et al. (2009)], ultrafast optical signal processing [Goda and Jalali (2013); Goda et al. (2009)], and dense wavelength demultiplexing communication systems [Xiao and Weiner (2004); Supradeepa et al. (2008)], to name a few. The heart of such systems is a spatial dispersive device. Typical devices are optical prisms, diffraction gratings, arrayed-waveguide gratings (AWGs) [Saleh and Teich (2007); Goodman (2004)], Bragg gratings [Kashyap (2009)], phasers [Caloz et al. (2013)], LWAs [Caloz et al.

(2011)], and virtual image phased arrays (VIPAs) [Shirasaki (1996)]. Among these several spatially dispersive devices, diffraction gratings and LWAs emerge as two most practical components to be employed for spectrum analysis applications and will be described next.

5.3.1 Diffraction Gratings

A diffraction grating is a spatial dispersion element that is commonly used in optics for spectrum analysis of broadband signals [Goodman (2004)]. Let us first consider a complex transmittance function, relating the input and output waveform distributions, of a thin sinusoidal diffraction grating of Fig. 5.3:

$$t_g(x, y) = 1 + m\cos(2\pi\zeta y), \tag{5.5a}$$

where

$$\frac{1}{\zeta} = \Lambda: \quad \text{grating period.} \tag{5.5b}$$

Now consider a plane wave propagating along the z-axis given by

$$\psi(x, y, z; t) = \mathrm{Re}\{e^{-j(kz-\omega t)}\}, \tag{5.6}$$

so that $\psi(x, y, 0_-; t) = 1$, where the explicit time dependence $e^{j\omega t}$ is dropped for convenience. The wave just after the grating can then be expressed as:

$$\psi(x, y, 0_+) = t_g(x, y; \omega)\,\psi(x, y, 0_-) = [1 + m\cos(2\pi\zeta y)] \tag{5.7}$$

Considering the lens transmittance function $t_1(x, y; \omega)$, wave output after the lens is given by

$$\psi(x, y, l_+) = [1 + m\cos(2\pi\zeta y)]\exp\left[-j\frac{\pi}{\lambda d}(x^2 + y^2)\right], \tag{5.8}$$

where d is the focal length of the lens. The output wave, in the focal plane, is

$$\psi(x, y, d) = \psi(x, y, l_+) * h(x, y), \tag{5.9}$$

where $h(x, y)$ is the free-space impulse response. Using the paraxial-wave approximations, the output wave can finally be written as:

$$|I(x, y; z = d)| = |\psi(x, y; l_+) * h(x, y)|^2$$

$$= \left| [1+m\cos(2\pi\zeta y)]) \exp\left[-j\frac{\pi}{\lambda d}(x^2+y^2)\right] * \frac{e^{jkd}}{j\lambda d}\exp\left[j\frac{\pi}{\lambda d}(x^2+y^2)\right]\right|^2$$
$$= \left|\int \exp\left(-j\frac{2\pi x}{\lambda d}X\right)dX \int [1+m\cos(2\pi\zeta y)]\exp\left(-j\frac{2\pi y}{\lambda d}Y\right)dY\right|^2 \qquad (5.10)$$
$$= \frac{|\delta(x)|}{(\lambda d)^2}\left[\delta(y)+\frac{m}{2}\delta(y+\zeta\lambda d)+\frac{m}{2}\delta(y-\zeta\lambda d)\right],$$

where $\delta(\cdot)$ is the Dirac-delta function. This equation consists of three terms where $\delta(y)$ corresponds to the 0th–order diffraction and the other two terms, which are symmetric about the origin, are the $\pm 1^{st}$–order diffraction terms. It is clear from the equation above that the location of intensity maxima of the $\pm 1^{st}$ diffraction order depends on the signal wavelength (or the frequency ω). Selecting one of these orders, we obtain the frequency-to-space mapping relation of this system where each frequency ω is mapped onto a specific point $[0, y(\omega)]$ on the output plane according to the relation:

$$y(\omega) = \zeta\left(\frac{2\pi c}{\omega}\right)d, \quad \text{or} \qquad (5.11a)$$

$$\theta(\omega) = \tan^{-1}\left[\frac{2\pi\zeta}{k_0}\right] = \tan^{-1}\left[\frac{2\pi}{\Lambda}\frac{1}{k_0}\right]. \qquad (5.11b)$$

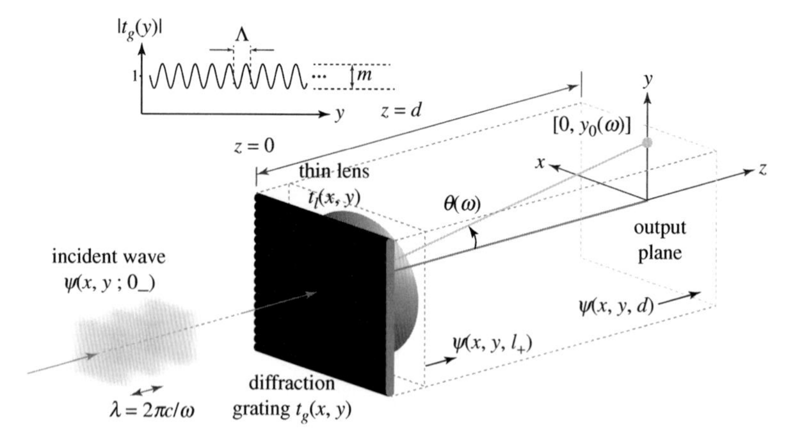

Figure 5.3 Thin sinusoidal grating excited with a normally incident plane wave.

This is the 1D frequency scanning along the y-axis using a thin sinusoidal diffraction grating. While a thin sinusoidal grating is taken here as an example, a large variety of diffraction gratings exist, which are used for spectral analysis using their higher-order diffraction patterns.

5.3.2 Leaky-Wave Antennas

A diffraction grating spectrally decomposes a pulsed wavefront in space. Another related component common at microwave and mm-wave frequencies is an LWA. LWAs are a class of antennas that use a traveling wave on a guiding structure as the main radiating mechanism [Hessel (1969); Tamir (1969); Oliner and Jackson (2007)]. These antennas are capable of producing narrow beams, with the beam-width limited by the size of the structure. LWAs support a fast wave on the guiding structure, where the phase constant $\beta(\omega)$ is less than the free-space wavenumber k_0. The leaky wave is, therefore, fundamentally a radiating type of wave, which radiates power continuously as it propagates on the guiding structure. Typical implementations of an LWA are in slot-waveguides and printed metamaterial transmission line structures [Caloz et al. (2011); Caloz and Itoh (2006)].

An LWA resembles a diffraction grating in two main respects:

(1) LWAs are broadband structures and can support pulsed waves with finite signal bandwidths $\Delta\omega$.

(2) The radiation angle θ of an LWA depends on the signal frequency ω, as illustrated in Fig. 5.4.

On the other hand, they differ from each other in two respects:

(1) An equivalent 0th-order diffraction in an LWA is frequency dependent unlike that in the case of a diffraction grating, where higher-order diffraction orders are used for frequency discrimination.

(2) While a diffracting grating is excited with a 2D wavefront in space, an LWA is fed at a single point in space, as illustrated in Fig. 5.4.

A typical 1D periodic LWA structure may be seen as a uniform structure supporting a slow non-radiating wave, with $\beta_0(\omega) > k_0$, that has been periodically modulated in the longitudinal (y)-

direction. The periodic modulation generates an infinite number of space harmonics with propagation constant $\beta_n(\omega) = \beta_0(\omega) + 2\pi n/p$ where p is the unit cell period and n is an integer [Hessel (1969)]. Although the main ($n = 0$) space harmonic is a slow wave, one of the space harmonics (usually $n = -1$) is designed to be a fast wave, so that $-k_0 < \beta_{-1} < k_0$, and hence this space harmonic is a radiating wave. With recent developments in the field of LWAs inspired from metamaterial concepts, $n = 0$ fundamental space harmonic also radiates inside the fast-wave region enabling a full-space frequency scan including broadside radiation, as will be explained in the next subsection [Caloz and Itoh (2006); Otto et al. (2012); Otto et al. (2014)].

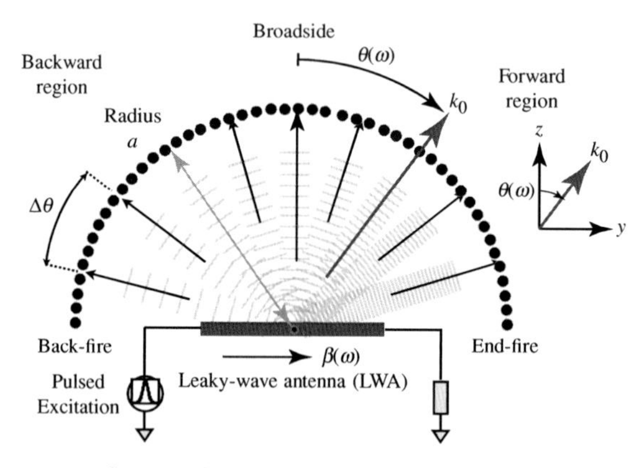

Figure 5.4 Spatial-spectral decomposition of a broadband temporal signal using an LWA.

An intuitive derivation of the radiation characteristics of an LWA can be made from a simple wave-propagation argument. Consider the LWA of Fig. 5.4, supporting a guided-wave mode $\psi(x) = \psi_0 e^{-j\beta(\omega)y}$. If n^{th} space harmonic is the radiating leaky-wave mode, the field in the air region above the aperture ($z > 0$) is given by $\psi(x,z) = \psi_0 e^{-j\beta_n x} e^{-jk_z z}$, satisfying $k_0^2 = \beta_n^2 + k_z^2$, where k_z is the wavenumber along the z-axis. The angle $\theta(\omega)$ of the resulting leaky-wave radiation is then given by:

$$\sin\theta(\omega) = \left[\frac{\beta_n(\omega)}{k_0}\right] = \left[\frac{\beta_0(\omega) + 2\pi n/p}{k_0}\right] = \left[\frac{2\pi}{\lambda_{g,n}(\omega)}\frac{1}{k_0}\right], \quad (5.12a)$$

which for small angles around $\theta = 0°$ can be written as:

$$\theta(\omega) \approx \tan^{-1}\left[\frac{2\pi}{\lambda_{g,n}(\omega)}\frac{1}{k_0}\right]. \tag{5.12b}$$

This small angle approximation is made to establish a link between an LWA and a diffraction grating operated under paraxial conditions. Compared to the frequency-scanning relation of (5.11) of a grating, the above LWA scanning relation has the same form. Therefore, an LWA essentially operates as a diffraction grating, or vice versa.

5.3.3 Composite Right/Left-Handed Transmission Lines

Among several types of LWAs reported in the literature, one structure particularly stands out due to its unique radiation properties. It is a metamaterial CRLH transmission line LWA [Caloz and Itoh (2006)], which radiates in its $n = 0$ space harmonic, as compared to $n = -1$ space harmonic in other conventional LWA structures, and is capable of full-space frequency scanning from backfire to endfire, including broadside.

The general properties of a CRLH structure can be established based on a lumped circuit unit cell model. An ideal left-handed (LH) transmission line consists of a series capacitance and a shunt inductance, and subsequently exhibits anti-parallel phase and group velocities [Ramo et al. (1994)]. However, such a line does not exist in nature because of the presence of parasitic series inductance and shunt capacitance, which are responsible for right-handed (RH) contributions. To take into account these effects, Caloz et al. [Caloz and Itoh (2006)] developed the concept of a CRLH transmission line, which acts as an LH transmission line at low frequencies and RH transmission line at high frequencies.

The CRLH artificial transmission line is composed of RH elements (L_R, C_R) and LH elements (L_L, C_L), as shown in Fig. 5.5(a), and is characterized by the following dispersion relation [Caloz and Itoh (2006)]:

$$\beta(\omega) = \frac{1}{p}\cos^{-1}\left(1 - \frac{\chi}{2}\right),$$
$$\text{with} \quad \chi = \left(\frac{\omega}{\omega_R}\right)^2 + \left(\frac{\omega_L}{\omega}\right)^2 - \kappa\omega_L^2, \tag{5.13}$$

where $\kappa = L_L C_R + L_R C_L$, $\omega_R = 1/\sqrt{L_R C_R}$, $\omega_L = 1/\sqrt{L_L C_L}$ and p is the unit cell size or period, and by the Bloch impedance

$$Z_B = Z_L \sqrt{\frac{(\omega/\omega_{se})^2 - 1}{(\omega/\omega_{sh})^2 - 1} - \frac{\omega_L}{2\omega}\left[\left(\frac{\omega}{\omega_{se}}\right)^2 - 1\right]^2}, \tag{5.14}$$

where $\omega_{se} = 1/\sqrt{L_R C_L}$, $\omega_{sh} = 1/\sqrt{L_L C_R}$ and Z_L is the load impedance. Depending on the relative values of the LH and RH contributions, this transmission line can be unbalanced or balanced, *i.e.*, exhibiting a gapless transition between the LH and RH bands.

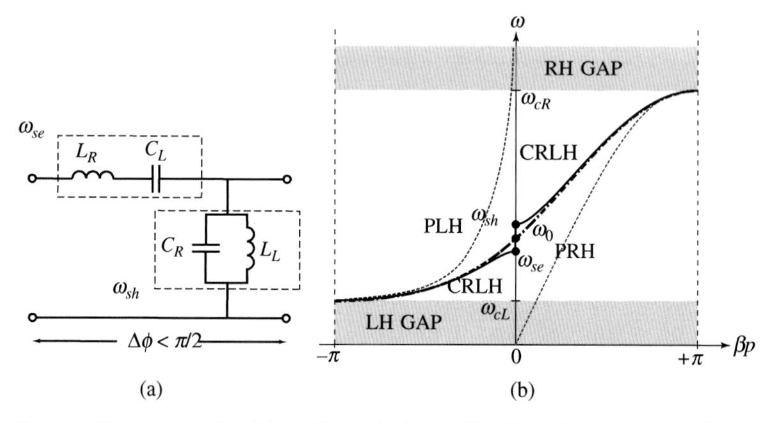

Figure 5.5 Composite right/left-handed (CRLH) MTMs fundaments. (a) Unit cell transmission model. (b) Dispersion diagram [Caloz and Itoh (2006)].

The typical dispersion relation of a balanced CRLH transmission line, where ϕ is the phase shift across the structure, curves are shown in Fig. 5.5b. As seen in (5.13), the CRLH transmission line offers a certain degree of dispersion (phase) control via the CRLH parameters L_R, C_R, L_L, and C_L. Thanks to its dispersive properties and subsequent design flexibility, the CRLH transmission line provides low-loss, compact, and planar dispersion-engineered solutions, avoiding frequency limitations, complex fabrication, cryogenics, circulators, or amplifiers. Moreover, the CRLH transmission line's operational frequency and bandwidth are dependent only on a single unit cell's RH/LH capacitor and inductor values [Caloz and Itoh (2006)]. Thus, a compact CRLH transmission line can be designed to operate at high frequencies while also exhibiting wide bandwidth.

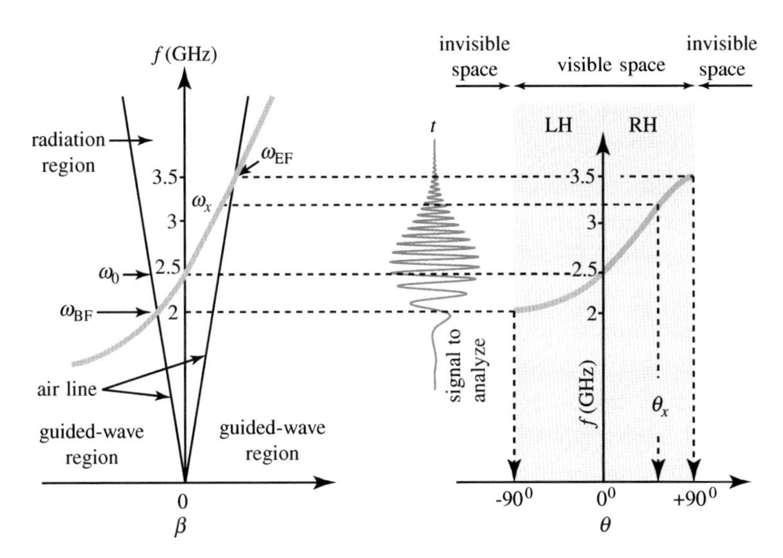

Figure 5.6 Frequency-space mapping associated with temporal-spatial dispersion for the CRLH LWA employed in the RTSA. Reprinted with permission from Caloz et al., 2013, Copyright 2013, IEEE.

A CRLH transmission line can also be operated in a radiative leaky-wave mode when open to free-space, since the CRLH dispersion curve penetrates into the fast-wave region, $\omega \varepsilon$ [ω_{BF}, ω_{EF}]. The resulting LWA radiates from backfire (θ = –90°) to end-fire (θ = +90°), including broadside (θ = 0°) as frequency is scanned from ω_{BF} (where $\beta = -k_0$) to ω_{EF} (where $\beta = +k_0$) [Liu et al. (2002); Caloz and Itoh (2006)], following the scanning law of (5.12a), which is plotted in Fig. 5.6.

5.4 LWA-Based Real-Time Spectrum Analyzers

Consider now some signal processing applications such as a general Fourier transforming operation: short-time Fourier transform (STFT) or a spectrogram, from which simpler operation of determining the overall spectral contents is implicitly obtained. This operation is useful for characterizing non-stationary signals, which are typical of most of today's ultra-wideband (UWB) systems, such as radar, security and instrumentation, and electromagnetic interference and compatibility (EMI/EMC), displaying rapid spectral variations in time [Thummler and Bednorz (2007)]. In order to effectively observe such signals, both time information and spectral information are simultaneously needed.

The STFT belongs to the general class of joint time-frequency representation. Joint time-frequency representations are 2D plots of a signal where a 1D signal is represented as an image in a time-frequency plane, with the signal energy distribution coded in the color-scale levels of the image. The joint time-frequency representation of a given signal thus provides information on the temporal location of the signal's spectral components, which depends on the temporal/spectral structure of this signal [Cohen (1989)]. Such analysis not only provides an intuitive insight into the transient behavior of the signals, but also completely characterizes their frequency, phase, and amplitude responses. Thus, the joint time-frequency representation is a highly informative tool for real-time spectrum analysis. Various numerical techniques exist to compute the joint time-frequency representations, with STFT/Spectrogram and Wigner-Ville distributions being the most common [Cohen (1989)].

The STFT (Spectrogram) of a signal $x(t)$ is calculated using

$$S(\tau,\omega)=\left|\int_{-\infty}^{\infty}x(t)g(t-\tau)e^{-j\omega t}dt\right|^2 \tag{5.15}$$

where $g(t)$ is a gate function. Numerous RTSAs are currently utilized for various applications. At microwaves, RTSAs are generally based on a digital STFT. For UWB signals with ultrafast transients, small-gate duration is required in the STFT for high time resolution. Such small-gate durations inherently lead to a long acquisition time and large acquisition bandwidth. Thus, the STFT process requires heavy computational resources and large memory buffers. These restrictions severely affect the system functionalities and limit its performance to restricted acquisition bandwidth, thereby preventing its application to UWB systems.

In optics, joint time-frequency analysis is often achieved by frequency-resolved optical gating (FROG) systems, which are used to measure and characterize ultrashort optical pulses. A FROG system is an analog implementation of the STFT employing a self-gating process achieved by a nonlinear second harmonic generating crystal

[Trebino (2002)]. Another optical system capable of performing joint time-frequency analysis is the acoustic spectrum analyzer reported in [Lee and Wight (1986b)], where Bragg cell plays a similar role as the nonlinear crystal in the FROG systems. In acoustics, real-time spectrum monitoring is achieved using acoustic filters for speech signal analysis [Wood and Hewitt (1963)]. Whereas the optical and acoustic spectrum analyzers mentioned above are capable of handling and analyzing very broadband signals, the digital RTSAs available at microwaves are unfortunately restricted to bandwidths generally too narrow for practical broadband applications.

With this background in context, the objective of this section is to describe a system that achieves a spectrogram of an arbitrary non-stationary signal using analog means without resorting to digital computation, in the spirit of the R-ASP paradigm.

5.4.1 One-Dimensional Real-Time Spectrum Analyzer

As established in Section 5.3, an LWA may be seen as a microwave counterpart of a diffraction grating. Just like a diffraction grating is used in optics for spectrally decomposing a broadband signal in space, an LWA can also be analogously used to spectrally analyze a broadband microwave signal. This equivalence inspires us to devise an RTSA as shown in Fig. 5.7 for analog computation of STFT of a broadband signal. This system is based on the following three successive operations:

(1) Spatial-spectral decomposition using an LWA to discriminate the frequency components of the signal.
(2) Probing and monitoring of the time variation of each frequency component. Probing is achieved by antenna receivers, while monitoring is performed by envelope demodulation.
(3) Post-processing, including analog/digital conversion, data processing, and display.

In the first step, an LWA first spectrally decomposes the non-stationary signal in space. While there are several choices for the antenna, metamaterial-based CRLH LWA is of particular usefulness as it offers three distinct benefits: (i) full-space radiation from backfire to endfire, including broadside in the fundamental mode, offering a simple and real-time frequency-space separation mechanism; (ii) frequency and bandwidth scalability, allowing to handle UWB signals; (iii) simple and compact design and implementation.

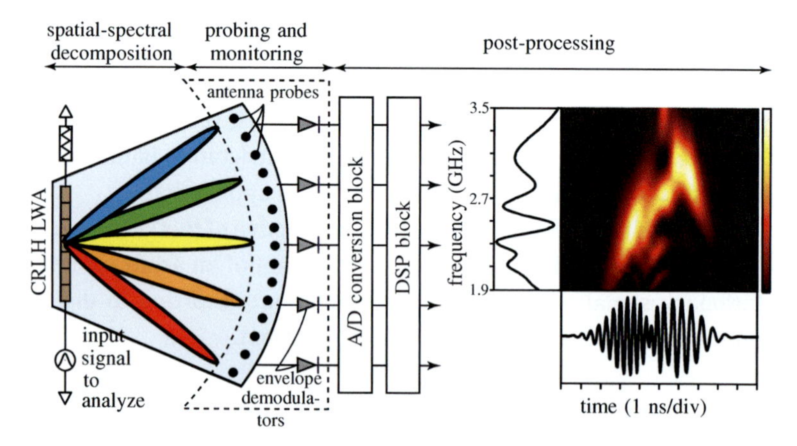

Figure 5.7 LWA-based real-time spectrum analyzer. Reprinted with permission from Gupta et al., 2009, Copyright 2009, IEEE and Caloz et al., 2013, Copyright 2013, IEEE.

As mentioned in Section 5.3.3, the typical dispersion curve of a CRLH transmission line always penetrates into the fast-wave region, resulting in leaky-wave radiation [Liu et al. (2002); Caloz and Itoh (2006)]. Therefore, according to the scanning law of (5.12a), if the CRLH LWA is excited by a pulse signal, the various spectral components of the signal radiate in different directions at any particular instant. Thus, the CRLH LWA performs a real-time spectral-to-spatial decomposition of the signal following the beam-scanning law of the LWA, thereby discriminating the various spectral components present in the testing signal, as indicated in the left of Fig. 5.7.

Once the signal has been spatially decomposed in space by the CRLH LWA, the various frequency components need to be probed and monitored in real time. For this purpose, n antenna probes are arranged circularly in the far-field around the antenna at the positions $r(a, \theta_n = n\Delta\theta)$, where a is the observation distance from the center of the antenna and $\Delta\theta$ is the angular separation between two observation points. For broadband applications, the far-field distance is given by $d_{\text{ff}} \approx 2l^2/\lambda_{\text{min}}$, where λ_{min} is the smallest wavelength in the operating frequency range. At each time instant, the different probes, based on their angular location in space θ_n, receive the different frequencies, thereby achieving real-time frequency–space mapping of the signal propagating across the LWA. The voltages induced along the antenna probes are subsequently envelope-demodulated and monitored as a function of time to track the temporal evolution of the different spectral components of the input signal.

Finally, the envelope-demodulated induced voltages across the antenna probes are next converted into digital format via A/D converters, combined, and then post-processed, as shown in Fig. 5.7. After digitizing and combining the envelope-demodulated voltage waveforms from the antenna probes, an energy distribution function $g(\theta, t)$ is obtained, where θ is the radiation angle. If the beam-scanning law was linear, the resulting spectrogram $S(\omega, t)$ would be directly proportional to this function $g(\theta, t)$ and would, therefore, be immediately available. Because the beam-scanning law of the CRLH LWA (5.12a) is a nonlinear function of frequency, a post-processing operation is required to compensate for this effect. Instead of uniform angular spacing, the antenna probes can also be placed non-uniformly corresponding to uniform increment in frequency, thus avoiding an extra operation of spectrogram linearization. However, in order to avoid spatial crowding and overlapping in the forward region where the CRLH LWA exhibits a larger fast-wave bandwidth, uniformly spaced probes were employed, which is also practically more convenient. Nonetheless, this spectrogram linearization step requires minimal computational resources and can be easily done at the end of the post-processing stage.

The typical full-wave computed spectrograms for diverse range of non-stationary signals are shown in Fig. 5.8 using the RTSA setup of Fig. 5.7. It can be seen that the instantaneous spectral features of complicated chirped signals are faithfully reproduced, thereby validating the usage of LWA for analog computation of STFT. More details about the experimental prototypes and measured results are available in [Gupta et al. (2009)].

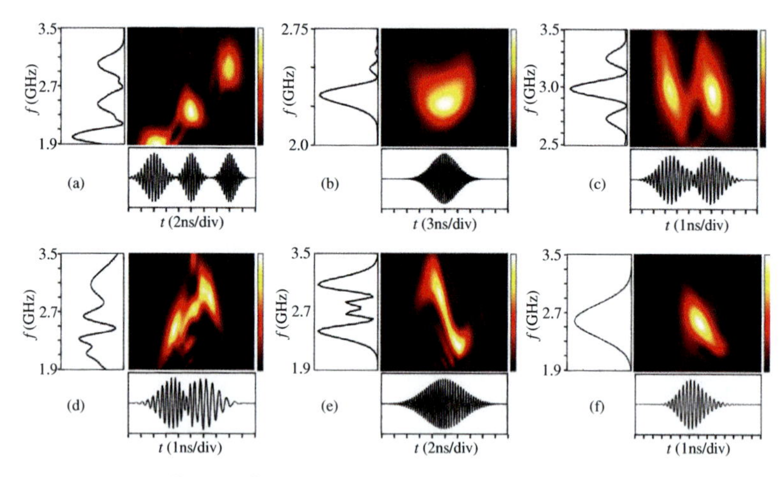

Figure 5.8 Full-wave (CST Microwave Studio) spectrograms. (a) Multiple modulated Gaussian pulses. (b) Nonlinear cubicly chirped Gaussian pulse. (c) Doubly negative chirped Gaussian pulses. (d) Oppositely chirped Gaussian pulses. (e) Self-phase modulated pulses. (f) Dispersed pulse through a CRLH transmission line. Reprinted with permission from Gupta et al., 2009, Copyright 2009, IEEE.

5.4.2 RTSA Features and Time-Frequency Resolution Tradeoff

This analog microwave RTSA system based on LWA offers several benefits and advantages compared to the conventional real-time analysis systems. Some of its key features are as follows:

(1) It is a completely analog system. Since no digital computation is required to generate the RTSA, the only digital processing is done at the post-processing stage, thus requiring neither fast processors nor large memory. Moreover, since the measurements are single-shot (no memory buffers), the systems operate, therefore, in a real-time mode.

(2) The proposed RTSA is a frequency-scalable system. The CRLH LWA may be designed at any arbitrary frequency to meet the requirements of the specific applications [Caloz and Itoh (2006)]. Moreover, the length of the CRLH lines (periodic in nature) can be tailored to obtain practically any lengths by cascading the required number of cells and engineering the footprint of the CRLH unit cell to obtain the desired directivity.

Therefore, it is suitable for a wide variety of signals, from microwaves potentially up to mm-wave frequencies.

(3) The proposed RTSA is inherently broadband or UWB, with bandwidth controllable by proper design of the CRLH transmission line. As experimentally demonstrated in [Nguyen et al. (2007)], the CRLH line can easily attain a bandwidth of 100% using differential metal-insulator-metal (MIM) structure. It can, therefore, handle ultrashort pulses and transient signals in the order of few nanoseconds to a few hundred picoseconds.

(4) The proposed RTSA is based on far-field probe configurations. However, since the far-fields of the antenna may be deduced from its near-field measurements, by standard near-to-far-field transformations, the system may be made more compact by using a near-field probing mechanism with additional postprocessing.

One important aspect of the STFT is the fundamental tradeoff between the time and frequency resolution achieved in the computed spectrograms. These spectrograms suffer from the well-known fundamental uncertainty principle limitation [Cohen (1989)]

$$\Delta t \Delta f \geq \frac{1}{2}, \tag{5.16}$$

where Δt is the gate duration and Δf is the bandwidth of the gated signal. There is an inherent tradeoff between time resolution and frequency resolution in spectrograms, where increasing time resolution decreases frequency resolution, and vice versa. This can be understood by considering the STFT integral of (5.15). The function of the gate signal $g(t)$ in the STFT integral is to time sample the testing signal $x(t)$, which is then Fourier transformed to observe the spectral contents inside this time sample. The time resolution is thus defined by the width of this gate signal, which can be increased by decreasing the gate signal duration. However, a small time duration signal has a larger spectrum, and thus the actual instantaneous frequency of the testing signal is smeared by the large spectrum of the narrow gated signal. Consequently, a low frequency resolution is achieved. In a similar fashion, a longer gate signal provides a high-frequency resolution, but at the cost of lower time resolution. This is a characteristic of all spectrograms. While a gate function is explicitly present in an STFT integral and thus in all digital approaches to compute the spectrogram, one may ask: What is the gate function in the CRLH LWA RTSA?

Whereas a digital STFT time-gates and stores the signal to perform the digital Fourier transform operation, the analog RTSA works in a real-time mode by gating the spatial profile of the signal through the aperture of the LWA. The LWA aperture is the gate function, where the physical length of the LWA represents a space gating mechanism that controls the time resolution of the resulting spectrograms. The shorter the antenna, the better the sampling of signal spatial profile and, therefore, the better the time resolution of the spectrogram, and vice versa. The space gate is thus effectively equivalent to a time gate of $\Delta t = l/v_g$, where v_g is the group velocity along the LWA. On the other hand, the physical length of the aperture controls the directivity of the LWA and, therefore, also controls the frequency resolution of the generated spectrogram. The longer the antenna, the better the directivity and, therefore, the better the frequency resolution of the resulting spectrogram, and vice versa. Due to this fundamental tradeoff between the time and frequency resolution, the antenna length is critical. A given analog RTSA will have a specified minimum time resolution directly related to this length. In a practical circumstance, a commercial analog RTSA could offer CRLH LWA of different lengths, which could be switched by the user to accommodate different signal bandwidths.

5.4.3 Spatio-Temporal 2D RTSA

Since the 1D RTSA spectrally decomposes a signal along one dimension of space, the maximum angular extent of the detection array is fixed to half-space, *i.e.*, 180°. For a fixed distance a between the observation location and the LWA, the total scanning path length is (πa). If the minimum spacing between the detectors is δd due to practical detector size constraints, the total number of detectors and thus the measurement of unique frequency samples is given by $\pi a/\delta d$.

While the 1D RTSA systems have been extensively used for frequency discrimination along a single dimension, dominantly in optics, there has always been a great demand to increase their frequency sensitivity and resolution by exploiting a second dimension of space. For this reason, various 2D real-time frequency analysis systems have been recently reported in the literature, primarily at optical frequency ranges. The first system of this type was introduced in a patent filed by Dragone and Forde filed in 1999, and used an arrayed-waveguide grating, a diffraction grating, and focusing lenses [Dragone and Ford (2001)]. Several related systems have been proposed since then, all utilizing free-space propagation between two diffractive elements to achieve the same effect [Metz et al. (2014); Xiao and Weiner (2004); Yang et al. (2004)]. However, while offering enhanced resolution and sensitivity, these 2D spectrum analyzers are bulky systems. While these systems have been proposed and demonstrated at optics, their microwave counterparts have never been suggested. Based on this context, one may ask: Can the 1D microwave RTSA based on LWAs be extended to a 2D RTSA? The answer is Yes.

The general objective of a 2D spectral decomposition system is illustrated in Fig. 5.9. The temporal frequencies (ω_m) of an input signal to analyze [$v_{\text{in}}(t)$] are mapped, in real-time, onto different spatial frequencies ($\text{k}_m = (\phi_m, \theta_m)$) over a 2D region of space.

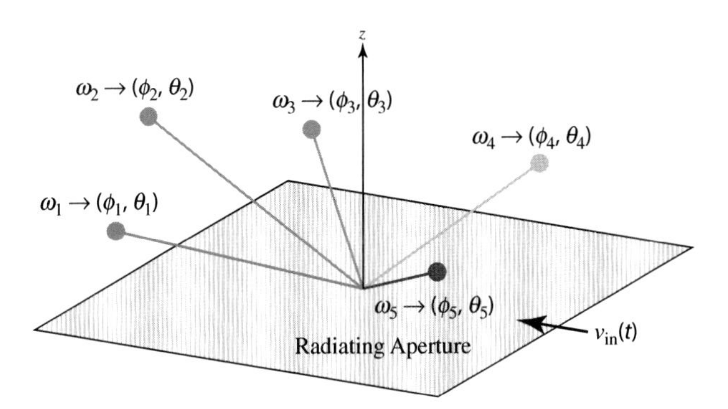

Figure 5.9 Illustration of a real-time temporal-to-spatial frequency decomposition of a broadband signal in 2D spatial coordinates (θ, φ). Reprinted with permission from Gupta and Caloz, 2015, Copyright 2015, IEEE.

To recall, the conventional 1D LWA of Section 5.5 decomposes the signal into its spectral components in space, where each frequency component is radiated along a specific direction of space according to the scanning law $\theta_{\text{MB}}(\omega) = \sin^{-1}[\beta(\omega)/c]$ [Caloz et al. (2011)]. The dispersion relation $\beta(\omega)$ of the radiating structure dictates the mapping between temporal and spatial frequencies, and thus a 1D

LWA acts as an analog RTSA along one dimension of space [Gupta et al. (2009)].

To extend the operation of spatial-spectral decomposition from 1D to 2D, consider the system shown in Fig. 5.10. [Gupta and Caloz (2015)]. It consists of an array of M LWAs, where each LWA is excited by a unique phaser [Caloz et al. (2013)], as shown in Fig. 5.10a. Let us also assume that each of the LWAs is a CRLH, covering the signal bandwidth (ω_{start}, ω_{stop}) [Caloz et al. (2011)]. This antenna array naturally provides spectral decomposition along the y-axis from forward to backward region, including broadside, according to $\beta(\omega) = \omega/\omega_R - \omega_L/\omega$ [Gupta et al. (2009); Caloz et al. (2011)].

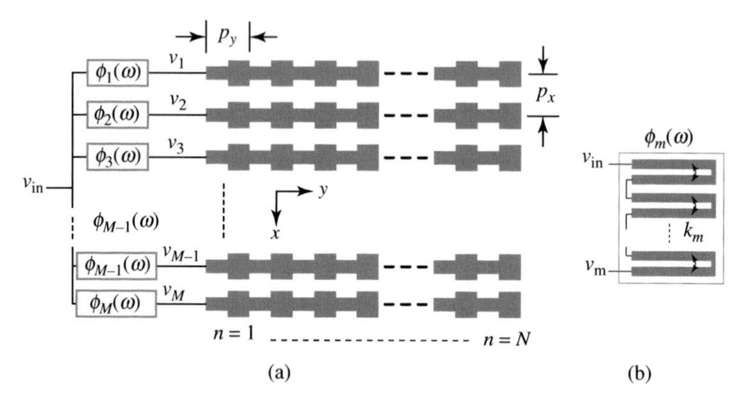

Figure 5.10 Proposed structure for 2D spectral decomposition, consisting of an array of LWAs fed by phasers. (a) Antenna array layout. (b) One of the phasers, which is here a cascaded C-section phaser. Reprinted with permission from Gupta et al., 2013, Copyright 2013, John Wiley and Sons and Gupta and Caloz, 2015, Copyright 2015, IEEE.

In this system, the frequency discrimination along the x-axis is achieved using a dispersive feeding network wherein the mth antenna in the array is fed with a phaser, providing a frequency-dependent phase shift $\phi_m(\omega)$. First assume, for the sake of understanding, that the LWAs do not scan with frequency and only radiate along broadside ($\theta = 0°$ in the $\phi = 90°$ plane). Consider the radiation at a certain frequency ω_0 in the $\phi = 0°$ plane (or along the x-axis) in the following three cases:

(1) When all the LWAs are fed in phase, *i.e.*, $|\phi_{m+1}(\omega_0) - \phi_m(\omega_0)| = 2k\pi$, where k is an integer: In this case, frequency ω_0 points to broadside, *i.e.*, $\theta = 0°$.

(2) When $\phi_{m+1}(\omega_0) - \phi_m(\omega_0) > 0$: In this case, ω_0 points in the forward direction, *i.e.*, $90° > \theta_{MB} > 0$.

(3) When $\phi_{m+1}(\omega_0) - \phi_m(\omega_0) < 0$: In this case, ω_0 points in the backward direction, *i.e.*, $-90° < \theta_{MB} < 0$.

Therefore, an appropriate choice of the feeding phasers enables a full-space frequency scanning along the *x*-axis. In addition, if the above three phase conditions are satisfied at multiple frequency points within the specified bandwidth, as will be shown shortly to be the case with cascaded C-section phasers, the backward-to-forward frequency scan will periodically repeat for each frequency sub-band. Finally, combining the *x*-axis scanning with the conventional *y*-axis scanning of the antenna provides the sought 2D frequency scanning in space.

Let us look at an example. Consider a cascaded C-section phaser as the dispersive feed element to the array, as shown in Fig. 5.10b [Gupta et al. (2013)]. This phaser consists of the series connection of C-sections, which are formed using a coupled-line coupler with a coupling coefficient *k*. The corresponding transmission phase is given by $\phi_i(\omega) = -2q_0 \tan^{-1}\left[\sqrt{(1+k_i)/(1-k_i)}\tan\left((\pi\omega)/(2\omega_\lambda)\right)\right]$, where q_0 is the number of C-sections in the phaser and ω_λ is the frequency at which the coupled lines are quarter-wavelength long. The radiation characteristics of the LWA array of Fig. 5.10a can then be computed taking into account the above dispersive network and CRLH LWA dispersion relation.

Figure 5.11 shows the computed array factor [Stutzman and Thiele (1997)] results for different dispersion characteristics of the feeding network [Gupta and Caloz (2015)]. First, a non-dispersive network is assumed by using $k = 0$ as shown in the first row of Fig. 5.11. In this trivial case, the LWA array exhibits conventional 1D scanning along *y*-axis, as all the antennas are fed in phase. Next, a small amount of dispersion in $\phi(\omega)$ is introduced by using $k \neq 0$, as shown in the second row of Fig. 5.11. Consequently, the frequency-scanning plane is rotated to about $\phi = 45°$ plane. In this case, the frequency point ω_A, where $|\phi_{m+1}(\omega_A) - \phi_m(\omega_A)| = 0$, corresponds to the broadside frequency in the $\phi = 0°$ plane. The most interesting case is when the dispersion characteristics of the phaser exhibit a periodic response in frequency due to its commensurate nature, within the radiating band of the LWA, as shown in the third row of Fig. 5.11. In this case, a periodic frequency scanning is achieved between the left and right half of the $(\theta - \phi)$ plane, thereby accommodating the same frequency band in a larger spatial area. Since the overall scanning path length Δl is larger than πa, a larger number of detectors can be placed in the region above the antenna, thereby achieving a larger frequency sampling of the signal spectrum.

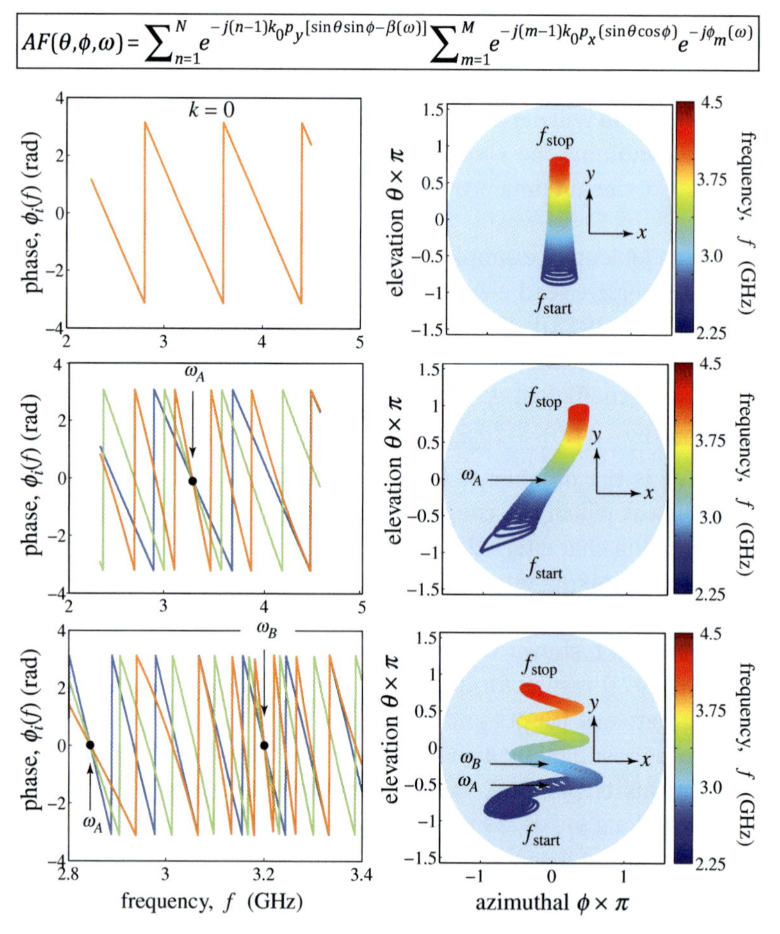

Figure 5.11 Array factor computed results corresponding to the structure in Fig. 5.10 with $M \times N = 20 \times 30$ showing the dispersion profile of the phaser feeding elements at the input of each LWA and the 3 dB contour plots of the 2D radiation patterns in the (θ, ϕ) plane. Reprinted with permission from Gupta and Caloz, 2015, Copyright 2015, IEEE.

5.5 Metasurface-Based Spatial 2D RTSA

The 2D RTSA is a perfect example of combining temporal dispersion of a phaser with the spatial dispersion of an LWA to achieve a larger frequency resolution in a test signal spectrum. With the engineered phase response of the phaser feeding network, the overall scanning path length in this system can be conveniently extended to achieve multiple cycles of frequency scanning in the $\phi = 0°$ plane so as to maximally populate the (θ, ϕ) region of space. While this system requires inter-operability between guided-wave phasers and the spatial dispersion element (here a LWA), a purely spatial 2D RTSA can also be developed based on the principles described in the previous sections. Compared to the 2D RTSA where the test signal is fed at a single point in space, the 2D spatial RTSA decomposes a pulsed wave spectrally in space upon incidence on the dispersive element.

5.5.1 Conventional 2D Spectral Decomposition

A functional schematic of a conventional system for 2D spectral decomposition in a 2D Spatial RTSA is illustrated in Fig. 5.12, which employs two dispersive elements separated in space. Consider a broadband pulse:

$$\psi(x, y, z; t) = \text{Re}\{\Psi(x, y; t)e^{-j(kz-\omega_0 t)}\}, \tag{5.17}$$

where ω_0 is the carrier frequency, wavenumber $k = 2\pi/\lambda = 2\pi\omega/c$ and $\Psi(x, y; t)$ is the pulse envelope of spectral bandwidth $\Delta\omega = (\omega_{\text{start}} - \omega_{\text{stop}})$. At the input of the first dispersion element, $\psi(x, y, 0_-; t) = \text{Re}\{\Psi(x, y; t)\}$, where the time dependence $e^{j\omega_0 t}$ is dropped for convenience.

This system works in two stages. In the first stage, dispersion element #1 spectrally decomposes $\psi(x, y, 0_-; t)$ along the x-axis first at the first image plane at $z = z_1$. In the second stage, the dispersion element #2 then decomposes the wave $\psi(x, y, z_1; t)$ along the y-axis. The spectrally decomposed wave is then finally focused onto the focal plane of the lens, so that each frequency ω (or wavelength A) of the input is focused at a unique point $[x_0(\omega), y_0(\omega)]$ on the output plane. The first dispersion element is typically either an AWG or a VIPA, which includes multiple free-spectral ranges (FSR) $\Delta\omega_{\text{FSR}}$ within the operation bandwidth $\Delta\omega$ of the system, so that $x_0(\omega) = x_0(\omega + n\Delta\omega_{\text{FSR}})$, where n is an integer. This existence of multiple FSRs within the operating bandwidth leads to the so-called spectral shower, as shown in the right of Fig. 5.12. The total length of this system is $\Delta z = (z_1 + z_2 + d)$, which can, in principle, be reduced by placing the lens directly after the second dispersion element [Goodman (2004)], so that $\Delta z_{\min} = (z_1 + d)$.

The question one may ask is: Is it possible to realize a spectral shower using just a single dispersion element, so that the system length is reduced to $\Delta z = d$, as illustrated in Fig. 5.13? This question is addressed here by proposing a metasurface as a single dispersive element that operates on the broadband signal $\psi(x, y, 0_-, t)$ to produce a unique frequency-to-space decomposition between its frequency components ω and spatial coordinates (x_0, y_0) in the output plane $z = d$, and thereby acting as a 2D spatial phaser.

Metasurfaces generally consist of non-uniform spatial arrays of subwavelength scattering particles, with sizes much smaller than the operating wavelengths and provide unprecedented flexibility in controlling wavefronts in space, time or both space and time [Kildishev et al. (2013); Holloway et al. (2012); Liu and Zhang (2013); Yu and Capasso (2014); Aieta et al. (2014); Minovich et al. (2015)]. The sought metasurface for 2D RTSA is characterized by a complex transmittance function, $t_m(x, y; \omega)$, which directly operates on the input and produces a spectrally resolved output at the image plane $z = d$.

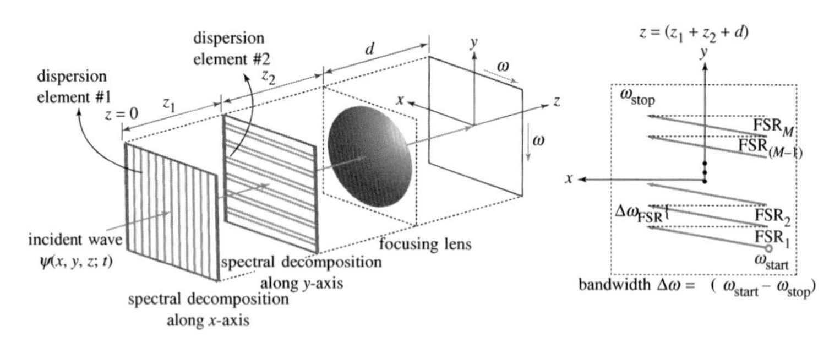

Figure 5.12 A conventional free-space optical system to spectrally decompose a broadband $\psi(x, y, z; t)$ in two spatial dimensions, using two dispersive elements.

5.5.2 Metasurface Transmittance

The complex transmittance function $t_m(x, y; \omega)$ of a metasurface phaser decomposing the frequency contents of an incident electromagnetic pulsed wave can be established by considering the arrangement of various dispersion elements in Fig. 5.14. It consists of a series cascade of a thin linear wedge with transmittance $t_w(x, y)$ and a wavelength-dependent refractive index $n(\omega)$, a diffraction grating with transmittance $t_g(x, y)$, and a focusing lens with transmittance $t_l(x, y)$ and focal length d. The thin wedge and the lens are placed directly after the diffraction grating, and any refractions within the wedge are assumed to be negligible. Also, paraxial-wave propagation is assumed for simplifying the analytic expressions. It will be shown that the cascade of these three elements is functionally equivalent to the metasurface phaser of Fig. 5.13.

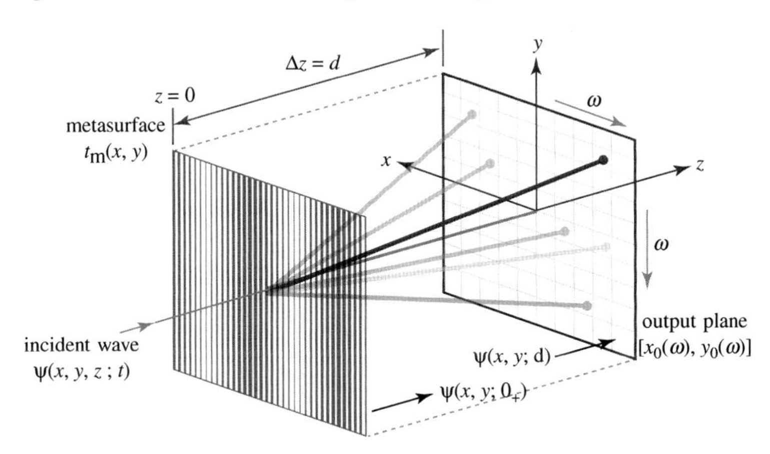

Figure 5.13 A metasurface as a spatial phaser spectrally decomposing a broadband pulse $\psi(x, y, z; t)$ along two dimensions, where each frequency u is mapped to a specific coordinate point $[x_0(\omega), y_0(\omega)]$ on the image plane.

Let us first consider a thin wedge of thickness $d(x) = a_0(x + w_0)$, where a_0 and w_0 are constants. The corresponding transmittance function is given by:

$$\begin{aligned} t_w(x,y;\omega) &= \exp[-jkn(\omega)d(x)] \\ &= \underbrace{\exp[-jkn(\omega)a_0w_0]}_{t_0(\omega)}\exp[-jkn(\omega)a_0x]. \end{aligned} \quad (5.18)$$

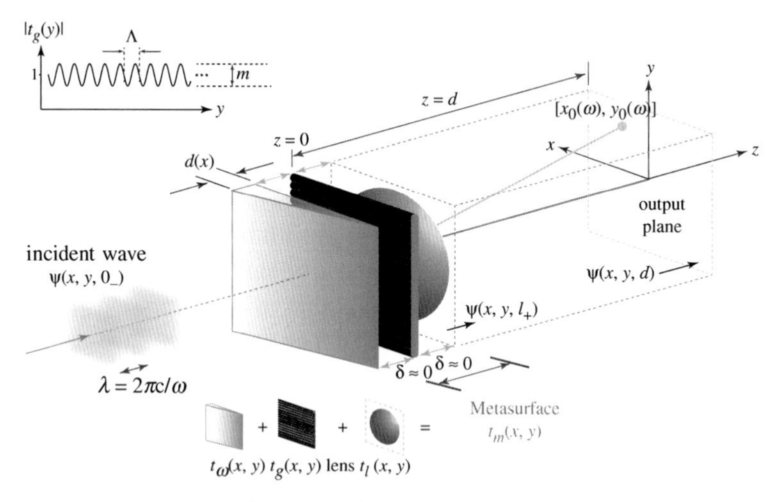

Figure 5.14 An equivalent setup for mapping a temporal frequency ω to a specific point on the output plane ω to $[x_0(\omega), y_0(\omega)]$.

Next, the complex transmittance functions of a thin sinusoidal diffraction grating are given by (5.5):

$$t_g(x, y) = 1 + m\cos(2\pi\zeta y). \tag{5.19}$$

Consider a plane wave propagating along the z-axis:

$$\psi(x, y, z; t) = \mathrm{Re}\{e^{-j(kz-\omega t)}\}, \tag{5.20}$$

so that $\psi(x, y, 0_-; t) = \psi(x, y, 0_-) = 1$, where the explicit time dependence $e^{j\omega t}$ is dropped for convenience. The wave just after the grating can then be expressed as:

$$\psi(x, y, 0_+) = t_w(x, y; \omega) t_g(x, y; \omega)\, \psi(x, y, 0_-)$$

$$= t_0(\omega) \exp[-jkn(\omega)a_0 x]\, [1 + m\cos(2\pi\zeta y)]$$

$$= t_0(\omega)\left\{\underbrace{\exp[-jkn(\omega)a_0 x]}_{0^{\text{th}}\text{order}} + m\,\underbrace{\exp[-jkn(\omega)a_0 x]\cos(2\pi\zeta y)}_{\pm 1^{\text{st}}\text{order}}\right\} \tag{5.21}$$

The first term inside the brackets on the RH side of (5.21) corresponds to the 0th diffraction order, whose location in the output plane is not sensitive to ω along the y-axis of the grating [Goodman (2004)], and thus will be ignored. Considering the lens transmittance function $t_1(x, y; \omega)$, wave output after the lens is given by

$$\psi(x,y,l_+) = mt_0(\omega)\exp[-jkn(\omega)a_0x]\cos(2\pi f_0 y)\exp\left[-j\frac{\pi}{\lambda d}(x^2+y^2)\right], \tag{5.22}$$

which, after free-space propagation, leads to the field intensity:

$$I(x,y,f) = m^2\omega^2\frac{|\delta\{x+a_0n(\omega)d\}\delta(y-2\pi\zeta cd/\omega)|}{2(dc)^2}, \tag{5.23}$$

where only one of the diffraction orders $\delta(y - 2\pi\zeta cd/\omega)$ is kept due to symmetry. This equation finally gives the frequency-to-space mapping relation of this system where each frequency ω is mapped onto a specific point (x_0, y_0) on the output plane according to the relation:

$$x_0(\omega) = a_0n(\omega)d, \tag{5.24a}$$

$$y_0(\omega) = \zeta\left(\frac{2\pi c}{\omega}\right)d. \tag{5.24b}$$

Equation 5.24b is the conventional 1D scanning achieved using a diffraction grating only, *i.e.*, (5.11). On the other hand, the frequency scanning along the x-axis directly depends on the refractive index, $n(\omega)$, of the wedge. Therefore, a precise control over the refractive index of the wedge and the spatial period of the diffraction grating provides a mechanism to achieve the specified unique frequency-to-space decomposition between frequency and space, thereby enabling the sought 2D spectral decomposition in the output plane. The setup of Fig. 5.14 consisting of a thin dispersive wedge, a diffraction grating, and a focusing lens may be replaced by a single metasurface with a complex transmittance function

$$t_\mathrm{m}(x,y;\omega) = \overbrace{\exp[-jka_0n(\omega)x]}^{\text{dispersive wedge}}\underbrace{\exp\left[-j\frac{\pi\omega}{dc}(x^2+y^2)\right]}_{\text{focusing lens}}\overbrace{\exp(j2\pi\zeta y)}^{\text{diffraction grating}}. \tag{5.25}$$

where the diffraction grating transmittance function is written in complex exponential form to avoid symmetric diffraction orders.

Such a metasurface will focus on the spectral contents of a pulsed wave following the frequency-to-space mapping of (5.24) at the focal plane located at $z = d$, by directly operating on the input field as:

$$\psi(x,y,d;\omega) = \underbrace{\mathcal{F}_t\left[\psi(x,y,0_-;t)\right]t_{\mathrm{m}}(x,y;\omega)}_{\text{metasurface output}} * \overbrace{h(x,y)}^{\text{free space}} \tag{5.26a}$$

$$\psi(x,y,d;t) = \mathcal{F}_t^{-1}\{\psi(x,y,d;\omega)\}. \tag{5.26b}$$

where $\mathcal{F}_t(\cdot)$ and $\mathcal{F}_t^{-1}(\cdot)$ are the Fourier transform and inverse Fourier transform operators in the time domain. This finally completes the synthesis of the metasurface phaser for 2D spectrum analysis.

5.5.3 Numerical Examples

Let us consider a wedge that exhibits Lorentz dispersion with multiple resonances at frequencies $\omega_{r,i}$ with line width $\Delta\omega$, $i \in [1, \mathrm{M}]$. The corresponding susceptibilities are given by

$$\chi(\omega) = \sum_{i=1}^{M} \frac{\omega_{r,i}^2}{(\omega_{r,i}^2 - \omega^2) + j\omega\Delta\omega}. \tag{5.27}$$

The refractive active index is related to χ by $n(\omega) = \sqrt{1+\chi(\omega)}$. Following (5.27), Fig. 5.15. shows a typical Lorentz dispersion response showing the real and imaginary parts of the susceptibilities for the cases of a single resonance (M = 1) and multiple resonance (M = 3). The resonance region is also associated with the strongest absorption as required by Kramers–Kronig relations [Jackson (1998)]. Engineering the locations of the resonant frequencies $\omega_{r,i}$ thus allows to tailor the dispersion response of such a medium over a broad frequency range.

Let us assume a Gaussian monochromatic beam illumination $\psi(x, y, 0_-)$ in the system of Fig. 5.13, where the metasurface has the complex transmittance Eq. 5.25. In the numerical examples herein, the paraxial propagation constraint assumed in theoretical formulation is suppressed and a full transfer function is considered, *i.e.*, $\tilde{H}(k_x, k_y) = e^{-j(k_x^2+k_y^2)^{1/2}z}$ in (5.26) to compute the system output $\psi(x, y, d)$ [Goodman (2004)]. The field distribution just after the metasurface is shown in Fig. 5.16 along the physical parameters used in the setup. The choice of these parameters is somewhat arbitrary but represents quantities that can be easily realized with current technologies.

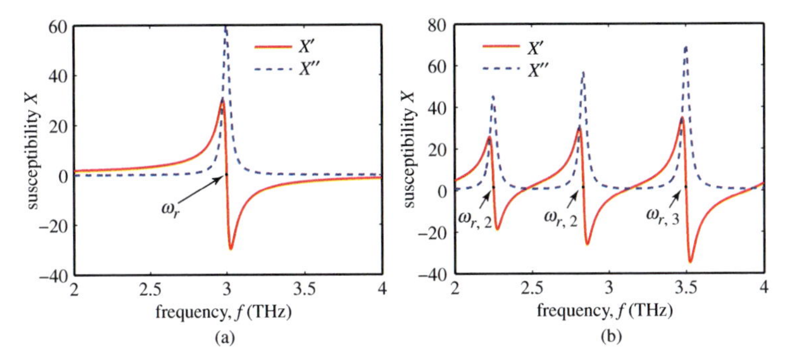

Figure 5.15 Lorentz susceptibility functions for (a) single resonance (M = 1), and (b) multiple resonances (M = 3). $\Delta\omega = 2\pi$ (0.05 THz).

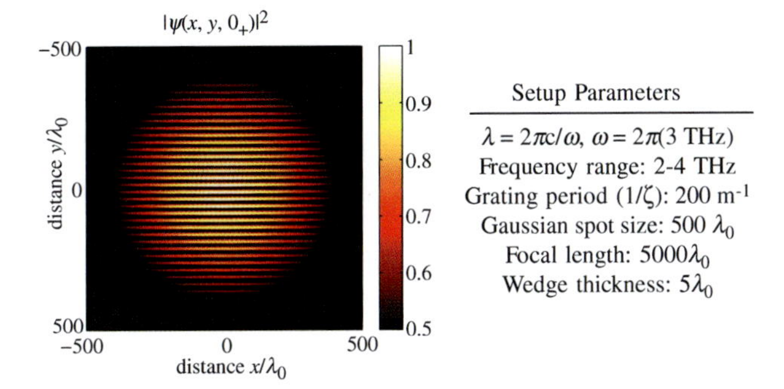

Figure 5.16 Normalized wave intensity $|I(x, y, 0_+)|$ corresponding to a Gaussian beam after the metasurface (left), and the parameters used in the numerical examples (right). Modulation index of the grating is assumed to be m = 1.

First consider a trivial case when the metasurface transmittance function (5.25) has no wedge contribution. In this case, the system behaves as a conventional 1D frequency scanned system where the diffraction grating does the conventional spectral decomposition along one dimension (y-axis here). Figure 5.17a shows the intensity distribution at the image plane for several frequencies covering the specified bandwidth. In this case, the intensity beam spots move toward the center with increasing frequencies along the y-axis following the scanning law (5.24b). Next, a non-dispersive wedge is introduced with $n(\omega) = n_0$, and consequently the axis of scanning is translated along the x-axis to $x_0 = a_0 n_0 d$ following the scanning law (5.24a), still maintaining the 1D scan. This behavior is shown in Fig. 5.17b.

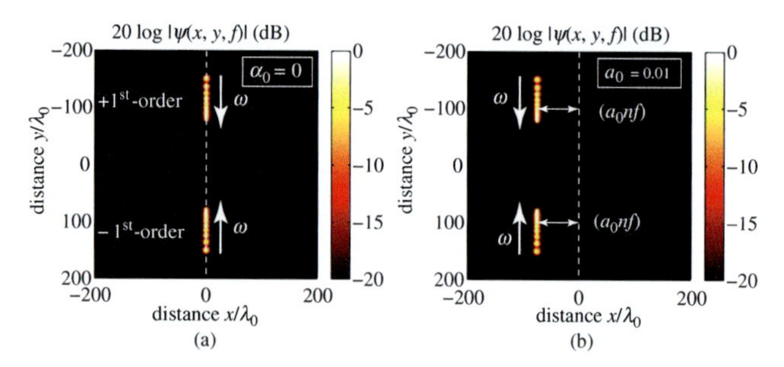

Figure 5.17 Normalized wave intensity $|I(x, y, d)|$ obtained using (5.26) (a) showing 1D frequency scanning in the output plane for uniformly spaced frequencies within the specified bandwidth in (a) the absence of a wedge contribution in the metasurface transmittance function, and (b) in the presence of a non-dispersive wedge contribution in the metasurface transmittance function with refractive index $n = 3$.

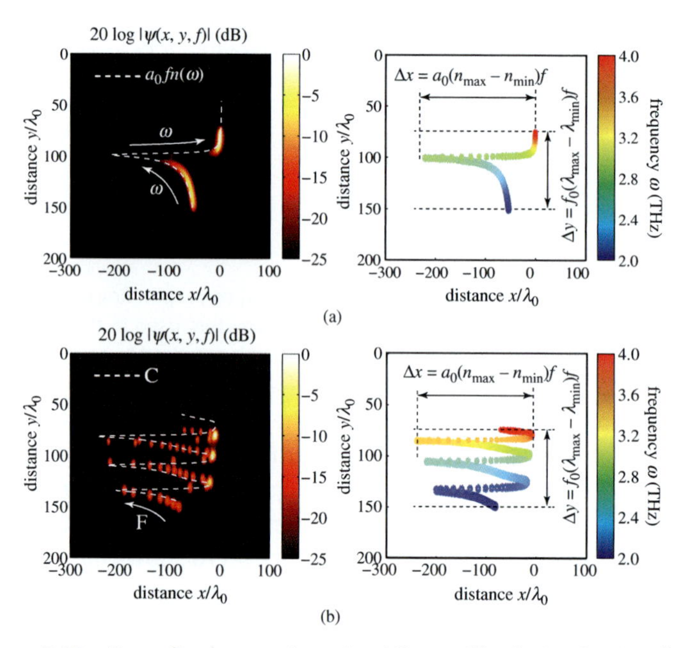

Figure 5.18 Normalized wave intensity $|I(x, y, d)|$ obtained using (5.26) showing 2D frequency scanning in the output plane for uniformly spaced frequencies within the specified bandwidth in the presence of a dispersive wedge contribution in the metasurface transmittance function exhibiting (a) a single resonance of Fig. 5.15a, and (b) multiple resonances of Fig. 5.15b. Left figures show the superimposed beam spots of different frequencies, and the right figures show the corresponding 3 dB contour plots.

The most interesting case is when the metasurface consists of a dispersive wedge contribution, following the Lorentz response of Fig. 5.15. Figure 5.18 shows the output intensity distribution for this case when a single resonant and a multi-resonant dispersion response are considered. As expected, the frequency spots are moving along both x- and y-directions, as frequency ω is changed, following the scanning law of (5.24). The 3 dB contour plots show the exact trajectory of the frequencies in the image plane, in both single- and multi-resonant cases, the frequency scanning closely follows the Lorentz dispersion response of the metasurface. The region of frequency scanning is $\Delta x = a_0(n_{\max} - n_{\min})d$ and $\Delta y = \zeta(\lambda_{\max} - \lambda_{\min})d$. The number of spatial scanning periods is equal to the number of resonant points M in the Lorentz response. This successfully demonstrates the principle of frequency scanning in two dimensions using a metasurface.

5.6 Summary

This chapter has laid the conceptual foundations for performing ultrafast real-time spectrum analysis using broadband metamaterials. The principle is based on the R-ASP paradigm, which is founded upon temporal and spatial phasers discriminating spectral components of a broadband signal in time and space, without resorting to digital computations. The chapter focused on three different systems for real-time spectrum analysis: (1) A 1D RTSA based on CRLH transmission lines to decompose a broadband signal along one dimension of space. By monitoring the far-field radiation and exploiting the unique spectra-to-space mapping in LWAs, not only the spectral contents of the signal be determined, but a more general operation of STFT be performed for complete characterization of non-stationary signals. (2) A spatio-temporal 2D RTSA combining temporal and spatial phasers to perform spectral decomposition in two dimensions of space to achieve higher-frequency resolution in the RTSA system, and (3) a spatial 2D RTSA using metasurface phasers to spectrally decompose an incident pulsed wavefront onto two dimensions of space. These three systems cover the different aspects of temporal and spatial dispersion engineering for ultrafast real-time spectrum analysis applications, and particularly suited for mm-wave and terahertz highspeed systems, where digital computation is not readily available. This is still a topic of current research and more advancements are expected to be seen in the near future.

References

Abielmona, S., Gupta, S. and Caloz, C. (2007). Experimental demonstration and characterization of a tunable crlh delay line system for impulse/continuous wave, *IEEE Antennas Wirel. Propagat. Lett.*, **17**(12), pp. 864–866.

Abielmona, S., Gupta, S. and Caloz, C. (2009). Compressive receiver using a CRLH-based dispersive delay line for analog signal processing, *IEEE Trans. Microw. Theory Tech.*, **57**(11), pp. 2617–2626.

Abielmona, S., Nguyen, H. V. and Caloz, C. (2011). Analog direction of arrival estimation using an electronically-scanned CRLH leaky-wave antenna, *IEEE Trans. Antennas Propagat.*, **59**(4), p. 14081412.

Aieta, F., Kats, M. A., Genevet, P. and Capasso, F. (2014). Achromatic metasurface optical components by dispersive phase compensation, arXiv:1411.3966.

Caloz, C. (2009). Perspectives on EM metamaterials, *Materials Today*, **12**(3), pp. 12–30.

Caloz, C., Gupta, S., Zhang, Q. and Nikfal, B. (2013). Analog signal processing: A possible alternative or complement to dominantly digital radio schemes, *Microw. Mag.*, **14**(6), pp. 87–103.

Caloz, C. and Itoh, T. (2006). *Electromagnetic Metamaterials, Transmission Line Theory and Microwave Applications*, Wiley-IEEE Press.

Caloz, C., Jackson, D. R. and Itoh, T. (2011). *Leaky-Wave Antennas*, McGraw Hill.

Campbell, C. K. (1989). *Surface Acoustic Wave Devices and Their Signal Processing Applications*, Academic Press.

Cohen, L. (1989). Time-frequency distributions—a review, *Proceedings of the IEEE*, **77**(7), pp. 941–981.

Dragone, C. and Ford, J. (2001). Free-space/arrayed-waveguide router, US Patent 6,263,127.

Goda, K. and Jalali, B. (2013). Dispersive fourier transformation for fast continuous single-shot measurements, *Nature Photonics*, **7**, pp. 102–112.

Goda, K., Tsia, K. K. and Jalali, B. (2009). Serial time-encoded amplified imaging for real-time observation of fast dynamic phenomena, *Nature*, **458**, pp. 1145–1149.

Goodman, J. (2004). *Introduction to Fourier Optics*, 3rd edn., Roberts and Company Publishers.

Gupta, S., Abielmona, S. and Caloz, C. (2009). Microwave analog realtime spectrum analyzer (RTSA) based on the spatial-spectral decomposition property of leaky-wave structures, *IEEE Trans. Microw. Theory Tech.*, **59**(12), pp. 2989–2999.

Gupta, S. and Caloz, C. (2009). Analog signal processing in transmission line metamaterial structures, *Radio Engineering*, **18**(2), p. 155167.

Gupta, S. and Caloz, C. (2015). Realtime 2-d spectral-decomposition using a leaky-wave antenna array with dispersive feeding network, in *IEEE AP-S Int. Antennas Propag. (APS)*, Vancouver, BC, pp. 29–30.

Gupta, S., Nikfal, B. and Caloz, C. (2011). Chipless RFID system based on group delay engineered dispersive delay structures, *IEEE Antennas Wirel. Propagat. Lett.*, **10**, pp. 1366–1368.

Gupta, S., Parsa, A., Perret, E., Snyder, R. V., Wenzel, R. J. and Caloz, C. (2010). Group delay engineered non-commensurate transmission line all-pass network for analog signal processing, *IEEE Trans. Microw. Theory Tech.*, **58**(9), pp. 2392–2407.

Gupta, S., Sounas, D. L., Nguyen, H. V., Zhang, Q. and Caloz, C. (2012). CRLH-CRLH C-section dispersive delay structures with enhanced group delay swing for higher analog signal processing resolution, *IEEE Trans. Microw. Theory Tech.*, **60**(21), pp. 3939–3949.

Gupta, S., Sounas, D. L., Zhang, Q. and Caloz, C. (2013). All-pass dispersion synthesis using microwave C-sections, *Int. J. Circ. Theory Appl.*, **42**(12), pp. 1228–1245.

Gupta, S., Zhang, Q., Zou, L., Jiang, L. J. and Caloz, C. (2015a). Generalized coupled-line all-pass phasers, *IEEE Trans. Microw. Theory Tech.*, **63**(3), pp. 1007–1018.

Gupta, S., Zou, L., Salem, M. A. and Caloz, C. (2015b). Bit-error rate (BER) performance in dispersion code multiple access (DCMA), IEEE International Symposium on Antennas and Propagation, Vancouver, British Columbia, Canada.

Hessel, A. (1969). General characteristics of traveling-wave antennas, Chapter 19 in *Antenna Theory*, Part 2, R. E. Collin and F. J. Zucker, Eds., McGraw-Hill, New York.

Holloway, C., Kuester, E. F., Gordon, J., O'Hara, J., Booth, J. and Smith, D. (2012). An overview of the theory and applications of metasurfaces: The two-dimensional equivalents of metamaterials, *Antennas and Propagation Magazine, IEEE*, **54**(2), pp. 10–35.

Jackson, J. D. (1998). *Classical Electrodynamics*, 3rd edn. Wiley.

Kashyap, R. (2009). *Fiber Bragg Gratings*, 2nd edn., Academic Press.

Kildishev, A. V., Boltasseva, A. and Shalaev, V. M. (2013). Planar photonics with metasurfaces, *Science*, **339**, 6125.

Lee, J. P. Y. and Wight, J. S. (1986a). Acoustooptic spectrum analyzer: Detection of pulsed signals, *Appl. Opt.*, **25**(2), pp. 193–198.

Lee, J. P. Y. and Wight, J. S. (1986b). Acoustooptic spectrum analyzer: detection of pulsed signals, *Applied Physics*, **25**(2), pp. 193–198.

Liu, L., Caloz, C. and Itoh, T. (2002). Dominant mode (DM) leaky-wave antenna with backfire-to-endfire scanning capability, *Electron. Lett.*, **38**(23), pp. 1414–1416.

Liu, Y. and Zhang, X. (2013). Metasurfaces for manipulating surface plasmons, *Applied Physics Letters*, **103**, 141101.

Matthaei, G., Young, L. and Jones, E. M. T. (1980). *Microwave Filters, Impedance-Matching Networks, and Coupling Structures*, Artech House.

Metz, P., Adam, J., Gerken, M. and Jalali, B. (2014). Compact, transmissive two-dimensional spatial disperser design with application in simultaneous endoscopic imaging and laser microsurgery, *Appl. Opt.*, **53**(3), pp. 376–382.

Minovich, A. E., Miroshnichenko, A. E., Bykov, A. Y., Murzina, T. V., Neshev, D. N. and Kivshar, Y. S. (2015). Functional and nonlinear optical metasurfaces, *Laser and Photonics Reviews*, **9**(2), pp. 195213.

Muriel, M. A., Azana, J. and Carballar, A. (1999). Real-time Fourier transformer based on fiber gratings, *Opt. Lett.*, **24**(1), pp. 1–3.

Nguyen, H. V. and Caloz, C. (2008). Composite right/left-handed delay line pulse position modulation transmitter, *IEEE Microw. Wireless Compon. Lett.*, **18**(5), pp. 527–529.

Nguyen, H. V., Yang, N. and Caloz, C. (2007). Differential bi-directional CRLH leaky-wave antenna in CPS technology, *Proc. Asia-Pacific Microwave Conference (APMC)*, Bangkok, Thailand.

Nikfal, B., Badiere, D., Repeta, M., Deforge, B., Gupta, S. and Caloz, C. (2012). Distortion-less real-time spectrum sniffing based on a stepped group-delay phaser, *IEEE Microw. Wireless Compon. Lett.*, **22**(11), pp. 601–603.

Nikfal, B., Gupta, S. and Caloz, C. (2011). Increased group delay slope loop system for enhanced-resolution analog signal processing, *IEEE Trans. Microw. Theory Tech.*, **59**(6), pp. 1622–1628.

Nikfal, B., Salem, M. and Caloz, C. (2013). A method and apparatus for encoding data using instantaneous frequency dispersion, US 62/002,978.

Oliner, A. A. and Jackson, D. R. (2007). Leaky-wave antennas, Chapter 11 in *Antenna Engineering Handbook*, 4th edn., J. L. Volakis (Ed.), McGraw-Hill, New York.

Otto, S., Al-Bassam, A., Rennings, A., Solbach, K. and Caloz, C. (2012). Radiation efficiency of longitudinally symmetric and asymmetric periodic leaky-wave antennas, *IEEE Antennas Wirel. Propagat. Lett.*, **11**(10), pp. 612–615.

Otto, S., Al-Bassam, A., Rennings, A., Solbach, K. and Caloz, C. (2014). Transversal asymmetry in periodic leaky-wave antennas for bloch impedance and radiation efficiency equalization through broadside, *IEEE Trans. Antennas Propagat.*, **62**(10), pp. 5037–5054.

Ramo, S., Whinnery, J. R. and Duzer, T. V. (1994). *Fields and Waves in Communication Electronics*, 3rd edn., Wiley.

Saleh, B. E. A. and Teich, M. C. (2007). *Fundamentals of Photonics*, 2nd Ed., Wiley-Interscience.

Schwartz, J. D., Azana, J. and Plant, D. (2006). Experimental demonstration of real-time spectrum analysis using dispersive microstrip, *IEEE Microw. Wireless Compon. Lett.*, **16**(4), pp. 215–217.

Shirasaki, M. (1996). Large angular dispersion by a virtually imaged phased array and its application to a wavelength demultiplexer, *Opt. Lett.*, **21**(5), pp. 366–368.

Steenaart, W. J. D. (1963). The synthesis of coupled transmission line all-pass networks in cascades of 1 to n, *IEEE Transactions on Microwave Theory and Techniques*, **11**(1), pp. 23–29.

Stutzman, W. L. and Thiele, G. A. (1997). *Antenna Theory and Design*, 2nd edn., Wiley.

Supradeepa, V. R., Huang, C.-B., Leaird, D. E. and Weiner, A. M. (2008). Femtosecond pulse shaping in two dimensions: Towards higher complexity optical waveforms, *Opt. Express*, **16**(16), pp. 11878–11887.

Tamir, T. (1969). Leaky-wave antennas, Chapter 20 in *Antenna Theory*, Part 2, R. E. Collin and F. J. Zucker, Eds., McGraw-Hill, New York.

Thummler, F. W. and Bednorz, T. (2007). Measuring performance in pulsed single devices: a multi-faceted challenge, *IEEE Antennas Wirel. Propagat. Lett.*, **50**(9), pp. 196–208.

Trebino, R. (2002). *Frequency-Resolved Optical Gating: The Measurement of Ultrashort Laser Pulses*, Springer.

Wood, D. E. and Hewitt, T. L. (1963). New instrumentation for making spectrographic pictures of speech, *J. Acoust. Soc. Am.*, **35**, pp. 1274–1278.

Xiao, S. and Weiner, A. (2004). 2-D wavelength demultiplexer with potential for > 1000 channels in the C-band, *Opt. Express*, **12**(13), pp. 2895–2902.

Yang, J., Jiang, X., Wang, M. and Wang, Y. (2004). Two-dimensional wavelength demultiplexing employing multilevel arrayed waveguides, *Opt. Express*, **12**(6), pp. 1084–1089.

Yu, N. and Capasso, F. (2014). Flat optics with designer metasurfaces, *Nature Materials*, **13**, pp. 13–150.

Zhang, Q., Gupta, S. and Caloz, C. (2012). Synthesis of narrow-band reflection-type phaser with arbitrary prescribed group delay, *IEEE Trans. Microw. Theory Tech.*, **60**(8), pp. 2394–2402.

Zhang, Q., Sounas, D. L. and Caloz, C. (2013). Synthesis of cross-coupled reduced-order dispersive delay structures (dds) with arbitrary group delay and controlled magnitude, *IEEE Trans. Microw. Theory Tech.*, **61**(3), p. 10431052.

Chapter 6

Broadband Performance of Lenses Designed with Quasi-Conformal Transformation Optics

Jogender Nagar, Sawyer D. Campbell, Donovan E. Brocker, Xiande Wang, Kenneth L. Morgan, and Douglas H. Werner
Department of Electrical Engineering, Penn State University, University Park, PA 16802, USA
jun163@psu.edu

6.1 Introduction

Gradient-index (GRIN) lenses have become an increasingly popular topic in modern optical system design. GRIN lenses possess spatially varying indices of refraction throughout the bulk material of the lens. By choosing a proper GRIN distribution, the power of a lens can be increased. It has also been shown that carefully selected GRIN distributions within a lens can also reduce aberrations including, but not limited to, chromatic and spherical aberrations [39, 16]. When applied to entire optical systems, GRIN technology could potentially lead to smaller and lighter optical devices by reducing the

Broadband Metamaterials in Electromagnetics: Technology and Applications
Edited by Douglas H. Werner
Copyright © 2017 Pan Stanford Publishing Pte. Ltd.
ISBN 978-981-4745-68-0 (Hardcover), 978-1-315-36443-8 (eBook)
www.panstanford.com

number of optical components needed to reach diffraction-limited performance. While GRIN lenses are currently hailed as an advanced and modern technology, the concept of GRIN lenses has actually existed for over a century [40, 61, 62]. One of the earliest GRIN lens examples was a spherical fish-eye lens proposed by J. C. Maxwell in 1854 [40], which is similar to the more modern Luneburg lens [61]. Interestingly, GRIN lenses have been found to exist naturally in the eyes of many living creatures. For example, the nucleus of the lens in the human eye has an index distribution with a Δn of approximately 0.03 [43, 19, 22] and whose origin is attributed to varying protein densities within the lens [43]. For the case of humans, the curvature and the GRIN of the eye aid in the focusing of light onto the retina so that humans can see a focused image of their surroundings. In order to make use of the potential of GRIN in human-made lenses, the index distribution of the GRIN must be carefully designed to obtain the desired focusing performance. When including traditional lens parameters such as radii of curvature, aspherical curvature terms, thickness, bandwidth, field of view (FOV), and other constraints of the system, it becomes apparent that a sophisticated methodology is needed to effectively design GRIN lenses. Transformation optics (TO) or transformation electromagnetics (TEM) is a method of designing electromagnetic and optical devices that exploit the form-invariance of Maxwells equations under coordinate transformations [54]. The method involves mapping a given geometric space onto a second, more desired geometric space via coordinate transformations, then realizing this geometric transformation through spatially varying material parameters. In order to physically implement the design, exotic materials are often required, including inhomogeneous and anisotropic materials. The necessary refractive indices can be zero or even negative, which require the use of certain types of metamaterials to physically realize in the design. Unfortunately, these metamaterials can be very difficult to manufacture and are often very narrowband and lossy [34]. However, if the transformation is a conformal map, the medium parameters will be strictly isotropic and inhomogeneous (graded-index), thus improving the design manufacturing feasibility. The application of a conformal mapping will yield an entirely dielectric material if we only consider a single polarization, resulting in potentially broadband designs with low loss [6]. However, it is impossible to analytically define a conformal map for any general transformation. A numerical method called quasi-conformal transformation optics (qTO) is a generalization of conformal mapping that uses grid generation tools to create maps between two bounded regions by solving Laplace's equations [31]. Without any specified boundary conditions, there are an infinite number of transformations that are not strictly conformal. However, through a judicious selection of boundary conditions [6, 31], we can generate a map with negligible anisotropy, which can be accurately realized through an inhomogeneous isotropic all-dielectric medium [27]. Some applications of qTO include electromagnetic cloaks, flat focusing lenses, virtual conformal arrays, and right-angle bends [28].

This chapter will focus on broadband lens designs enabled or inspired by qTO by exploring a wide variety of applications. Here, the term "broadband" is used to describe lenses that possess desirable performance over large FOVs and/or wide frequency ranges. First, the mathematics of TO and the numerical algorithm of qTO will be described. Refractive index profiles engineered through qTO will be referred to as qTO-derived inhomogeneous metamaterials, while those developed using TO will be called anisotropic metamaterials. Next, a series of broadband qTO-derived inhomogeneous metamaterials created from transforming the geometry of classical designs, including the Luneburg and Fresnel lenses, will be presented. Then, qTO will be used to transform wavefronts and design a multibeam lens antenna. One unfortunate downside of these designs is that there could be significant reflections due to discontinuities at the interfaces. To deal with this, the next section will describe a systematic design procedure using qTO for anti-reflective coatings. Another downside of these designs is that there could be significant dispersion in the optical regime and the traditional qTO procedure cannot be directly used to correct for this. An alternative method that can correct for dispersion, called wavefront matching (WFM), will then be described. While this method is extremely general, it relies on the performance of a numerical optimizer to be effective. The final section will discuss the effects of material dispersion on the performance of GRIN lenses and a set of classical as well as qTO-inspired corrections will be presented.

6.2 Mathematics of Transformation Optics

This section will detail the mathematical formulation of TO by first introducing the concept of conformal coordinate transformation mapping. Then, TO will be described by showing that a coordinate transformation that can be represented analytically can be reproduced through the use of spatially varying material parameters. If the transformation is conformal, the resulting media is isotropic. Finally, a numerical scheme for generating a nearly conformal map for an arbitrary coordinate transformation, called qTO, will be discussed.

6.2.1 Conformal Mapping

Laplace's equation is a very important general second-order partial differential equation with applications in electrostatics, low-speed fluid flows, and gravitational fields [4]. Given a scalar potential ϕ, Laplace's equation is represented as:

$$\Delta\phi = \nabla^2\phi = \left(\frac{\partial^2}{\partial x^2} + \frac{\partial^2}{\partial y^2}\right)\phi = 0 \tag{6.1}$$

where $\Delta = \nabla^2$ is the Laplace operator. Analytical functions are functions that can be represented by convergent power series, are infinitely differentiable, and can be used to map one complex space $z = x + jy$ to another complex space $f(z) = f'(x + jy) = x'(x, y) + jy'(x, y)$. These functions also obey the Cauchy–Riemann equations:

$$\frac{\partial x'}{\partial x} = \frac{\partial y'}{\partial y'} \quad \frac{\partial x'}{\partial y} = -\frac{\partial y'}{\partial x} \tag{6.2}$$

This implies that $f(z)$ has the same derivative $\partial f/\partial z$ independent of which direction in the complex plane dz is oriented. For two different orthogonal orientations of dz in the complex plane, we have:

$$\frac{df(z)}{dz}\bigg|_{dz=dx} = \frac{\partial x'}{\partial x} + j\frac{\partial y'}{\partial x}$$

$$\frac{df(z)}{dz}\bigg|_{dz=jdy} = \frac{\partial y'}{\partial y} - j\frac{\partial x'}{\partial y} \tag{6.3}$$

The Cauchy–Riemann equations directly follow by setting these two expressions equal to each other. The real and imaginary parts of any complex analytic function are automatically harmonic functions and thus are twice continuously differentiable functions, which satisfy Laplace's equations [42]. Conformal mappings use analytic complex functions and ensure angles and aspect ratios are preserved through coordinate transformation. This can be proved by decomposing dx and dy into its differential components in the transformed space, dx' and dy', using the coordinate systems shown in Fig. 6.1:

$$dx' = a\,dx\cos\theta_x,\ dy' = -a\,dx\sin\theta_x$$

$$dx' = b\,dy\cos\theta_y,\ dy' = -b\,dy\cos\theta_y \tag{6.4}$$

In order to satisfy the Cauchy–Riemann equations, the following conditions must be satisfied:

$$\theta_x = \theta_y,\quad a = b \tag{6.5}$$

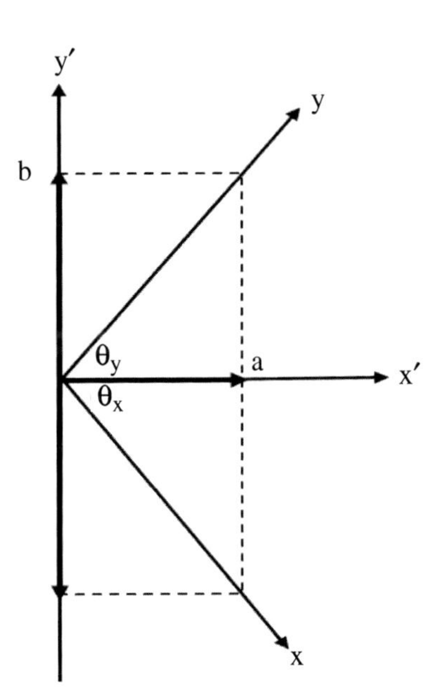

Figure 6.1 Coordinate systems used in transformations.

This proves that a conformal map leads to a locally orthogonal coordinate system that ensures the scaling of x is the same as the scaling of y. An important thing to note about conformal maps as they relate to TO is that they lead to media that are locally isotropic and non-magnetic, which is crucial for broadband performance [36]. Quasi-conformal transformation optics is a numerical algorithm that attempts to minimize anisotropy for a general transformation by solving equation 6.1 with carefully selected boundary conditions [6].

6.2.2 Transformation Optics

One of the most interesting properties of Maxwell's equations is that they are form-invariant under any general coordinate transformation. It is this property that gave rise to the field of TO. For example, consider the transformation from an unprimed coordinate system to a primed coordinate system $(x, y, z) \rightarrow (x', y', z')$, we have:

$$\nabla \times \mathrm{E} = -j\omega\mu\mathrm{H} \rightarrow \nabla' \times \mathrm{E}' = -j\omega\mu'\mathrm{H}'$$

$$\nabla \times \mathrm{H} = j\omega\varepsilon\mathrm{E} \rightarrow \nabla' \times \mathrm{H}' = j\omega\varepsilon'\mathrm{E}' \tag{6.6}$$

It is important to note that the curl in the right-hand side is primed, *i.e.*, it must be performed in the transformed space. Given a Jacobian A defined as:

$$A = \begin{bmatrix} \frac{\partial x'}{\partial x} & \frac{\partial x'}{\partial y} & \frac{\partial x'}{\partial z} \\ \frac{\partial y'}{\partial x} & \frac{\partial y'}{\partial y} & \frac{\partial y'}{\partial z} \\ \frac{\partial z'}{\partial x} & \frac{\partial z'}{\partial y} & \frac{\partial z'}{\partial z} \end{bmatrix}, \tag{6.7}$$

we can find the primed electromagnetic fields:

$$E' = (A^{-1})^{\mathrm{T}} E, \quad H' = (A^{-1})^{\mathrm{T}} H \tag{6.8}$$

In order to emulate the coordinate transformation, the required constitutive material parameters are given via the following relations:

$$\varepsilon' = \frac{A\varepsilon A^{\mathrm{T}}}{\det A}, \quad \mu' = \frac{A\mu A^{\mathrm{T}}}{\det A} \tag{6.9}$$

By using these equations, the primed electric fields of the transformed space are recreated in the original space by constructing a material whose constitutive parameters are prescribed by Eq. 6.9. If the transformation is a conformal map, the medium is isotropic.

As an example, consider a 2D transformation from (x, y, z) to (x', y', z):

$$A = \begin{bmatrix} \frac{\partial x'}{\partial x} & \frac{\partial x'}{\partial y} & 0 \\ \frac{\partial y'}{\partial x} & \frac{\partial y'}{\partial y} & 0 \\ 0 & 0 & 1 \end{bmatrix} \tag{6.10}$$

If the starting medium is isotropic (e.g., free space) and the transformation satisfies the Cauchy–Riemann equations, then the expressions for the constitutive parameters are very simple. For simplicity in notation, we define:

$$u \equiv \frac{\partial x'}{\partial x} = \frac{\partial y'}{\partial y}, \quad v \equiv \frac{\partial x'}{\partial y} = -\frac{\partial y'}{\partial x} \tag{6.11}$$

The Jacobian is given by:

$$A = \begin{bmatrix} u & v & 0 \\ -v & u & 0 \\ 0 & 0 & 1 \end{bmatrix} \tag{6.12}$$

whose determinant is

$$\det A = u^2 + v^2 \tag{6.13}$$

and the expression for the permittivity is given by:

$$\varepsilon' = \frac{A \varepsilon A^{\mathrm{T}}}{\det A}$$

$$= \frac{1}{u^2 + v^2} \begin{bmatrix} u & v & 0 \\ -v & u & 0 \\ 0 & 0 & 1 \end{bmatrix} \begin{bmatrix} \varepsilon & 0 & 0 \\ 0 & \varepsilon & 0 \\ 0 & 0 & \varepsilon \end{bmatrix} \begin{bmatrix} u & -v & 0 \\ v & u & 0 \\ 0 & 0 & 1 \end{bmatrix}$$

$$= \frac{1}{u^2+v^2}\begin{bmatrix} (u^2+v^2)\varepsilon & (uv-vu)\varepsilon & 0 \\ (uv-vu)\varepsilon & (u^2+v^2)\varepsilon & 0 \\ 0 & 0 & \varepsilon \end{bmatrix}$$

$$= \begin{bmatrix} \varepsilon & 0 & 0 \\ 0 & \varepsilon & 0 \\ 0 & 0 & \dfrac{\varepsilon}{u^2+v^2} \end{bmatrix} \tag{6.14}$$

Similarly, the primed permeability is given by:

$$\mu' = \begin{bmatrix} \mu & 0 & 0 \\ 0 & \mu & 0 \\ 0 & 0 & \dfrac{\mu}{u^2+v^2} \end{bmatrix} \tag{6.15}$$

For the TM^z polarization, where $E = E\hat{z}$ and $H \cdot \hat{z} = 0$, a non-magnetic, inhomogeneous isotropic dielectric could be used since μ_{zz} would have no impact and ε_{zz} would be the only parameter that has any polarizing effect.

An important thing to note is that for these equations to be valid, they need to be applied for all space. However, this is not possible in practice, as the area where the coordinate transformation takes place has to be physically restricted. If the fields are to be manipulated external to the designed media, discontinuities in the coordinate system across the outer boundary need to be introduced. Such a transformation features a material embedded in a finite region of space and is called a finite embedded coordinate transformation [45, 46]. But due to impedance mismatches, these designs may have reflections at the boundaries. An example that can be continuous across material boundaries is the well-known invisibility cloak [30, 31, 33]. The invisibility cloak can hide the presence of a PEC object such that an incident field illuminating the PEC and cloak will result in the same incident field outside the cloak. For such a design, it is required that the coordinate system be continuous across material boundaries, since reflections need to be eliminated. In contrast, there are designs that explicitly require discontinuities across material boundaries. One such example is a flat focusing lens [29, 37]. Traditional homogeneous lenses require curvature for focusing, but lenses with large curvatures can be difficult to manufacture. A flat focusing lens can be accomplished by mapping a curved surface to a flat planar surface which embeds the refractive properties of the surface into the bulk of the lens through a TO medium. To focus an incident plane wave to a point, the flat lens must modify the phase distribution of the electric field as it passes through the medium, so the field exiting the lens must be different from the incident field. Therefore, the transformation must be discontinuous and reflections will occur. Such reflections are not unique to TO lenses; homogeneous lenses are susceptible to such reflections as well and both designs can have their reflections mitigated through the introduction of anti-reflective coatings, which will be discussed in Section 6.3.4.

While none of the designs discussed so far involve sources in the transformation regions, charges and currents may be transformed in a similar way to the constitutive material properties, using what is called an optical source transformation, as described by Kundtz et al. [25].

$$J' = \frac{AJ}{\det A}, \quad \rho' = \frac{A\rho}{\det A} \tag{6.16}$$

Using these equations, a source distribution can be geometrically transformed without changing its radiation characteristics. Zhang et al. [38] used an elliptical transformation to map a line current to a spherical surface current. When combined with the appropriate medium, the radiation characteristics are similar to the original line current. This can easily be extended to mapping a linear phased array to a conformal surface, as described by Popa et al. [44].

These equations provide an analytical method of determining the permittivity and permeability tensors to achieve a certain geometrical transformation. To physically achieve these material parameters, anisotropic metamaterials must be utilized, which can be lossy and resonant. To achieve isotropic material parameters, the transformation must satisfy the Cauchy–Riemann equations, and it can be very difficult to derive an analytical expression for such a transformation. A more general numerical algorithm called qTO can be employed to find the required transformation between two general geometries.

6.2.3 Quasi-Conformal Transformation Optics

Quasi-conformal transformation optics is a generalization of TO that can, in principle, yield designs that are broadband and low loss [36]. Finding an exact analytical transformation that represents a conformal map for any general transformation can be extremely difficult and in many cases impossible. qTO is a numerical algorithm that finds the required transformation while also attempting to minimize the anisotropy. This can lead to a dielectric-only design in 2D. In 3D, a medium designed through qTO is inherently anisotropic and a magnetic medium is required. Returning to Eq. 6.10, the Jacobian was expressed as:

$$A = \begin{bmatrix} \frac{\partial x'}{\partial x} & \frac{\partial x'}{\partial y} & 0 \\ \frac{\partial y'}{\partial x} & \frac{\partial y'}{\partial y} & 0 \\ 0 & 0 & 1 \end{bmatrix} \tag{6.17}$$

If the initial unprimed material parameters are isotropic, we have:

$$\varepsilon' = \varepsilon \frac{AA^{\mathrm{T}}}{\det A}, \quad \mu' = \mu \frac{AA^{\mathrm{T}}}{\det A} \tag{6.18}$$

Here the restrictions on the Cauchy–Riemann equations are relaxed and the requirement for them to be strictly satisfied is removed. If the original space is assumed to be vacuum, the final result for the permittivity is given by

$$\varepsilon' = \varepsilon \frac{1}{\frac{\partial x'}{\partial x}\frac{\partial y'}{\partial x} - \frac{\partial y'}{\partial y}\frac{\partial x'}{\partial y}}$$

$$\cdot\begin{bmatrix} \frac{\partial x'^2}{\partial x}+\frac{\partial x'^2}{\partial y} & \frac{\partial x'}{\partial x}\frac{\partial y'}{\partial x}+\frac{\partial x'}{\partial y}\frac{\partial y'}{\partial y} & 0 \\ \frac{\partial x'}{\partial x}\frac{\partial y'}{\partial x}+\frac{\partial x'}{\partial y}\frac{\partial y'}{\partial y} & \frac{\partial y'^2}{\partial x}+\frac{\partial y'^2}{\partial y} & 0 \\ 0 & 0 & 1 \end{bmatrix} \quad (6.19)$$

For an isotropic material, the off-diagonal components are required to be zero:

$$\frac{\partial x'}{\partial x}\frac{\partial y'}{\partial x}=-\frac{\partial x'}{\partial y}\frac{\partial y'}{\partial y} \quad (6.20)$$

A function $f(x, y)$, which represents the degree of anisotropy through space, can be defined as follows:

$$f(x,y)\frac{\partial x'}{\partial x}=\frac{\partial y'}{\partial y}, \quad -\frac{1}{f(x,y)}\frac{\partial y'}{\partial x}=\frac{\partial x'}{\partial y} \quad (6.21)$$

These are the well-known Beltrami equations [36]. Note that if $f = 1$ everywhere, these equations reduce to the Cauchy–Riemann equations, resulting in a completely isotropic material. The partial derivative of the first equation in 6.21 with respect to x is given by:

$$\frac{\partial}{\partial x}f(x,y)\frac{\partial x'}{\partial x}=\frac{\partial}{\partial x}\frac{\partial y'}{\partial y} \quad (6.22)$$

Schwarz's theorem states that if the mapping has continuous second partial derivatives, then the partial derivatives are commutative. After applying Schwarz's theorem, the expression becomes:

$$\frac{\partial}{\partial x}f(x,y)\frac{\partial x'}{\partial x}=\frac{\partial}{\partial y}\frac{\partial y'}{\partial x} \quad (6.23)$$

Finally, plugging in the second of Eq. 6.21 and moving all terms to the left-hand side give:

$$\frac{\partial}{\partial x}f(x,y)\frac{\partial x'}{\partial x}+\frac{\partial}{\partial y}f(x,y)\frac{\partial x'}{\partial y}=0 \quad (6.24)$$

Similarly, the partial derivative of the second equation in 6.21 with respect to x is given by:

$$\frac{\partial}{\partial x}\frac{1}{f(x,y)}\frac{\partial y'}{\partial x}+\frac{\partial}{\partial y}\frac{1}{f(x,y)}\frac{\partial y'}{\partial y}=0 \tag{6.25}$$

If the mapping is perfectly isotropic ($f = 1$), these equations reduce to Laplace's equations for both x' and y':

$$\left(\frac{\partial^2}{\partial x^2}+\frac{\partial^2}{\partial y^2}\right)x'=0$$

$$\left(\frac{\partial^2}{\partial x^2}+\frac{\partial^2}{\partial y^2}\right)y'=0 \tag{6.26}$$

Quasi-conformal transformation optics consists of attempting to numerically solve the functional 6.21 given the boundaries of the original domain and the desired domain while simultaneously trying to minimize $|f(x, y) - 1|$ as well. The corners of the boundaries are absolutely fixed, but the transformation points are allowed to "slip" along the boundary of the desired domain [6]. With such a scheme, the resulting material will be purely dielectric and nonmagnetic. The transformed permittivity is given by:

$$\varepsilon'=\varepsilon\frac{1}{\left(\frac{\partial x'}{\partial x}\right)^2+\left(\frac{\partial x'}{\partial y}\right)^2} \tag{6.27}$$

Using Eq. 6.5 results in:

$$\varepsilon'=\varepsilon\frac{1}{a^2} \tag{6.28}$$

This leads to the very intuitive physical interpretation that reducing the area of the transformed cell will result in a larger refractive index [31]. Examining the transformed refractive index $n' = \varepsilon'^2 = a^{-1}n$ shows that the refractive index is scaled directly by the scaling factor of each locally orthogonal grid cell. This makes sense because the optical path length (OPL) $\int n{\cdot}dl$ must be invariant over the transformation. So if the size of a physical device is reduced, the refractive index must increase to maintain the same OPL within the transformed region.

As an example, consider flattening a convex-plano lens. The original lens is shown in Fig. 6.2a, and the desired flat lens is shown in Fig. 6.2b. The grid in Fig. 6.2b is perfectly orthogonal, while the grid in Fig. 6.2a was generated by solving Laplace's equation with slipping boundary conditions and attempting to maintain local orthogonality with approximately square mesh cells. This transformation then approximately satisfies the Cauchy–Riemann equations and is, therefore, approximately a conformal map. To verify this, each of the tensor components of the resulting permittivity are shown in Fig. 6.3. As can be seen, the ε_{xz}, ε_{yz}, ε_{zx}, and ε_{zy} components are all exactly zero, a result of the 2D transformation. Meanwhile, ε_{xy} and ε_{yx} are nearly zero over the entire lens but reach a magnitude of 0.12 near the corners of the front of the lens. This occurs because the grid is not completely locally orthogonal with constant scaling in those regions. The most important component is ε_{zz} as it determines the index distribution used when physically manufacturing the device. As seen in Fig. 6.3i, the ε_{zz} component and thus the GRIN distribution have a negative nearly parabolic radial profile, which is necessary to provide the power in a flat converging lens. Figure 6.4 shows the permittivity based on Eq. 6.28, which assumed that the transformation was conformal. The distribution is nearly identical to that of ε_{zz}, but the magnitudes are slightly off, owing to the imperfect grid created. This is typically the refractive index distribution used as a starting point when physically manufacturing these devices. Due to the fact that this is an imperfect numerical approximation and that the transformation is embedded in a finite region of space, the refractive index distribution may need to be slightly tweaked to yield the desired performance.

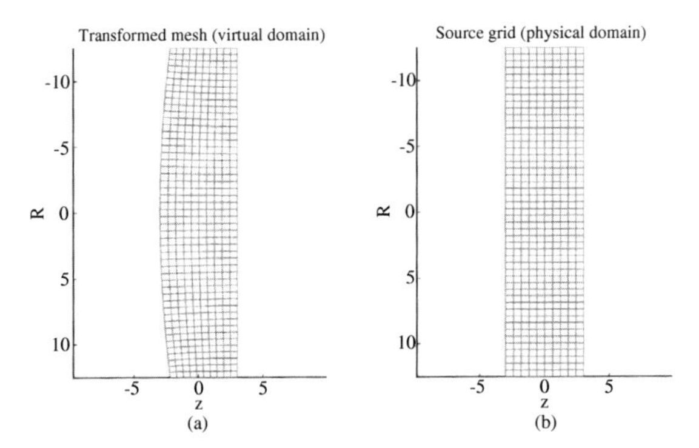

Figure 6.2 Example of coordinate transformation (a) original convex-plano lens, (b) desired flat lens.

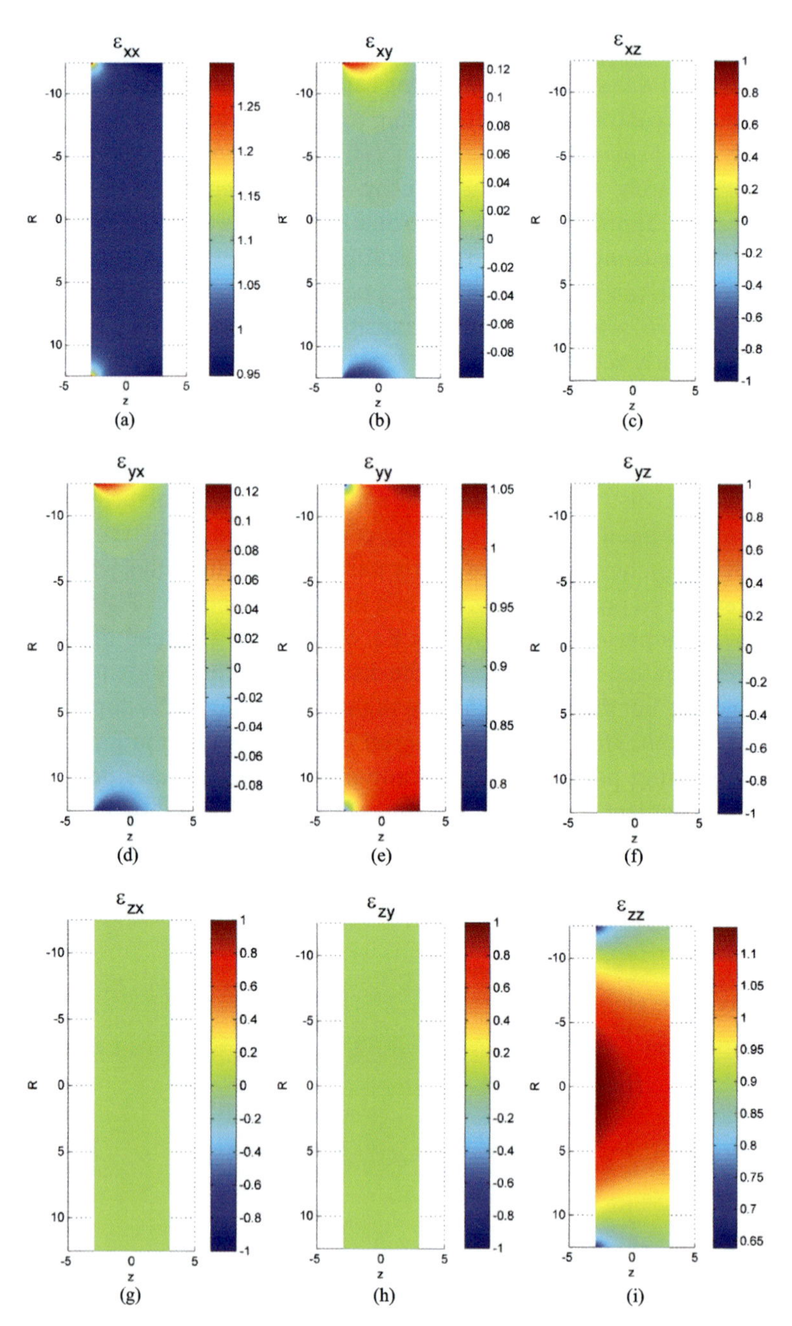

Figure 6.3 Permittivity tensor of convex-plano to flat transformation.

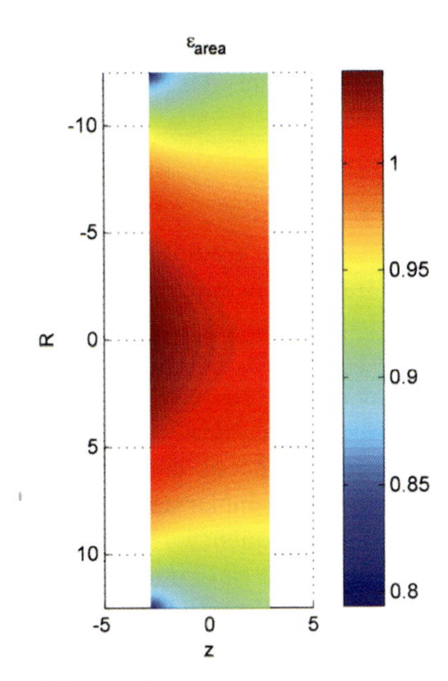

Figure 6.4 Permittivity based on areas.

6.3 Examples of qTO-Derived Lenses Inspired by Classical Designs

This section will describe several qTO-derived inhomogeneous metamaterials that are either explicit transformations of classical lenses or inspired by classical ideas. First, broadband wide-angle designs explicitly derived from the Maxwell, Luneburg, and Fresnel lenses will be described. After that, an inhomogeneous multibeam lens antenna and a GRIN anti-reflection coating (ARC) design procedure will be detailed by studying how these designs were done in the past and using qTO to generate the appropriate material profile.

6.3.1 Broadband Wide-Angle Lenses Derived from Refractive Lenses

In 1854, James Clerk Maxwell analytically determined the refractive index of a spherical lens [40] such that any point source on the spherical surface will focus to the opposite point on the sphere, as shown in Fig. 6.5a. The refractive index distribution of this so-called Maxwell lens for a sphere of radius R is given by:

$$n(r) = \frac{n_0}{1 + \left(\frac{r}{R}\right)^2} \tag{6.29}$$

As can be seen in Fig. 6.5b, a point source radiating spherical waves on one surface transforms into a plane wave in the center of the lens. This fact can be exploited by using only half of the Maxwell fish-eye (commonly referred to as an HMFE) to create highly directive lens antennas, as first discovered by Fuchs et al. [13] and shown in Fig. 6.5c. Fuchs et al. physically realized the HMFE through a series of concentric homogeneous dielectric shells and showed very good agreement for the co-polarized E- and H-plane radiated far-field patterns at a single frequency. The spherical surfaces make it difficult to fabricate, but qTO can be used to flatten the lens. To illustrate the process, the analytical study by Aghanejad et al. [1] will be briefly described. As shown in Fig. 6.6b, a direct implementation of numerically solving Laplace's equations to flatten the HMFE leads to refractive indices less than 1, requiring metamaterials that are typically highly resonant and lossy (we will call this design the dispersive design). To overcome this, the regions with refractive indices less than 1 are simply replaced by 1, as shown in Fig. 6.6b (this is referred to as the non-dispersive design). Figure 6.7 shows the fields from a COMSOL MultiPhysics simulation for the dispersive design, the non-dispersive design, and a benchmark horn antenna, all operating at 18 GHz. Figure 6.8 compares the far-field patterns. As can be seen, both the dispersive and non-dispersive designs are highly directional. The non- dispersive design has higher sidelobes but maintains approximately the same gain and half-power beamwidth as the dispersive design. To illustrate the broadband performance of this proposed design, the variations in gains, half-power beamwidths, and S_{11} over a frequency range of 15–21 GHz are shown in Fig. 6.9. As can be seen, the gain monotonically increases while the half-power beamwidth stays constant and S_{11} remains below -9 dB over the frequency range of interest. This design shows the theoretical performance of a broadband, directive, qTO-derived flat lens. The device was not manufactured but could be physically realized through non-resonant metamaterials or dielectric rods with spatially varying radii [59].

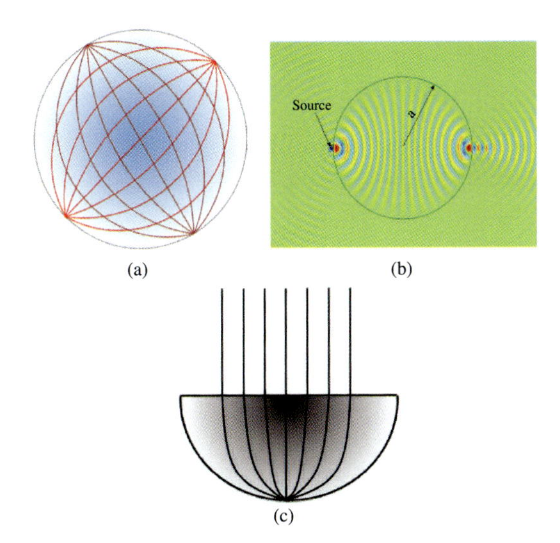

Figure 6.5 Maxwell's fish-eye lens. (a) Ray traces. Image by Catslash from Wikimedia Commons. Used under CC-BY-SA 3.0. https://upload.wikimedia.org/wikipedia/commons/d/d5/Maxwellsfish-eyelens.svg (b) Field distribution. Figure (b) reprinted with permission from Ref. 1, Copyright 2012, IEEE. (c) Half Maxwell's fish-eye.

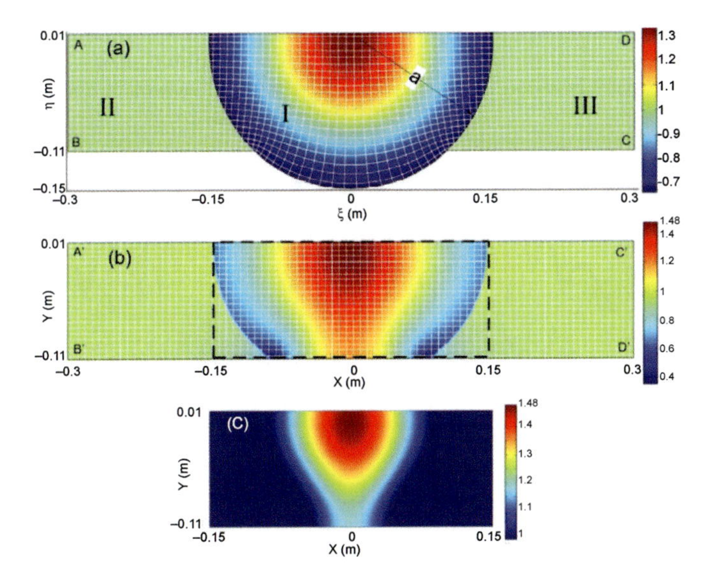

Figure 6.6 Flattening the half Maxwell's fish-eye (a) original HMFE, (b) flattened dispersive lens, (c) flattened non-dispersive lens. Reprinted with permission from Ref. 1, Copyright 2012, IEEE.

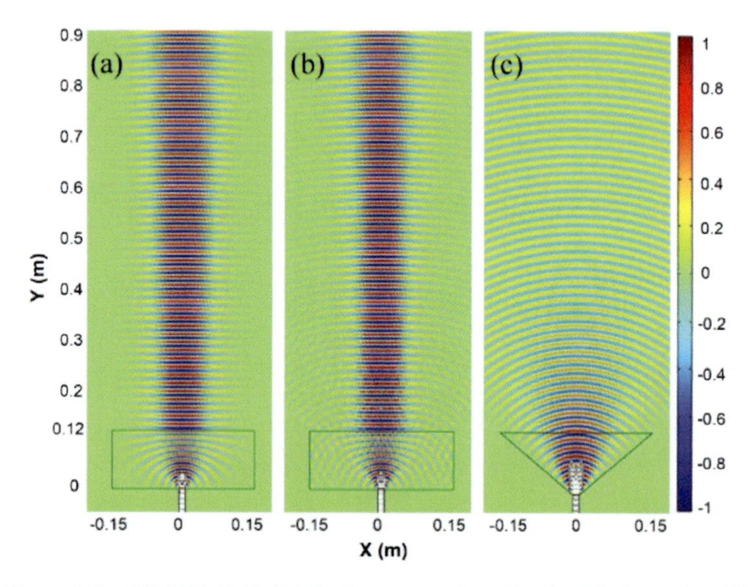

Figure 6.7 COMSOL field distribution comparison for the (a) dispersive, (b) non-dispersive, and (c) horn designs. Reprinted with permission from Ref. 1, Copyright 2012, IEEE.

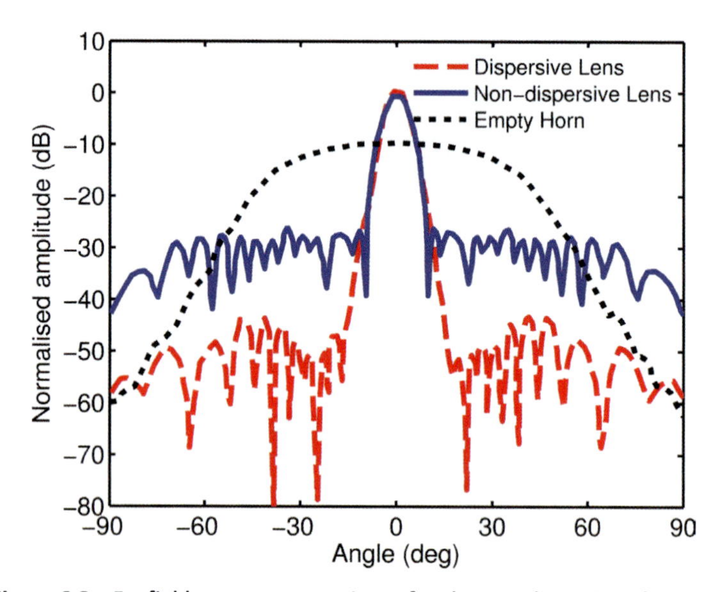

Figure 6.8 Far-field pattern comparisons for the non-dispersive, dispersive, and horn designs. Reprinted with permission from Ref. 1, Copyright 2012, IEEE.

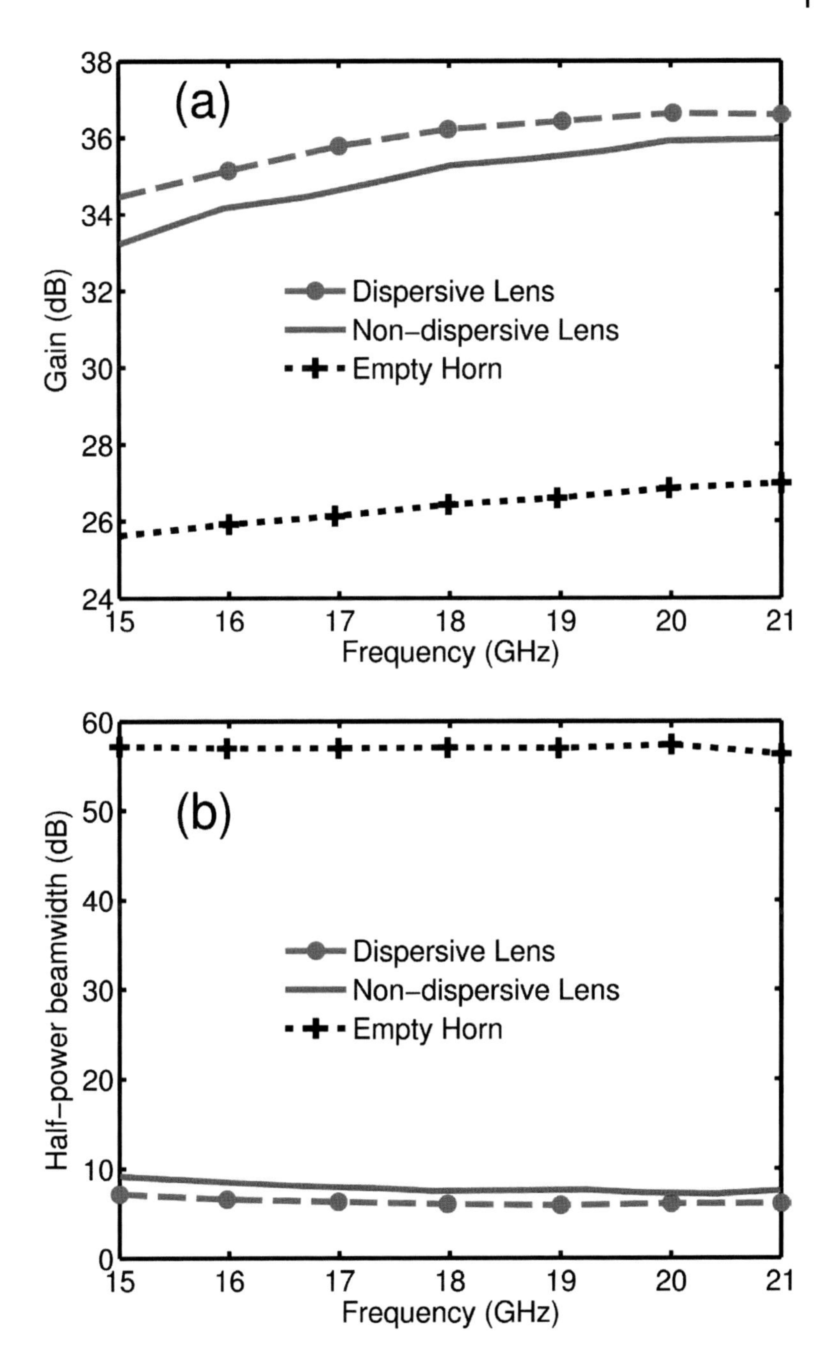
(a)
Gain (dB)
38
36
34
32
30
28
26
24
15
16
17
18
19
20
21
Frequency (GHz)
Dispersive Lens
Non–dispersive Lens
Empty Horn
(b)
Half–power beamwidth (dB)
60
50
40
30
20
10
0
15
16
17
18
19
20
21
Frequency (GHz)
Dispersive Lens
Non–dispersive Lens
Empty Horn

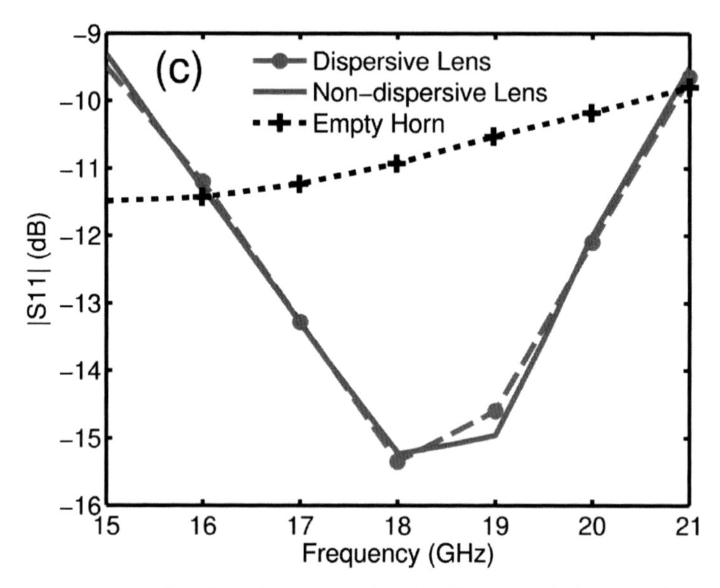

Figure 6.9 Broadband performance of the half Maxwell's fish-eye, (a) gain versus frequency, (b) half-power beamwidth versus frequency, and (c) S_{11} versus frequency. Reprinted with permission from Ref. 1, Copyright 2012, IEEE.

Another classic lens design was discovered by Rudolf Luneburg in 1944. The refractive index profile of this spherical lens was designed such that an incident plane wave from any direction will focus to a point on the surface of the lens [61], as shown in Fig. 6.10.

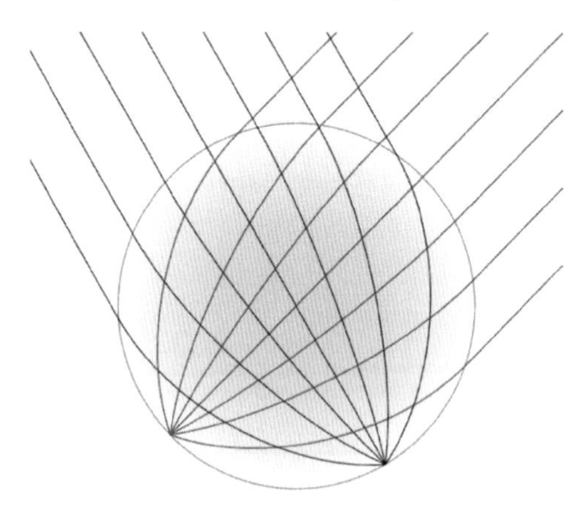

Figure 6.10 Luneburg lens. Image by Catslash from Wikimedia Commons. Used under CC-BY-SA 3.0. https://upload.wikimedia.org/wikipedia/commons/2/2f/Luneburg_lens.svg.

The refractive index for this so-called Luneburg lens of radius R is given by:

$$n(r) = \sqrt{2 - \left(\frac{r}{R}\right)^2} \tag{6.30}$$

Despite its incredible imaging properties, the Luneburg lens is not commonly used in optical applications since it requires a detector that conforms to the surface of the sphere, while most detectors in practice are planar. In 2008, D. Schurig proposed an analytical transformation that flattens the Luneburg lens [52]. The mathematical description of the transformation is given as follows:

$$\begin{aligned} \rho' &= \rho \\ \phi' &= \phi \\ z' &= \frac{1}{2}\left(z + \sqrt{R^2 - \rho^2}\right) \end{aligned} \tag{6.31}$$

A resulting ray trace is depicted in Fig. 6.11a. The required permittivity tensor is anisotropic and includes offdiagonal terms. This tensor can be expressed in an orthogonal diagonalizing basis by simply rotating the coordinate system, and the resulting diagonal permittivity tensor (which is the same as the permeability tensor) is:

$$\varepsilon = \mu = n(R)\begin{bmatrix} 1/\eta & 0 & 0 \\ 0 & 2 & 0 \\ 0 & 0 & \eta \end{bmatrix} \tag{6.32}$$

where

$$\eta = \frac{5R^2 - 4\rho'^2 - R\sqrt{9R^2 - 8\rho'^2}}{4(R^2 - \rho'^2)} \tag{6.33}$$

As expected, the final material properties are diagonalized, but still anisotropic. The resulting anisotropic permittivity distribution is provided in Fig. 6.11b.

Such a device could only be physically realized with anisotropic metamaterials. In 2009, Kundtz and Smith used qTO to derive a GRIN lens that can be implemented with simple dielectrics [26]. The numerical transformation performed is shown in Fig. 6.12 for both the virtual space (a) and physical space (b). The dashed yellow line

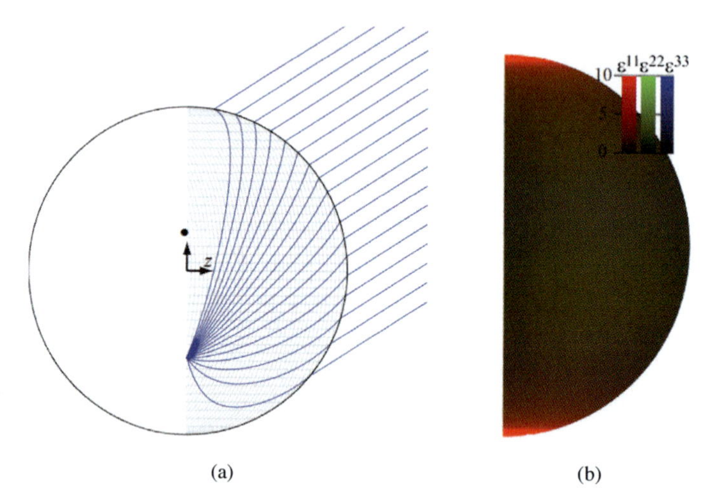

Figure 6.11 Flattening of Luneburg lens. (a) Coordinate transformation, (b) permittivity distribution. Reprinted from Ref. 52, Copyright 2008, IOP Publishing under Creative Commons Attribution license.

represents the focal surface, which gets mapped from a spherical surface to a flat surface via slipping Neumann boundary conditions. The red line represents the front surface of the flat lens; Dirichlet boundary conditions are employed to ensure the final design does not have reflections. The choice of $Y = 1.4$ for the upper boundary of the lens is relatively arbitrary, but it needs to be far enough away from the yellow line to ensure the conformal module is approximately one over the entire map, thereby minimizing anisotropy. The resulting refractive index distribution is shown in Fig. 6.13a. In the physical realization of this design, refractive indices less than 1 are set to 1.08, resulting in an index range of 1.08–4.0. Because this range is so large, the device was divided into two regions, as shown in Fig. 6.13b, a center region where $n < 2$ and an outer region where $n > 2$. In the center region, the distribution was broken up into small squares with a constant bulk effective permittivity. In each square region, the permittivity is realized by a metallic I-beam lithographically patterned on an FR4 substrate, as shown in Fig. 6.13d [58]. The dimensions required for a given permittivity at a certain point were found through a standard electromagnetic parameter retrieval method [53]. Over the frequency range of interest (7–15 GHz), I-beam metamaterials with refractive indices greater than two would be very large and yield significant spatial dispersion. Therefore, those regions where $n > 2$ are physically realized by patterning copper strips on an FR4 substrate, as shown in Fig. 6.13c. The resulting index changes over the frequency range by up to 10% in the outer regions and up to 3% in the inner region. To measure the device, a dielectric waveguide was used to approximate a point source at various points of the focal surface. This should result in plane waves directed at different angles. Figure 6.14 shows the experimental results for a variety of angles and frequencies, showcasing the broadband performance of the device.

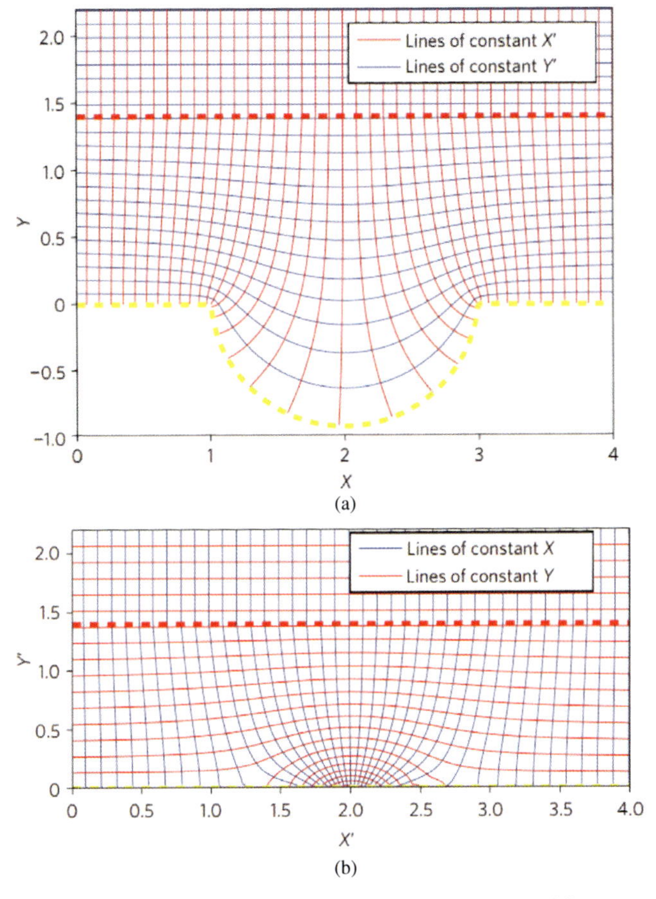

Figure 6.12 qTO solution for the flattened Luneburg lens in (a) virtual space and (b) physical space. Reprinted by permission from Macmillan Publishers Ltd: *Nature Publishing Group*, Ref. 26, Copyright 2010.

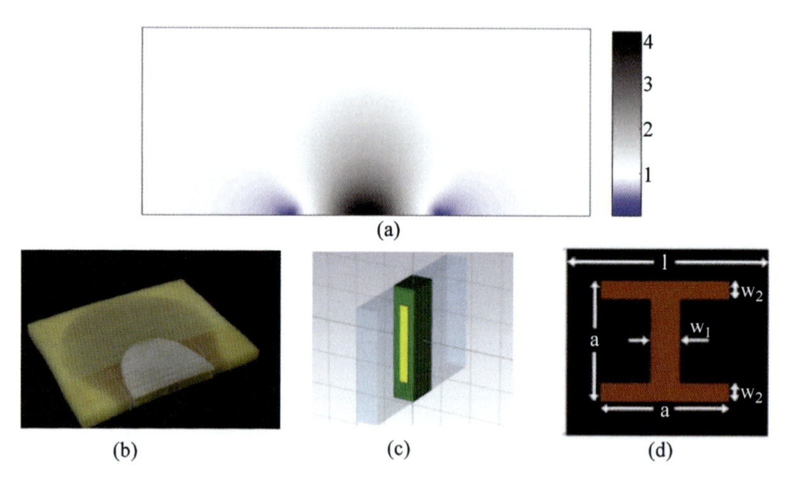

Figure 6.13 Flattened Luneburg (a) refractive index distribution; (b) physical device showing two regions, a center region where $n > 2$ and two outer regions where $n < 2$; (c) unit cell used to realize the index distribution where $n > 2$; (d) unit cell used to realize the index distribution where $n < 2$. Reprinted by permission from Macmillan Publishers Ltd: *Nature Publishing Group*, Ref. 26, Copyright 2010.

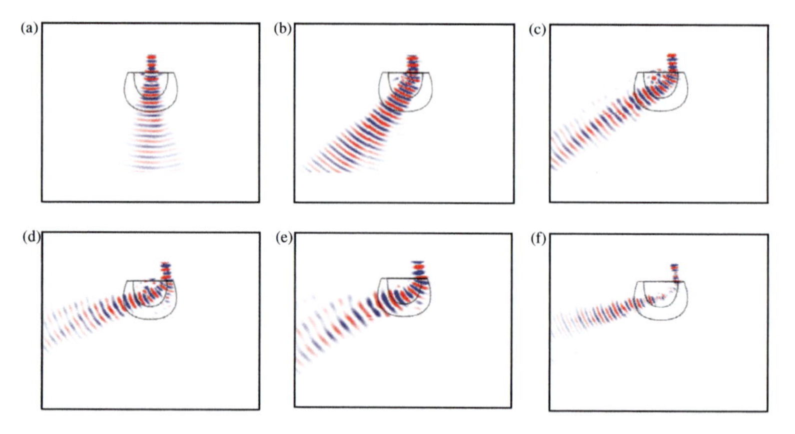

Figure 6.14 Experimental results for flattened Luneburg lens. Results are shown at 10 GHz for beams directed at incident angles of (a) 0°, (b) 35°, (c) 50°, and (d) off-axis at 50°. The broadband performance of the device is showcased by including off-axis 50° results at (e) 7 GHz and (f) 15 GHz. Reprinted by permission from Macmillan Publishers Ltd: *Nature Publishing Group*, Ref. 26, Copyright 2010.

While these results are extremely impressive at microwave frequencies, they likely would not scale well to optical frequencies where metals begin to behave like lossy dielectrics and less like conductors. For better scalability, an all-dielectric implementation would be preferred. This was done by Hunt et al. in 2011 by drilling sub-wavelength holes in a dielectric slab where the density of holes dictates the desired refractive index at a given location [21]. The same qTO transformation as [53] was employed, and regions where $n < 1$ were approximated as 1. The resulting refractive index profile of the inhomogeneous metamaterial is given in Fig. 6.15a, the density of holes in Fig. 6.15b, and the final manufactured design is shown in Fig. 6.15c. The device theoretically has the potential to scale well to optical frequencies, but it was constructed at microwave frequencies for ease of manufacturing and testing. A comparison between simulated results in COMSOL and measured results for a beam directed to 30° at 10 GHz is shown in Fig. 6.16. Similar to the previous designs, a point source at different positions of the flattened focal surface will yield perfectly planar wavefronts at different angles on planes containing the optical axis. On planes not containing the optical axis, there will be some distortion. To illustrate the FOV of the lens, ray traces with 33 rays were simulated through both the complete lens where no approximations have been made to the qTO-derived refractive index, and the approximated lens where $n < 1$ regions are replaced by $n = 1$. As can be seen in Fig. 6.17, the two designs agree very closely for on-axis and an incident angle of 10. At 20° and 30°, some of the rays travel through the $n < 1$ region, leading to a few stray rays. Still, the majority of the rays all focus to an extremely tight spot. Significant aberrations begin to appear at angles greater than 40°. This demonstrates that the exact qTO-derived index has an FOV of ±40°, while the approximated dielectric-only design has an FOV of ±30°. These results show the tremendous potential of obtaining manufacturable broadband lenses with a wide FOV by applying qTO to classic refractive lens designs such as the Maxwell and Luneburg lenses.

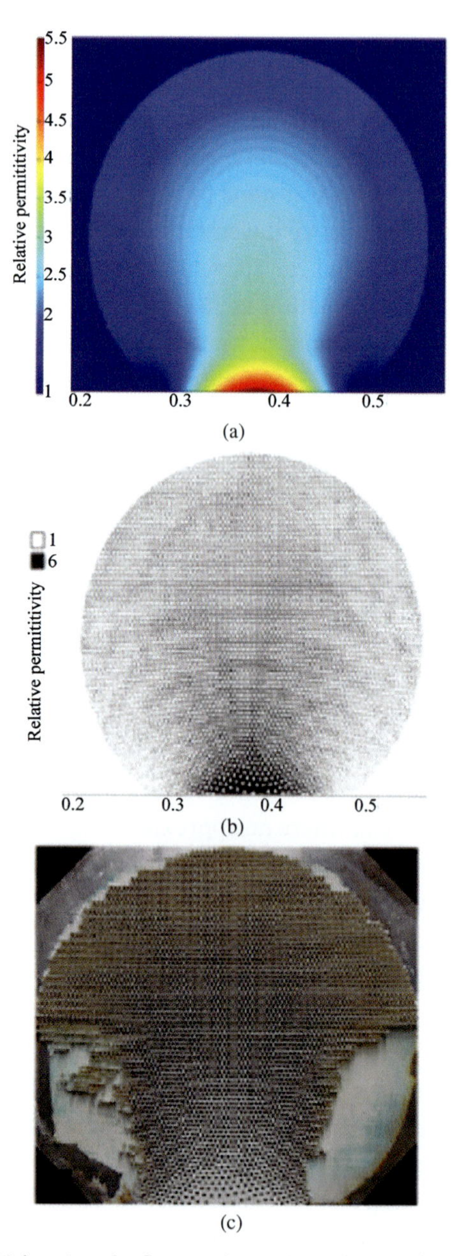

Figure 6.15 Dielectric-only flattened Luneburg lens. (a) Refractive index profile, (b) distribution of holes, and (c) manufactured design. Reprinted from Ref. 21, Copyright 2011, under Creative Commons Attribution license.

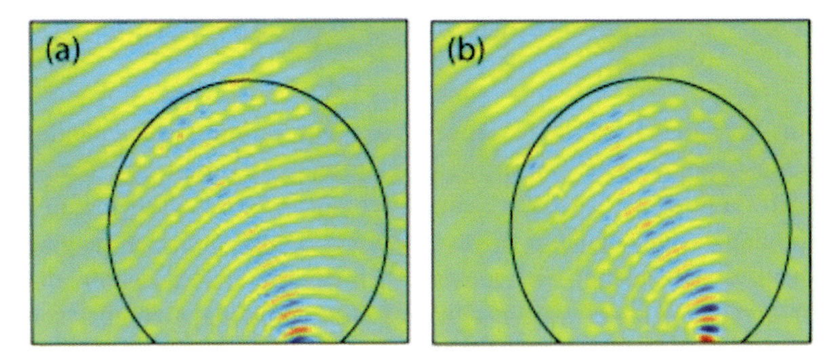

Figure 6.16 Results for dielectric-only flattened Luneburg lens: (a) simulation and (b) measurement. Reprinted from Ref. 21, Copyright 2011, under Creative Commons Attribution license.

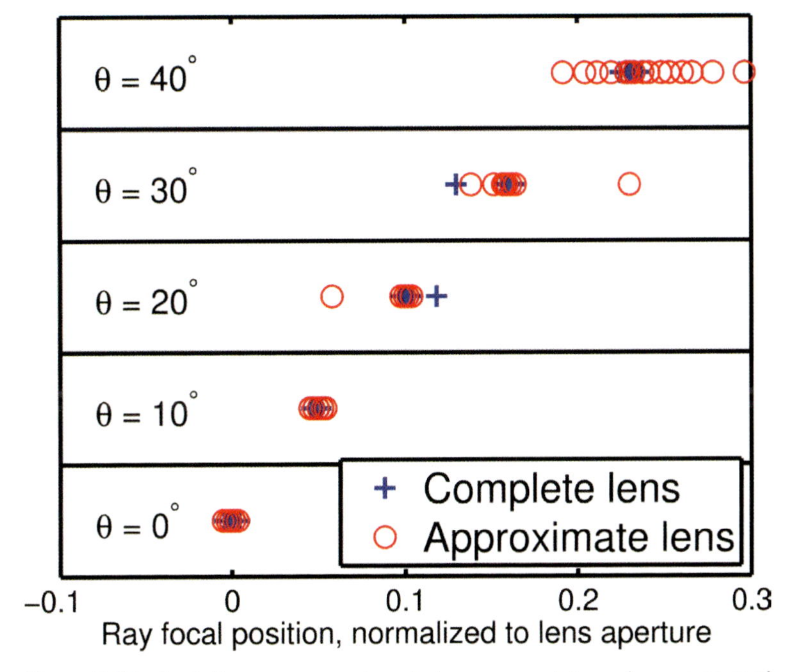

Figure 6.17 Spot diagram comparison between complete and approximated (where $n < 1$ regions are approximated by $n = 1$) at different incident angles. Reprinted from Ref. 21, Copyright 2011, under Creative Commons Attribution license.

6.3.2 Broadband Wide-Angle Lenses Derived from Diffractive Lenses

An alternative to refractive lenses is diffractive optics, which divides a lens into a set of concentric annular sections (often called zone plates) allowing the lens to be axially compressed, resulting in much thinner designs albeit at the expense of a reduction in the imaging quality. The Fresnel lens consists of a slightly convex center and an array of gradually steeper prisms on the outside. A comparison between a spherical Fresnel lens [35] and a refractive lens example is shown in Fig. 6.18. As can be seen, the overall thickness of the lens is drastically reduced, and simultaneously so is the lens mass and volume. As the name implies, diffractive lenses are based on the principle of wave diffraction instead of refraction, and therefore the design is inherently narrowband. Directly using qTO to transform the geometry is difficult due to the fact that the boundary of the Fresnel lens is discontinuous. This can lead to extremely large values of the refractive index around these discontinuities. To circumvent this problem, the Fresnel lens can be split into sections, as shown in Fig. 6.19.

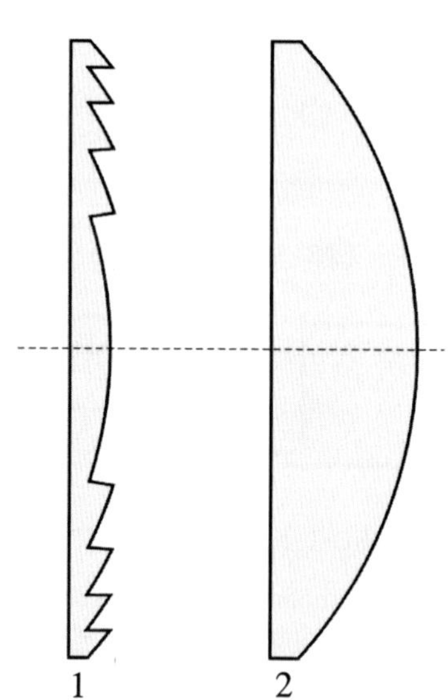

Figure 6.18 (a) Fresnel lens. (b) Plano-convex lens. Image by Pko from Wikimedia Commons. Used under Public Domain. https://upload.wikimedia.org/wikipedia/commons/e/e8/Fresnel_lens.svg.

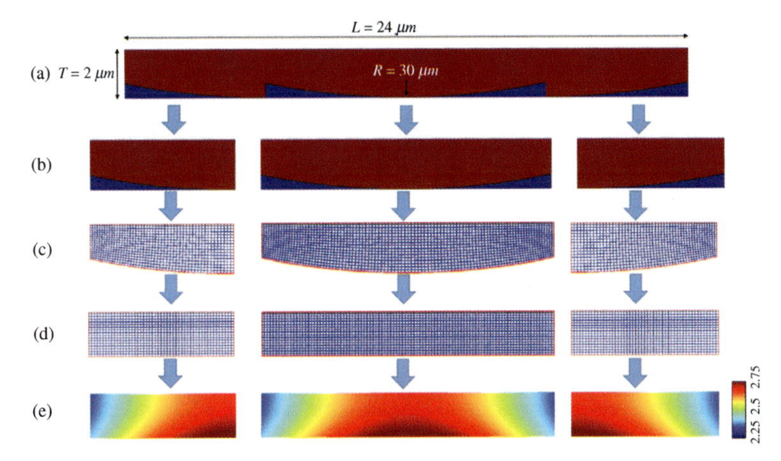

Figure 6.19 qTO transformation of the Fresnel lens: (a) original Fresnel lens, (b) splitting the lens into three sections, (c) mesh in the virtual domains, (d) mesh in the physical domains, and (e) resulting refractive index profiles.

The Fresnel lens under consideration has a refractive index of 2, thickness of 2 microns, and a length of 24 microns. For comparison, the corresponding refractive lens that would focus to the same spot would have a thickness of 2.37 microns, which means using the Fresnel lens results in a thickness reduction of 16%. For simplicity, only the first concentric ring outside the center section is included in the original Fresnel lens, as shown in Fig. 6.19a. This process could easily be extended to include multiple rings. As shown in Fig. 6.19b, the lens is split into three sections to eliminate the discontinuities in the geometry boundary. The curvilinear locally orthogonal grid generated in the virtual domain is shown in Fig. 6.19c, while the perfectly orthogonal grid of the physical domain is shown in Fig. 6.19d. As expected, there is perfect symmetry between the right and left halves of the refractive index profile, due to the fact that the original geometry is perfectly symmetric. The Δn is 0.5 with a range from 2.25 to 2.75. The thickness after flattening the lens is 1.77 microns. Because this was an embedded transformation, the refractive index profile that directly resulted from the qTO process on the center section was clipped on the edges. The most extreme values of the refractive index occur on the left and right sides because this is where the mesh cells are transformed the most. It has been found that removing these regions and then scaling the refractive index leads to a model that agrees with the performance of the original lens the most. For this model, regions of length 2.4 microns were removed from either side, the resulting refractive index profile was stretched to the original length of 12 microns, then it was scaled by a factor of 1.1. The final refractive index profiles are shown in Fig. 6.19e.

The results from an FDTD simulation with an on-axis incident TE-polarized plane wave with a magnitude of 1 V/m are shown in Fig. 6.20. The magnitude of the electric field at the final time step for the original Fresnel lens (located between $z = 57$ and $z = 59$ microns) is shown in Fig. 6.20a and for the flattened qTO-derived inhomogeneous metamaterial (located between $z = 57$ and $z = 58.77$ microns) is shown in Fig. 6.20b. As can be seen, the results are nearly identical. The focal length for this lens is 15 microns, and a peak can clearly be seen at 42 microns in both plots. There is considerable wave interference and diffraction, resulting in multiple peaks. To further test the validity of the qTO transformations, simulations were run where the lens is assumed to be mounted on a semi-infinite substrate with refractive index 2. The results are shown in Fig. 6.21a for the Fresnel lens and Fig. 6.21b for the qTO lens. The first peak is shifted to the same spot in both plots due to the presence of the substrate. Due to the fact that we performed qTO separately on different sections of the Fresnel lens, there are discontinuities in the index profile along the radial direction. This can lead to reflections for oblique incidence. Figures 6.22a,b show the results for the Fresnel and qTO lenses, respectively, for an incidence angle of 5°. Even at this incident angle, the electric field distributions in both cases match extremely well. If this method was used for a full Fresnel lens, like the one shown in Fig. 6.18a, the discontinuities in the lens surface would mean qTO would have to be applied separately for each concentric ring. However, modern Fresnel lenses typically consist of flat annular sections with different dielectric constants, resulting in a lens that corrects the phase of the incident wave at discrete locations [20]. A modern quarter-wave all-dielectric Fresnel lens is shown in Fig. 6.23. The radius of each Fresnel zone is computed at a specific design wavelength, meaning the phase correction of each zone will be too small or too large with variations in the wavelength. In contrast, refractive lenses are broadband but are typically curved and have a large volume.

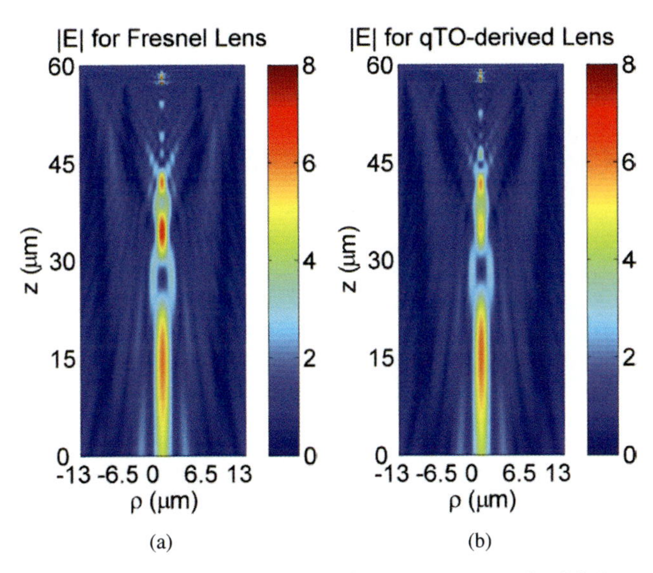

Figure 6.20 FDTD simulation results in free space on-axis for (a) the original Fresnel lens and (b) the flattened qTO-derived inhomogeneous metamaterial.

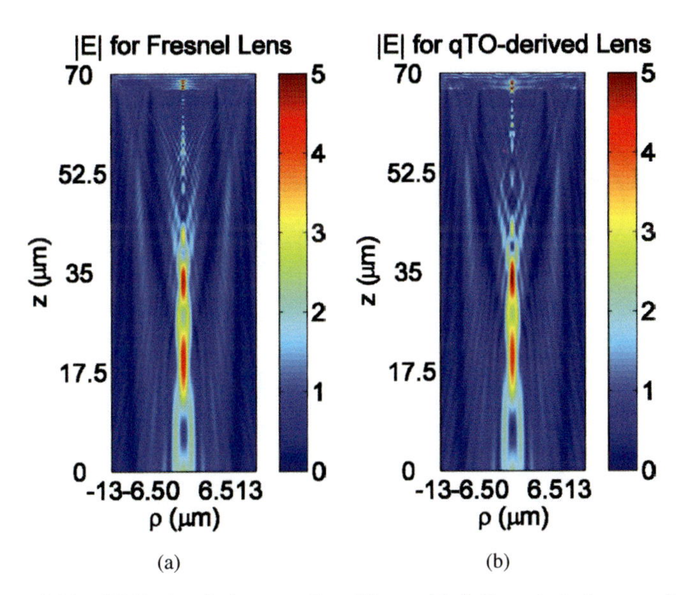

Figure 6.21 FDTD simulation results with semi-infinite substrate on-axis for (a) the original Fresnel lens and (b) the flattened qTO-derived inhomogeneous metamaterial.

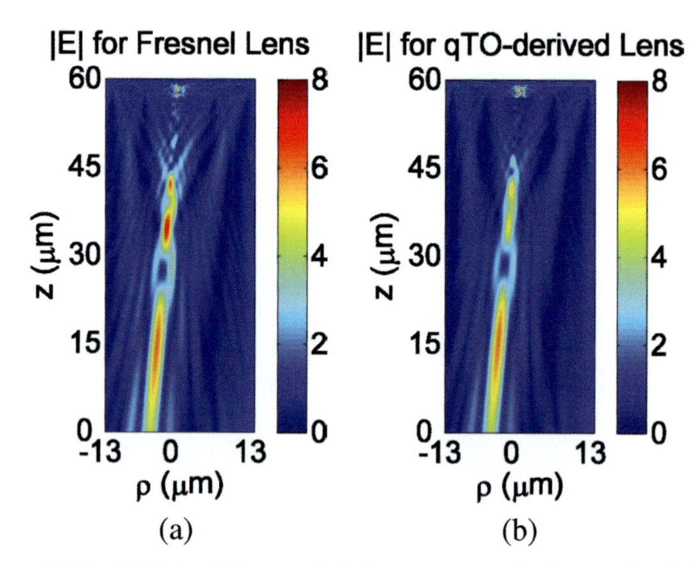

Figure 6.22 FDTD simulation results in free space at an incident angle of 5° for (a) the original Fresnel lens and (b) the flattened qTO-derived inhomogeneous metamaterial.

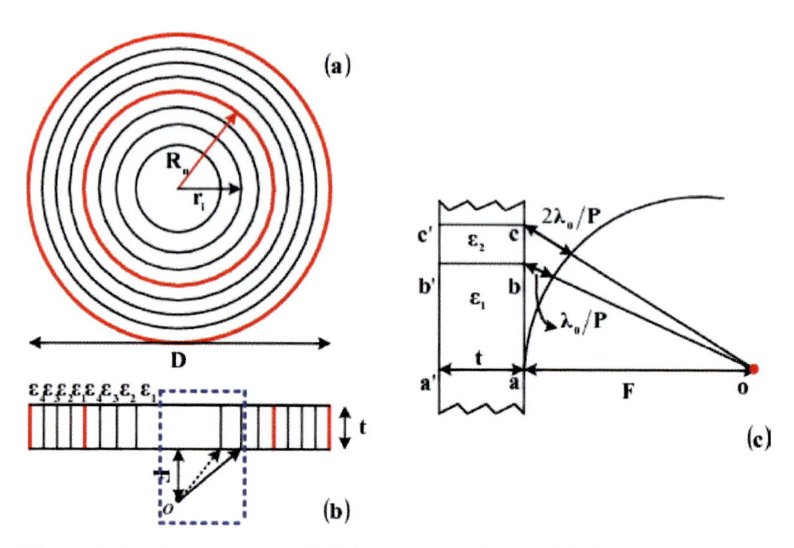

Figure 6.23 Quarter-wave all-dielectric Fresnel lens: (a) front view showing different radii for different zone plates, (b) side view showing different refractive indices for each layer, and (c) the spherical wavefront produced at a specific wavelength. Reprinted with permission from Ref. 64, Copyright 2011, The Optical Society.

Hao et al. in 2011 [64] designed a broadband zone plate lens based on qTO, which combines the small thickness and flat aperture of the diffractive lens with the broadband performance of the refractive lens by transforming a hyperbolic lens and taking inspiration from the quarter-wave Fresnel lens. Figure 6.24a shows the original hyperbolic lens along with a quasi-orthogonal grid in virtual space, which gets transformed through qTO to the flattened lens shown in Fig. 6.24b. The qTO process transforms the air region above the hyperbolic lens to a region labeled the "additional transformed layer" in Fig. 6.24b. This region consists of refractive indices less than 1, which must be physically realized through metamaterials, so this "additional transformed layer" was neglected for the design under consideration. The resulting permittivity distribution of Fig. 6.24b consists of 110 × 20 discrete blocks. This profile is then sampled at a lower resolution and a 22 × 4 block map is generated, as shown in Fig. 6.24d. Lastly, the 2D refractive index profile is rotated around its optical axis, resulting in the 3D zone plate lens shown in Fig. 6.24c. The resulting lens has the appearance of a conventional phase-correcting Fresnel lens, but as will be shown, it exhibits broadband performance. To verify the design, a full-wave finite-element simulation was performed using Ansys HFSS. Similar to the previous flattened Luneburg lenses, point sources are placed on the focal points and the resulting radiation pattern is highly directive, showcasing the converging ability of the lens. The radiation patterns of the original hyperbolic lens, the Fresnel lens, and the qTO-derived inhomogeneous metamaterial are shown in Fig. 6.25 for frequencies of (a) 20 GHz, (b) 30 GHz, the design frequency of the Fresnel lens, and (c) 40 GHz. All three lenses have nearly identical performance at 30 GHz, but the performance of the Fresnel lens is significantly degraded at 20 and 40 GHz, yielding much smaller directivities and very large sidelobes. In comparison, the qTO-derived inhomogeneous metamaterial shows extremely good agreement with the hyperbolic lens over the entire frequency range. The directivity of the Fresnel lens and qTO-derived inhomogeneous metamaterial are compared in Fig. 6.25d to further showcase the broadband performance of the proposed design. The permittivity range of the design is 2.62–15.02, allowing the zone plate lens to be constructed with standard dielectric materials instead of resonant metamaterials.

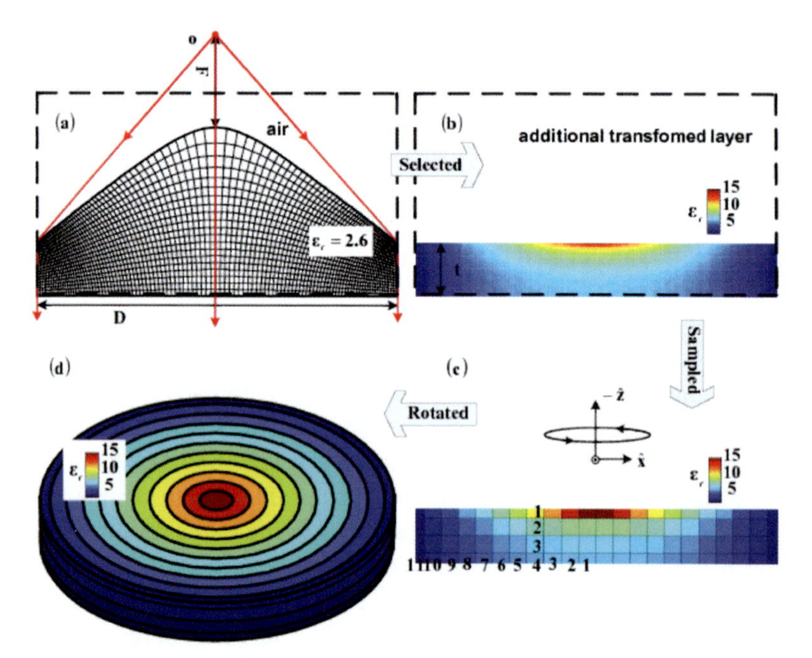

Figure 6.24 Depiction of the Fresnel lens transformation: (a) original coordinate system, (b) qTO-derived refractive index profile after flattening, (c) discretization of the refractive index, and (d) final zone plate lens. Reprinted with permission from Ref. 64, Copyright 2011, The Optical Society.

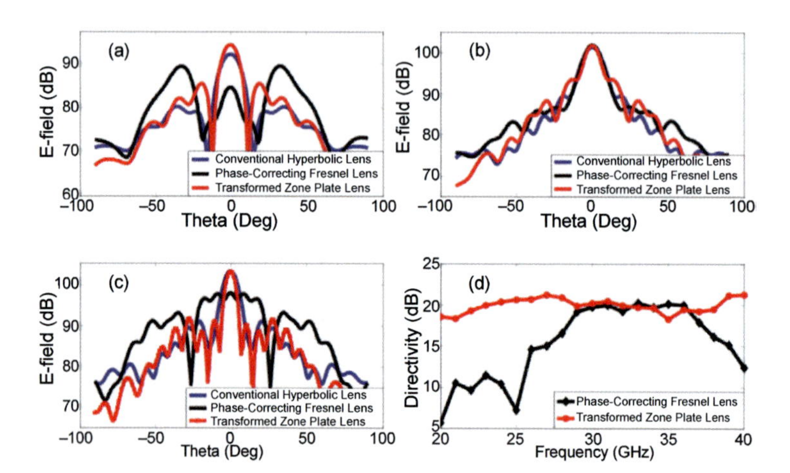

Figure 6.25 Far-field radiation patterns for hyperbolic, Fresnel, and zone plate lenses at (a) 20 GHz, (b) 30 GHz, (c) 40 GHz, and (d) directivity versus frequency. Reprinted with permission from Ref. 64, Copyright 2011, The Optical Society.

6.3.3 Broadband Directive Multibeam Lens Antennas

This section describes directive multibeam lenses that are not derived by explicitly transforming the geometry of a classical design. Instead, they are designed by relating the equi-phase wavefronts of theoretical sources with coordinate systems to effectively transform one source to another. Tichit et al. [57] described a coordinate transformation that changes an isotropically radiating source to an ultra-directive one. The appropriate form of the source transformations was given in Eq. 6.16. If a line source is oriented along z and radiating in a cylindrical space, this is accomplished by transforming the cylindrical equi-phase surfaces into planar surfaces, as shown in Fig. 6.26. An analytical transformation leads to anisotropic (though diagonal) material parameters, resulting in a design that is infeasible to fabricate. Another paper by Tichit [56] only considers the TM^z polarization, but the resulting design is magnetic and has permittivities less than one. This was physically realized through narrowband metamaterials.

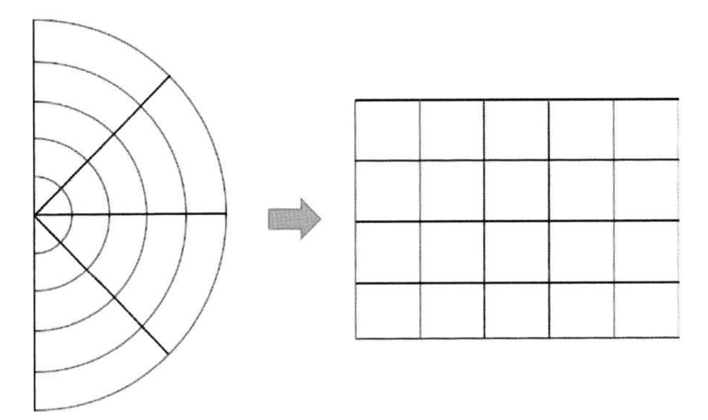

Figure 6.26 Coordinate transformation between an isotropic source and directional source.

Wu et al. [63] extended this basic concept to a multibeam lens antenna where there are multiple directive beams at different angles. Two designs are presented, one of which is homogeneous but anisotropic and can be realized by metamaterial inclusions, as detailed in Chapter 1. The other design is based on qTO and is inhomogeneous and dielectric only, providing a feasible path for a broadband and low-loss device. The design to be discussed will transform an isotropic source into one which exhibits quadbeam radiation. The coordinate transformation to realize this maps a circle of radius $R = 1$ into a square with side length $w = 1.4$, as shown in Fig. 6.27. The difference between this transformation and the previously described one for a single directive beam is that the circular equiphase surfaces are mapped into squares. This means that planar wavefronts will emerge from all four flat surfaces of the device, resulting in four directive beams. The mesh obtained through qTO is strongly orthogonal, which means anisotropy is minimized and this transformation can be physically realized to a large degree of accuracy with an isotropic medium, as discussed previously. The resulting refractive index varies from 1.0 to 1.9 and is shown in Fig. 6.28.

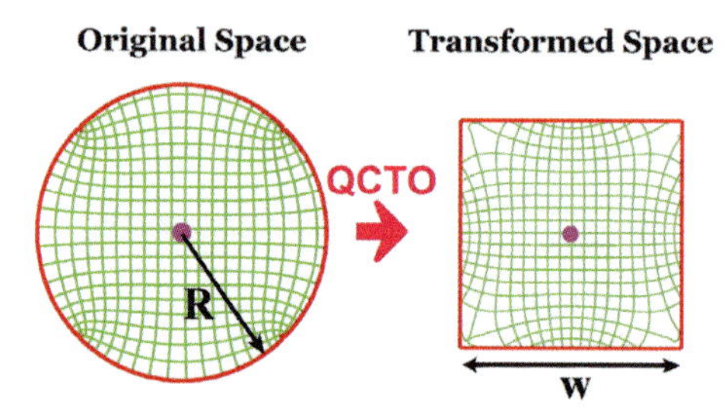

Figure 6.27 Coordinate transformation between an isotropic source and quadbeam antenna. Reprinted with permission from Ref. 63, Copyright 2013, IEEE.

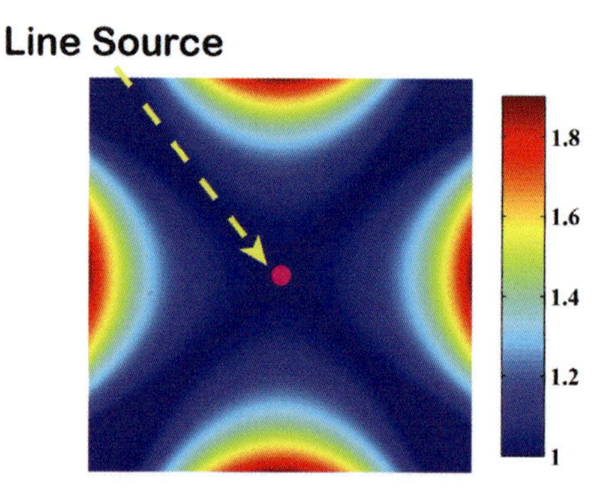

Figure 6.28 Refractive index distribution for quadbeam antenna. Reprinted with permission from Ref. 63, Copyright 2013, IEEE.

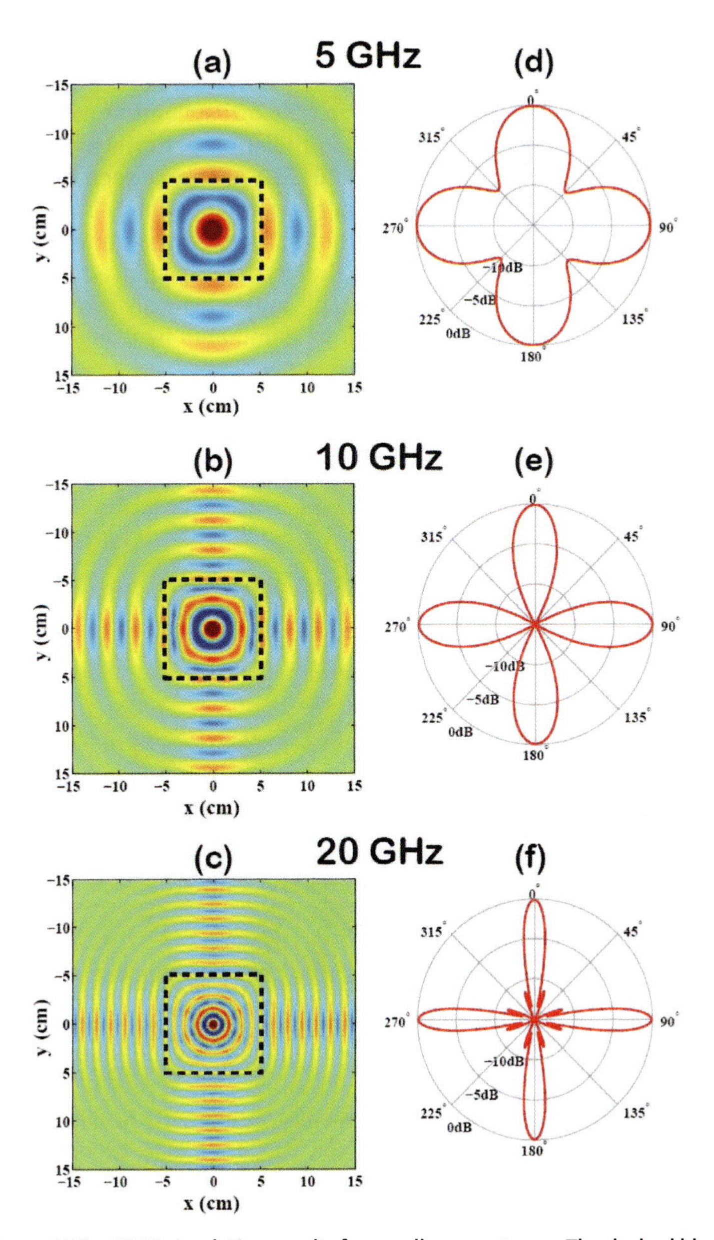

Figure 6.29 FDTD simulation results for quadbeam antenna. The dashed black lines indicate the boundaries of the lens. Snapshots of E_z are shown for (a) 5 GHz, (b) 10 GHz, and (c) 20 GHz. The corresponding far-field radiation patterns are also shown for (d) 5 GHz, (e) 10 GHz, and (f) 20 GHz. Reprinted with permission from Ref. 63, Copyright 2013, IEEE.

A 2D FDTD simulation was performed for a square lens measuring 10 cm on a side with an infinitely long current source in the center of the domain. The resulting electric field distribution at a given time and the corresponding far-field radiation patterns are shown in Fig. 6.29 for 5 GHz, 10 GHz, and 20 GHz. As can be seen, a higher frequency yields beams with larger directivity, which is a result of the changing electrical size of the radiator. As shown in (a)–(c), the wavefronts have small ripples due to the impedance mismatch between the device and free space. This most obviously manifests as sidelobes at 20 GHz (f). These can be eliminated by using the ARC design techniques discussed in Section 6.3.4.

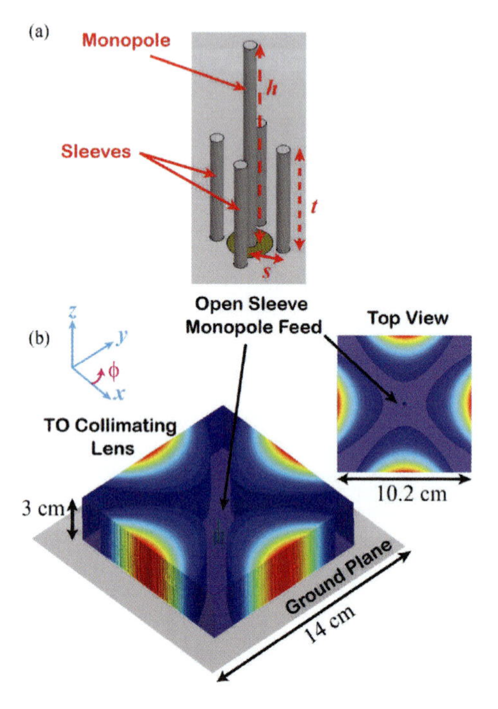

Figure 6.30 Design for more representative quadbeam lens antenna: (a) open-sleeve monopole feed and (b) dielectric index profile. Reprinted with permission from Ref. 63, Copyright 2013, IEEE.

A more representative simulation of the device as it would be physically fabricated was carried out using the finite-element solver in Ansys HFSS. To feed the antenna, a broadband open-sleeve monopole [24] that can be connected to a 50 Ω coaxial cable was used. The final design is shown in Fig. 6.30. The previous 2D design was extruded along the normal direction, and the design was discretized into homogeneous regions with curved boundaries. The resulting S_{11} is shown in Fig. 6.31. As can be seen, the monopole by itself shows a −10 dB bandwidth of 18%, while the open-sleeve monopole shows a −10 dB bandwidth of 53%. When the TO lens is added, the bandwidth is reduced slightly owing to a larger mismatch at the high end of the band. The radiation patterns are shown in Fig. 6.32. The open-sleeve monopole by itself shows an omnidirectional pattern in the H-plane and two main beams in the H-plane. When the lens is included, the H-plane patterns clearly show four highly directive beams with an increase in directivity as frequency increases. The enhancement in directivity is more than 6 dB at 7.5 GHz. Since the design was formulated in 2D assuming a certain polarization, the refractive index is constant along the z-axis, and therefore the H-plane patterns are only slightly perturbed by the presence of the lens. The beams with the lens tend to bend closer to the horizon, but the peak is closer to $\theta = 59° - 76°$ rather than $\theta = 90°$ as expected for perfectly planar wavefronts in the xz- and yz-planes. This is due to the fact that the original coordinate transformation was assumed to be 2D and because a finite ground plane was included in the simulation. To quantify this behavior more precisely, Fig. 6.33 shows the directivity versus frequency in both the H-plane and all of 3D space. The peak directivity in all of 3D space increases with frequency more or less monotonically, while the directivity in the H-plane has a significant ripple. This is due to the fact that the angle of the peak in the elevation plane changes with frequency. To remedy this situation, a 3D coordinate transformation could be employed, but the traditional qTO process requires a 2D transformation to yield an isotropic design for a given polarization.

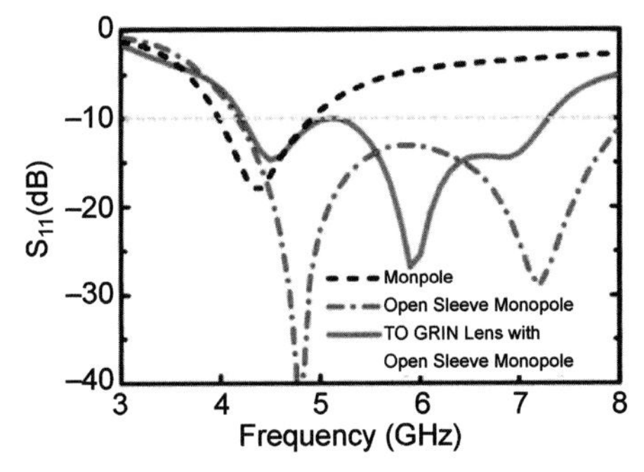

Figure 6.31 S_{11} versus frequency for a monopole, the open-sleeve monopole, and the lens fed by an open-sleeve monopole. Reprinted with permission from Ref. 63, Copyright 2013, IEEE.

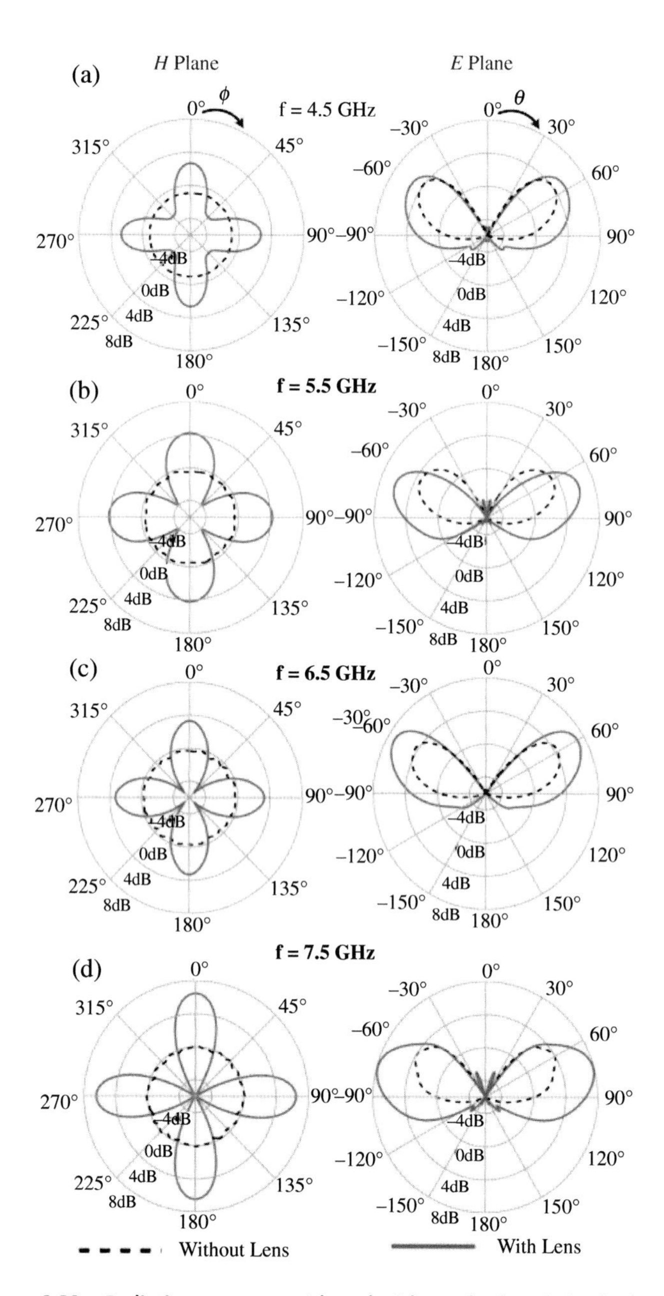

Figure 6.32 Radiation patterns with and without the lens in both the E- and H-planes at (a) 4.5 GHz, (b) 5.5 GHz, (c) 6.5 GHz, and (d) 7.5 GHz. Reprinted with permission from Ref. 63, Copyright 2013, IEEE.

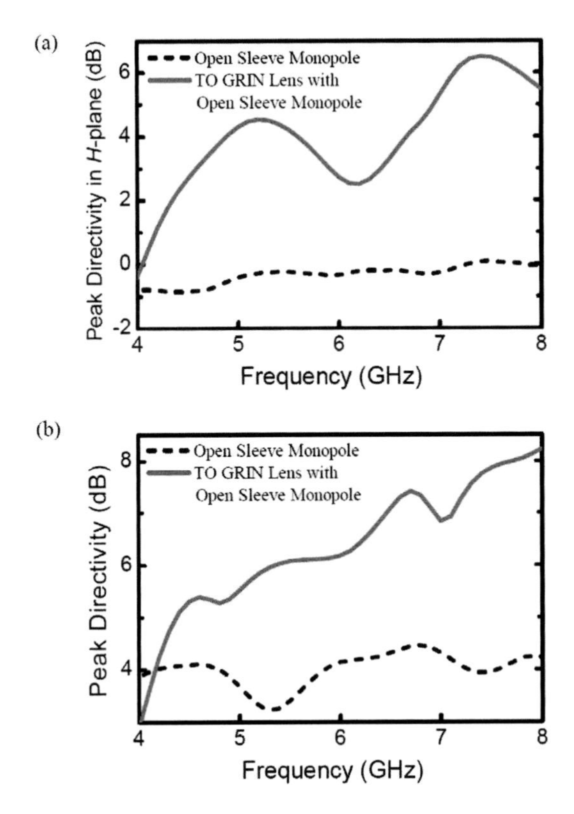

Figure 6.33 Directivity versus frequency in (a) the H-plane and (b) all of space. Reprinted with permission from Ref. 63, Copyright 2013, IEEE.

Another method of enhancing the directivity at $\theta = 90°$ is to replace the monopole and ground plane with a modified symmetric feed. Two symmetric feeds were explored: a dipole feed and a biconical feed. The S_{11} versus frequency response for both feeds with and without the lens is shown in Fig. 6.34. The dipole is extremely narrowband, exhibiting a similar response to the isolated monopole. Both the isolated biconical feed and the biconical feed with the lens show a very broadband response. Figure 6.35 displays the normalized radiation patterns of the qTO lens with biconical feed at various frequencies along the H- and E-planes. The H-plane cuts are similar to the open-sleeve monopole, but the E-plane cuts show a constant peak at $\theta = 90°$ over frequency. Finally, Fig. 6.36 shows the peak directivity versus frequency for the dipole and biconical feed, with and without the lens.

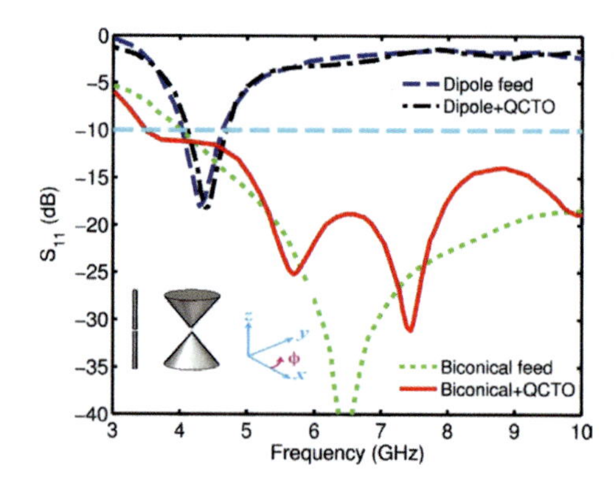

Figure 6.34 S_{11} versus frequency for a dipole and biconical feed, with and without the lens. Reprinted with permission from Ref. 63, Copyright 2013, IEEE.

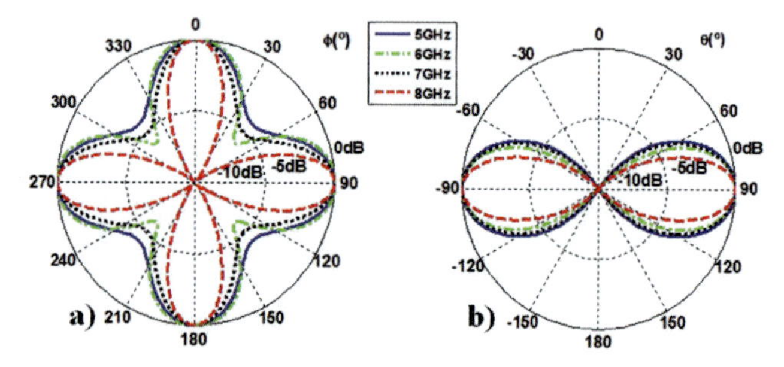

Figure 6.35 Radiation patterns of the lens and biconical feed in the (a) H-plane and (b) E-plane. Reprinted with permission from Ref. 63, Copyright 2013, IEEE.

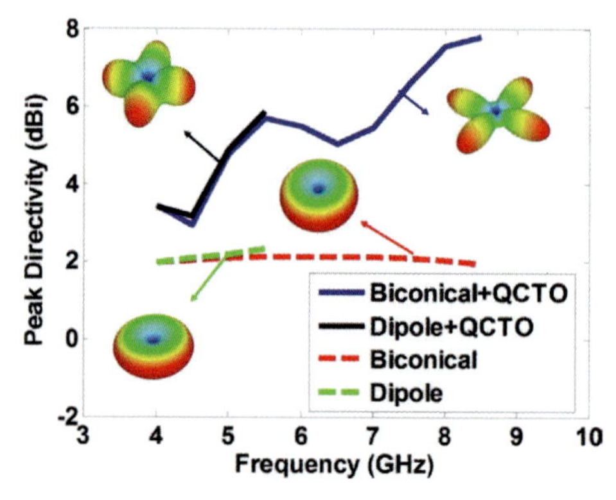

Figure 6.36 Directivity versus frequency in all of space for the dipole and biconical feed, with and without the lens. Reprinted with permission from Ref. 63, Copyright 2013, IEEE.

This section showcased the versatility of the qTO design procedure by including a source in the transformation. The resulting multibeam lens antenna is inhomogeneous and all-dielectric, resulting in good performance over a broad band.

6.3.4 Broadband qTO-Derived Anti-Reflective Coatings

The previous sections showed the potential of qTO-derived inhomogeneous metamaterials for improving traditional optical lenses and in developing multibeam sources. Unfortunately, due to the fact that there is a material discontinuity, there will inevitably be reflections and interference due to wave phenomena. Both classical lenses and lenses derived from coordinate transformations with discontinuities in the grid and Jacobian will exhibit reflections. Ray tracing software typically neglects reflected rays at surfaces and completely ignores wave interference, so a full-wave simulation will be used to fully model the effect of these reflections. In this chapter, a 2D finite element solver implemented in COMSOL will be used to simulate ARCs for GRIN lenses. While well-established for homogeneous lens systems [2, 47], ARC design is not fully understood for GRIN lens systems. By employing some of the qTO methodologies described previously, we can extend these classic ARC designs for use with GRIN lenses. One of the most popular and instructive examples is the quarter-wave coating shown in Fig. 6.37 [7]. By employing Fresnel's equations, we find that if the thickness of the ARC is $\lambda/4 = \lambda_0/(4n_1)$ for an incident wave at normal incidence, the wave reflected by the front surface will be exactly out of phase with the wave reflected by the back surface. For complete cancelation of the total reflected field, the in and out-of-phase reflected components need to have the same magnitudes. This condition is satisfied when $n_1 = \sqrt{n_0 n_s}$. The performance of this type of coating degrades for off-normal rays and becomes polarization dependent. Note also that due to the dispersive properties of materials, n_1 will vary with wavelength, and in addition, the optimal thickness is specified at one distinct wavelength, making this design very narrowband.

For common crown glass lenses (n = 1.52), the required refractive index for the quarter-wave coating is n = 1.23. This is a relatively low index that is difficult to achieve with optical materials. As a result, a common approach to lessen this restriction is to stack two quarter-wave coatings with a low-index coating first and a high-index coating second. The strong reflection from the first surface is canceled through two out-of-phase reflections. For crown glass, the required refractive indices are n = 1.38 for the low-index layer and n = 1.70 for the high-index layer. Broadband performance can be engineered into multilayer coatings. The design of these devices is analogous to traditional filter designs [2]. Some examples of quarter-wave multilayer coatings include the binomial design for maximally flat reflectivities and the Tschebyscheff design for an equi-ripple reflectivity over a wide bandwidth [2]. In addition to these traditional designs, GRIN- [7, 55] and nature-inspired coatings [12, 32] that are capable of wideband wide-angle performance have been developed. In this chapter, traditional ARC design techniques in conjunction with qTO will be employed for wideband performance [41]. For this study, a biconvex silicon $f/1$ micro-lens system with an aperture of $15\lambda_0$ and thickness of $2.14\lambda_0$ at a design wavelength λ_0 = 3.5 μm will be considered.

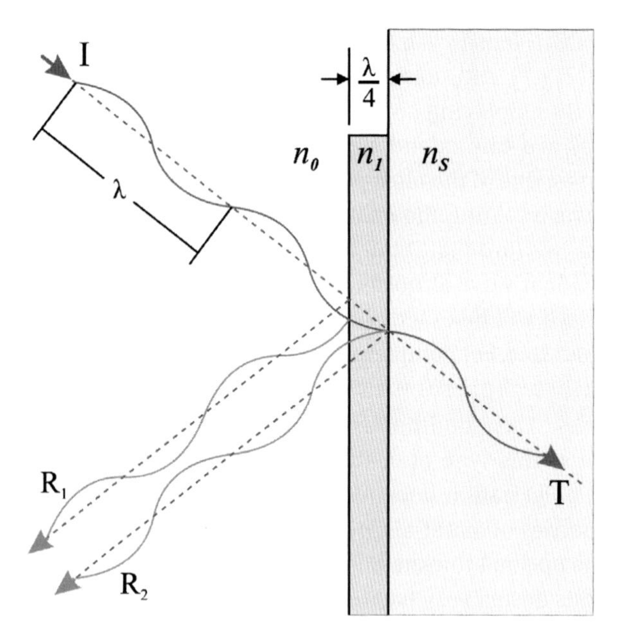

Figure 6.37 Quarter-wave anti-reflective coating. Image by Dr Bob from Wikimedia Commons. Used under CC-BY-SA 3.0. https://upload.wikimedia.org/wikipedia/commons/a/ad/Optical-coating-2.svg.

For a normally incident plane wave, the resulting normalized electric field distribution for the lens by itself is shown in Fig. 6.38a, while the lens with a quarter-wave coating is shown in Fig. 6.38b. The reflectance is reduced from 30% to almost 0% at the design wavelength. As shown in Fig. 6.39, the coating is extremely narrowband, with a reflectance < 2% bandwidth of only 29%.

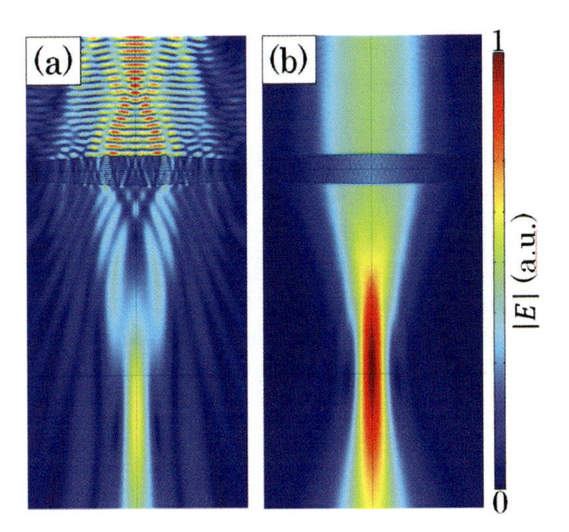

Figure 6.38 Electric field distributions for (a) the biconvex lens by itself and (b) the biconvex lens with a quarter-wave ARC. Reprinted with permission from Ref. 41, Copyright 2015, The Optical Society.

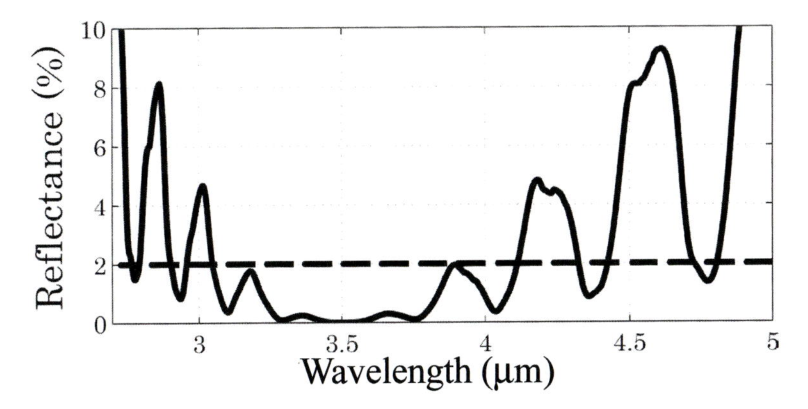

Figure 6.39 Reflectance versus wavelength for biconvex lens with a quarter-wave ARC. Reprinted with permission from Ref. 41, Copyright 2015, The Optical Society.

The biconvex lens is flattened using qTO, as shown in Fig. 6.40. Figure 6.41a shows that the flattened lens maintains the same focusing behavior as the biconvex lens, but the reflectance is increased to 50%. This is because the larger index of refraction leads to a larger impedance mismatch and higher reflections. It is not immediately obvious how to extend the traditional homogeneous quarter-wave ARC design to a GRIN lens since the refractive index varies over the front surface of the lens. The optimal refractive index and thickness were, therefore, determined through a parametric study. The optimal ARC index was found to be $n = 2.06$, which corresponds to $n = 4.24$, the mode of the refractive index profile over the front surface of the GRIN. The performance of the resulting system is depicted in Fig. 6.41b. This ARC again reduces the system reflectance to approximately 0% at the design wavelength, but, as with the biconvex lens system, this coating provides a narrow reflectance < 2% bandwidth of 24%, as shown by the blue line in Fig. 6.42.

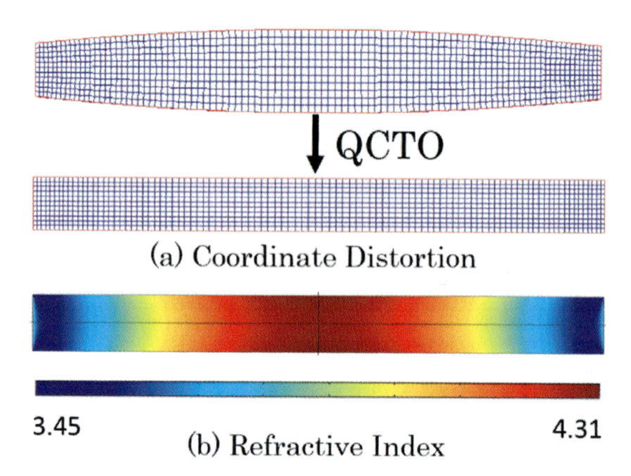

Figure 6.40 (a) Biconvex to flat transformation. (b) The resulting refractive index distribution. Reprinted with permission from Ref. 41, Copyright 2015, The Optical Society.

The homogeneous ARC for the GRIN lens performs slightly worse when compared with the ARC for the homogeneous biconvex lens. This is because the homogeneous ARC does not provide optimal matching across the entire gradient distribution of the lens interface. Hence, incorporating a GRIN distribution into the

ARC should enhance the performance of the system. To show this, a radial GRIN distribution is implemented with a refractive index corresponding to the geometric mean of the free-space index and the index gradient across the front surface of the lens. As shown by the red line in Fig. 6.42, this system's performance is nearly identical to the homogeneous ARC design.

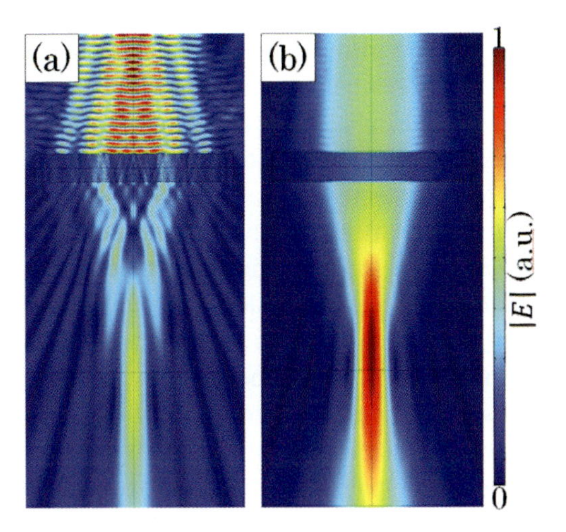

Figure 6.41 Electric field distributions for (a) the GRIN lens by itself and (b) the GRIN lens with a quarter-wave ARC. Reprinted with permission from Ref. 41, Copyright 2015, The Optical Society.

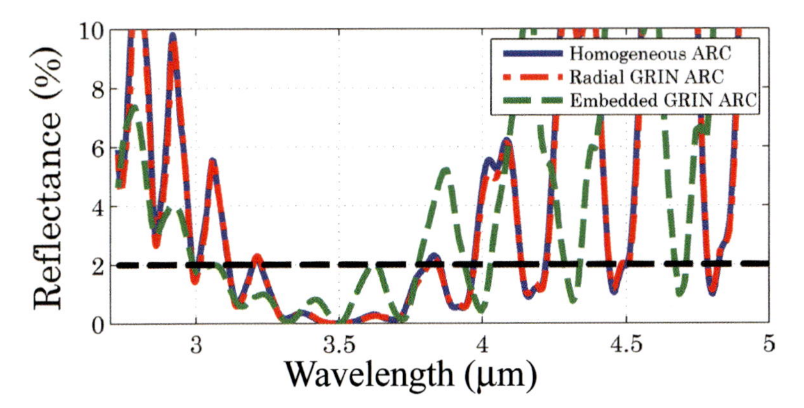

Figure 6.42 Reflectance versus wavelength comparison for homogeneous, radial GRIN, and embedded GRIN ARCs. Reprinted with permission from Ref. 41, Copyright 2015, The Optical Society.

A better idea is to first design a traditional quarter-wave ARC for the biconvex lens, then use qTO to flatten the homogeneous lens and the ARC concurrently, as shown in Fig. 6.43a. The resulting refractive index profiles are displayed in Fig. 6.43b, where the GRIN distributions corresponding to the lens and the ARC layers have been exploded to showcase the gradients more clearly. As before, this ARC is capable of achieving a reflectance of roughly 0% at the design wavelength. The primary performance distinction between this design and the previous implementations is its bandwidth, as shown by the green line in Fig. 6.42. The reflectance <2% bandwidth is approximately 24%, but the reflectance <4% bandwidth exceeds those of the homogeneous or radial ARC designs. The extra bandwidth is achieved at the expense of additional complexity in the ARC profile, with a refractive index that varies in both the radial and axial directions. Despite the modest improvement in bandwidth, the most promising feature of this approach is the fact that this methodology enables optical designers to streamline the rich field of filter theory and ARC design directly into the qTO process. Thus far, this approach has been used for single-layer coatings to show improved performance over a relatively narrow bandwidth. Next, these same design concepts will be extended to multilayer coatings for broadband operation.

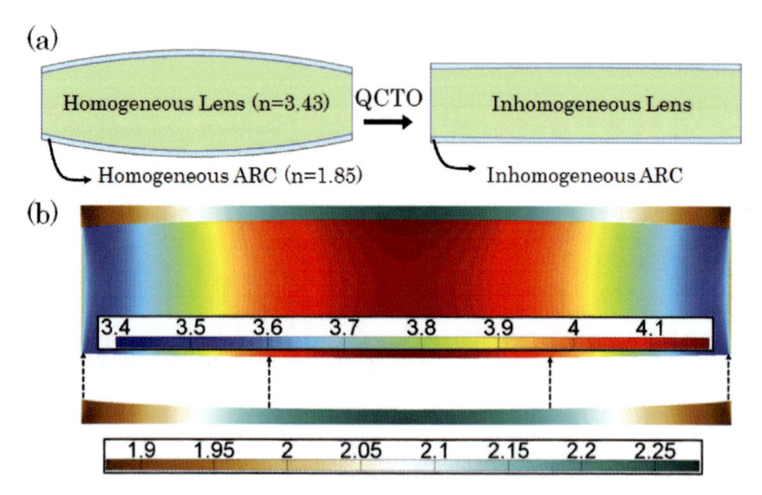

Figure 6.43 (a) qTO transformation for the lens-ARC system. (b) The resulting refractive index profile. Reprinted with permission from Ref. 41, Copyright 2015, The Optical Society.

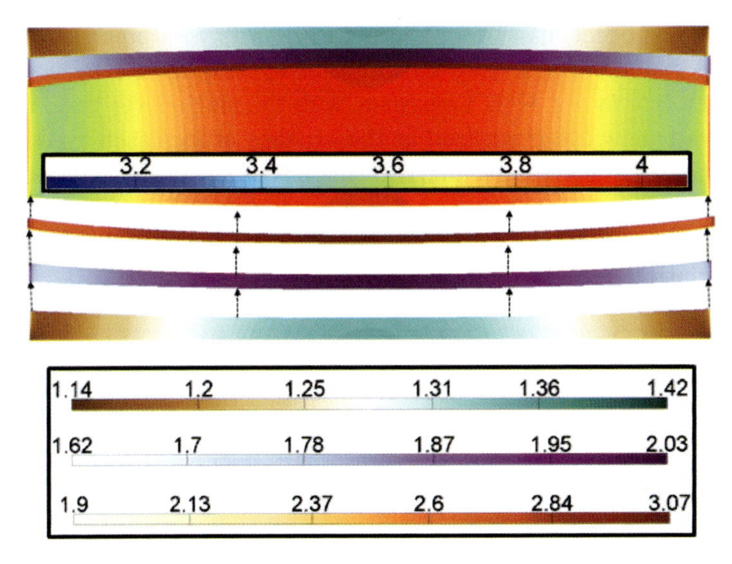

Figure 6.44 qTO transformation for the biconvex lens with a three-layer binomial ARC. Reprinted with permission from Ref. 41, Copyright 2015, The Optical Society.

To showcase the broadband potential of this design process, a three-layer homogeneous binomial ARC will now be considered. For comparison, a broadband ARC was first designed for the silicon biconvex lens, and the resulting reflectance versus wavelength is shown in Fig. 6.45. The bandwidth is increased from the single-layer case to 92%. For the GRIN lens, a binomial ARC was designed based on n = 4.24, the mode of the refractive index gradient over the front surface. The resulting ARC coatings are each quarter-wave thick and have refractive indices of 3.00, 1.87, and 1.17. As shown in Fig. 6.46, the multilayer homogeneous ARC applied to the GRIN lens achieves a bandwidth of 87%. As with the narrowband design, the GRIN lens system with homogeneous ARCs slightly underperforms the conventional biconvex lens system.

Now the embedded qTO approach will be used to transform the three-layer ARC and the biconvex lens concurrently. Using qTO, the lens system is flattened, resulting in the single GRIN region depicted in Fig. 6.44. As with the single-layer design, the ARC layers possess gradients in both the radial and axial directions. Each ARC layer is displayed in a separate color map to exaggerate the index distribution. This system outperforms both the original fully homogeneous system and the GRIN lens with homogeneous ARC system by achieving a bandwidth of 94%, as shown by the green line in Fig. 6.46. Consistent with the single-layer case, this extra bandwidth is realized at the expense of increased material composition complexity but with the advantage of flat interfaces. This section again showcased the tremendous potential of using qTO to improve optical devices, this time for broadband anti-reflective coatings.

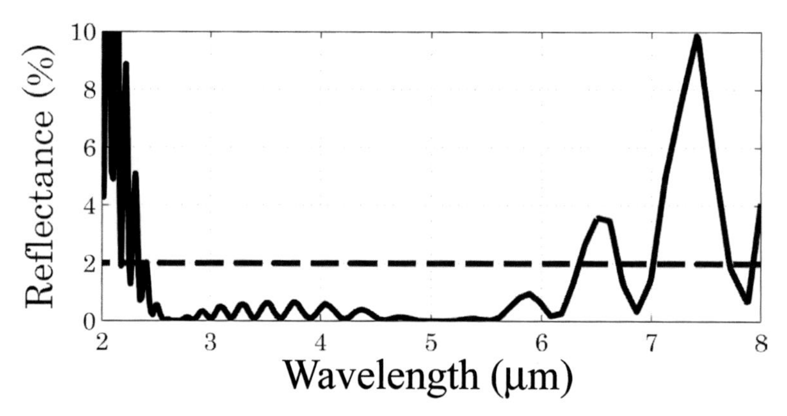

Figure 6.45 Reflectance versus wavelength for three-layer binomial ARC for homogeneous biconvex lens. Reprinted with permission from Ref. 41, Copyright 2015, The Optical Society.

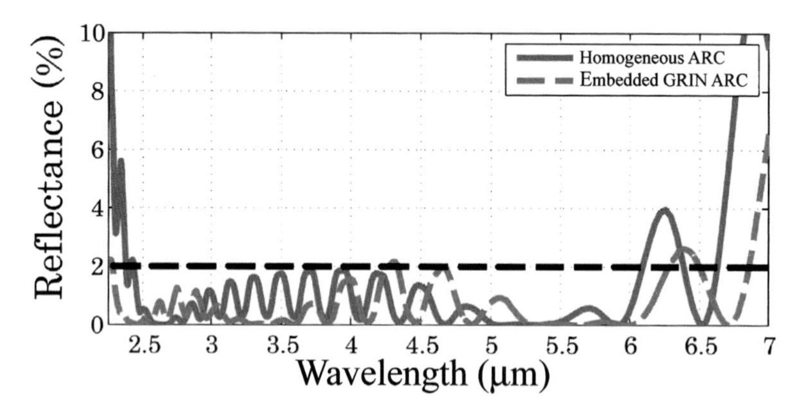

Figure 6.46 Reflectance versus wavelength comparison of the homogeneous and proposed embedded GRIN ARC designs. Reprinted with permission from Ref. 41, Copyright 2015, The Optical Society.

6.4 Wavefront Matching Method as an Alternative to qTO

In the previous sections, qTO was used with tremendous success to transform classical optics into inhomogeneous metamaterials, which analytically and experimentally showcased broadband performance at microwave frequencies. However, these devices likely would not perform as well at optical frequencies due to an increase in dispersion. Additional degrees of freedom are required to achieve chromatic correction, necessitating the need for doublets and triplets. This will be discussed in more detail in Section 6.5. Unfortunately, it is not straightforward to transform a cemented or air-spaced doublet or triplet into a continuously varying material. A direct application of qTO will lead to a refractive index distribution with a hard boundary. In a paper by Wang et al. on a Body of Revolution–Finite Difference Time Domain (BOR–FDTD) method [60], qTO was explicitly used to flatten a cemented doublet. The main focus of the paper is on the accuracy and efficiency of the simulation technique itself, and the example of the cemented doublet is used to showcase the robustness of the algorithm. Figure 6.47c shows the original cemented achromatic doublet, which consists of two lenses made of crown ($n_1 = 1.517$) and flint ($n_2 = 1.649$) glasses. This doublet is color corrected due to the different dispersion properties of the lenses. Figures 6.47a,b show the mesh grids created by qTO in the virtual and physical spaces, respectively. The resulting refractive index profile is shown in Fig. 6.47d. As can be seen, there is a hard interface in the profile, which will cause internal reflections. This could be mitigated through the use of the ARC techniques discussed in Section 6.3.4. For an incident TM-polarized plane wave in the $-\hat{z}$ direction (the optical axis), BOR–FDTD was employed to characterize the performance of these lenses.

Figures 6.48a,b show the electric field distribution in the *xz*-plane for the original cemented doublet and the qTO-derived flattened inhomogeneous metamaterial lens, respectively. The results are nearly identical, with the ideal focal point occurring at the same spot for both lenses. To further verify this, Figs. 6.48c,d show the electric field distributions on the *yz*-plane (the focal plane). Again, the results are nearly identical. This example shows the accuracy of the qTO process at a single frequency. However, the resulting flattened lens may not be color corrected depending on the dispersive properties of the materials considered. To mitigate this, a numerical optimization alternative to qTO called WFM will be presented, which can effectively transform any general geometry into a continuously varying inhomogeneous metamaterial. In addition, WFM is modular and can enable parallel optimization of a multi-element system. While this approach is extremely powerful and general, its performance depends on the optimization engine chosen. For this section, the Covariance Matrix Adaptation–Evolutionary Strategy (CMA–ES), a powerful global stochastic optimizer [15, 18], will be used. Section 6.5 introduces an approximate but analytical approach explicitly for color correction.

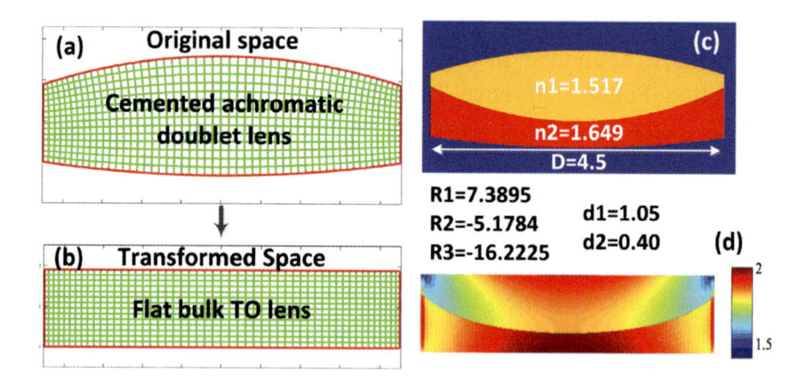

Figure 6.47 (a) The mesh grid of the achromatic homogeneous cemented doublet. (b) The mesh grid of the flattened qTO lens. (c) The geometry and refractive indices of the doublet. (d) The resulting qTO-derived refractive index profile of the GRIN lens. Reprinted with permission from Ref. 60, Copyright 2013, John Wiley and Sons.

The basic approach of qTO is to transform a given geometry into a more desired geometry that can be physically realized through an inhomogeneous material distribution. This is done using grid generation tools that mesh the input and output geometries into quadrilaterals and use the ratio of the areas to determine the refractive index in a given mesh cell. As a result, the Δn is governed by the transformation, and it cannot be easily minimized without introducing anisotropy. An alternative approach to using mesh cells is to represent the refractive index using a polynomial basis where the range can explicitly be restricted by constraining the coefficients appropriately. The most common GRIN distributions are radial, axial, and spherical. The first two can be easily represented in a polynomial using r^{2n} and z^n for $n = 1, 2, 3, \ldots$ where r is in the radial direction and z is the optical axis. The spherical GRIN distribution requires cross-terms to accurately represent in a polynomial basis. For generality, the refractive index profile in the desired geometry will be represented by:

$$\begin{aligned} n(r, z) &= n_0 + \alpha_1 r^2 + \alpha_2 r^4 + \alpha_3 z + \alpha_4 z^2 \\ &+ \alpha_5 r^2 z + \alpha_6 r^2 z^2 + \alpha_7 r^4 z + \alpha_8 r^4 z^2 \end{aligned} \tag{6.34}$$

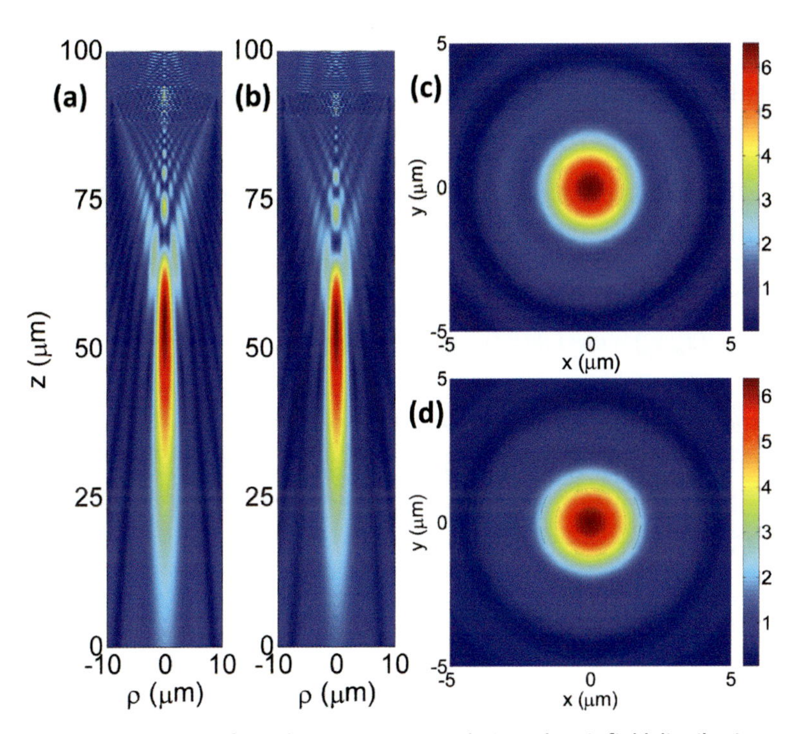

Figure 6.48 Results from the BOR–FDTD simulation. Electric field distributions in the *XZ*-plane for the (a) homogeneous doublet and (b) qTO-derived inhomogeneous metamaterial, and in the *XY*-plane (focal plane) for the (c) doublet and (d) GRIN lens. Reprinted with permission from Ref. 60, Copyright 2013, John Wiley and Sons.

Higher-order terms could be included, but this profile is general enough for the following examples. The WFM paradigm involves optimizing the α parameters of the GRIN lens to match the wavefront error (WFE) at the exit pupil of the original homogeneous optic. This allows the original lens to be effectively "transformed" since its behavior, including aberrations, will be reproduced exactly. The strategy essentially treats each optical element as a black-box transfer function, a claim that can be made more rigorously by relating the WFE with the optical transfer function (OTF). The WFE is the deviation of a wavefront from the desired spherical phase profile, which focuses perfectly to a single point. Figures 6.49a,b show examples of a non-perfect wavefront produced by a homogeneous doublet and a GRIN singlet with an as-of-yet unknown refractive index profile, respectively. The goal of WFM is optimizing the GRIN so these WFEs are the same. The cost function for this optimization will be the maximum difference in WFE, given by:

$$\text{cost} = \max_{P} \left[\left| W_{\text{orig}}(x_p, y_p) - W_{\text{GRIN}}(x_p, y_p) \right| \right] \tag{6.35}$$

In this expression, x_p and y_p are normalized pupil coordinates, W_{orig} and W_{GRIN} represent the WFE of the original and GRIN lens, respectively, and the max operator is over the exit pupil. Given a WFE W, the pupil function ϕ can be computed by:

$$\phi(x_p, y_p) = exp\,[jkW(x_p, y_p)] \tag{6.36}$$

Finally, the OTF H as a function of spatial frequencies f_x and f_y is expressed by an autocorrelation of the pupil function:

$$H(f_x, f_y) = \frac{\iint_{-1} \phi\left(x_p + \frac{f_x}{2}, y_p + \frac{f_y}{2}\right) \phi^*\left(x_p + \frac{f_x}{2}, y_p + \frac{f_y}{2}\right) dx_p dy_p}{\iint_{-1} |\phi(x_p, y_p)|^2 \, dx_p dy_p} \tag{6.37}$$

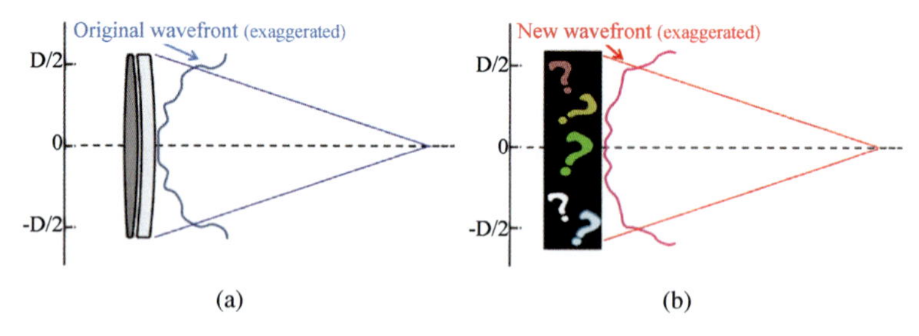

Figure 6.49 Example wavefront errors for (a) a homogeneous doublet and (b) a flat GRIN singlet.

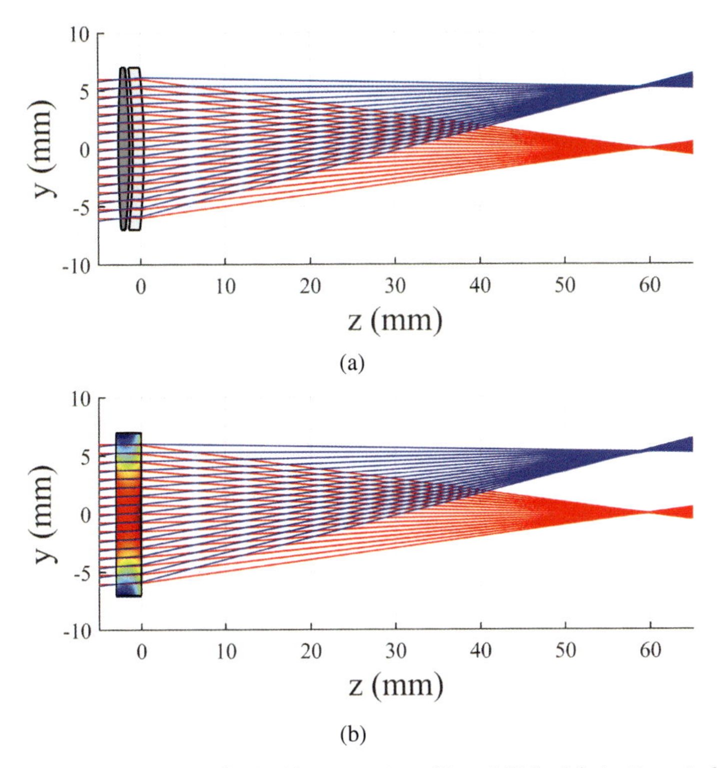

Figure 6.50 Ray traces for incidence angles of 0 and 5° for (a) the Fraunhofer doublet and (b) the GRIN replacement.

As a simple proof of concept, a Fraunhofer doublet [23] will be transformed using WFM. WFM for incidence angles of 0, 1, 2, and 5° was performed on the doublet to transform it to a flat inhomogeneous GRIN. Ray traces for incidence angles of 0 and 5 degrees for the homogenous doublet and the GRIN replacement are shown in Figs. 6.50a,b, respectively. Comparisons of the spot diagrams and WFE on the exit pupil are shown in Fig. 6.51. The agreement is excellent, showing that the flat GRIN lens can be used as a replacement for the homogeneous doublet. For brevity, only the spots and WFE for an incidence angle of 5° are shown, but the agreement is also very good for angles between 0 and 5. For this optimization, the full eight-term polynomial was used and the base index was chosen to be 1.8. For this optimization, the refractive index limits were unconstrained. As can be seen, this GRIN distribution is very complex with strong influence from the cross-terms. The major advantage of this approach over using qTO explicitly is that the refractive index profile is continuous, in contrast to the qTO-derived profile shown in Fig. 6.47. This leads to reduced reflections due to the lack of a hard boundary in the GRIN material itself.

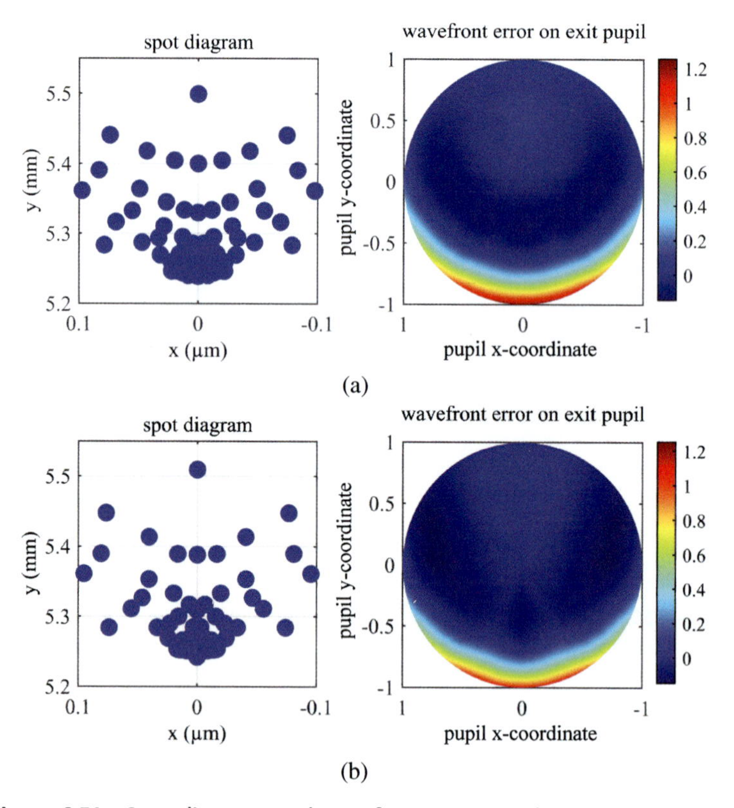

Figure 6.51 Spot diagrams and wavefront error on the exit pupil for an incidence angles of 5° for (a) the Fraunhofer doublet and (b) the GRIN replacement.

To verify the theoretical basis of the WFM method, the absolute value of the OTF of Eq. 6.37 was computed for both the homogeneous doublet and the GRIN replacement. This is called the modulation transfer function (MTF) and is a very important parameter in characterizing the behavior of an optical element [23]. Tangential and radial cuts of the MTF for four incidence angles are shown in Figs. 6.52a,b for the doublet and GRIN replacement, respectively. As can be seen, the MTF is nearly identical, verifying the one-to-one relationship between the WFE and the MTF.

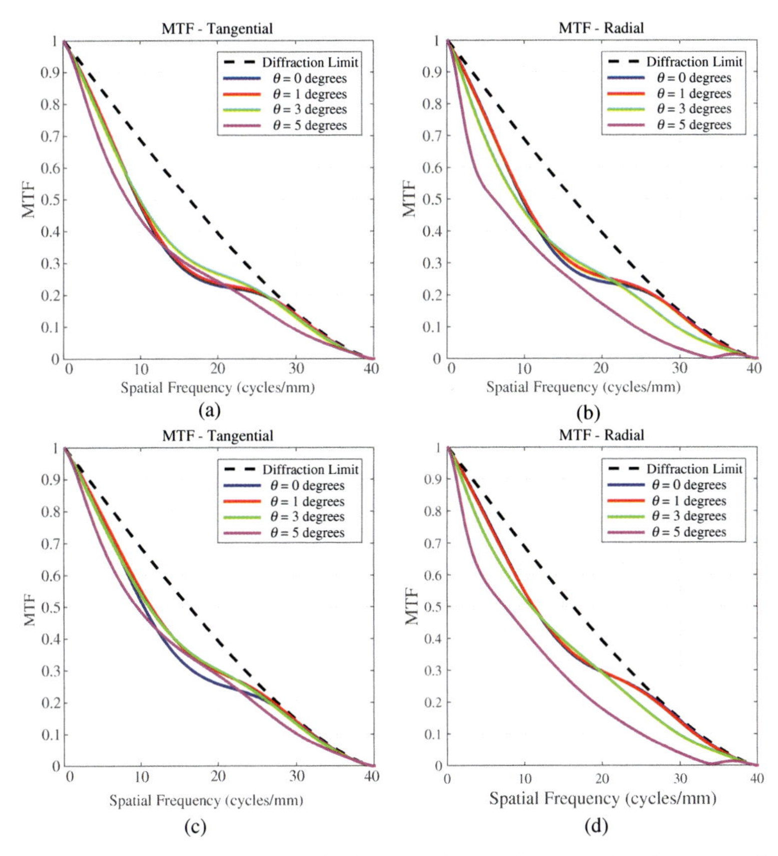

Figure 6.52 Modulation transfer functions for incidence angles of 0, 1, 3, and 5°. (a) Tangential and (b) radial cuts for the Fraunhofer doublet. (c) Tangential and (d) radial cuts for the GRIN replacement.

Another important advantage of WFM over TO is that it can be extended to include color correction. This can easily be done by extending the cost of Eq. 6.35 to include the WFE at multiple wavelengths. For this optimization, a doublet will again be matched, but this time the material will be constrained to be a chalcogenide [14] and the matching will be performed over three wavelengths in the mid-wave infrared, λ = 3 μm, 4 μm, and 5 μm. The original homogeneous doublet was designed using Si and Ge, two commonly used materials in this frequency range [8]. Because of the requirement for color correction, this GRIN replacement will have curved surfaces, a requirement that will be explained in Section

6.5. Ray traces for λ = 4 μm for three incidence angles, along with the spot diagram and OPD for λ = 4 μm and an incidence angle of 2° are shown in Figs. 6.53 and 6.54, again demonstrating very good agreement. One important application of the WFM idea is that it can be used to easily and efficiently create GRIN replacements for already existing multi-element homogeneous optical systems. Many optical systems include two or more homogeneous doublets, and WFM can be employed to replace each of these doublets with GRIN singlets, resulting in a system where optical alignment issues of closely spaced doublets will not be a problem. The major disadvantage of this method is that its effectiveness relies on the convergence properties of an optimization engine using the cost specified in Eq. 6.35, while qTO only relies on the numerical solution of Laplace's equation, mathematically a much more tractable problem. Fortunately, qTO can still be leveraged to understand the physics of color correction, as will be discussed in the next section.

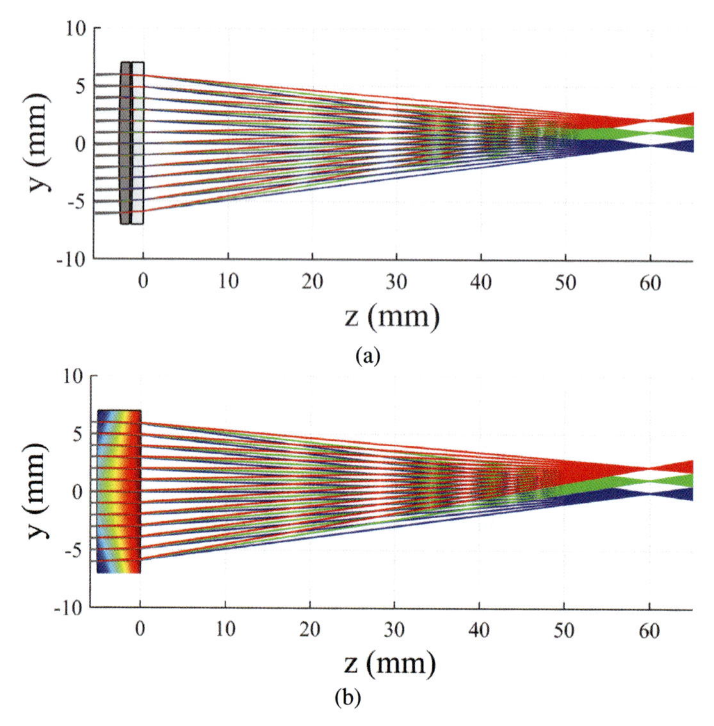

Figure 6.53 Ray traces for incidence angles of 0, 1, and 2° for (a) the Si–Ge doublet and (b) the GRIN replacement.

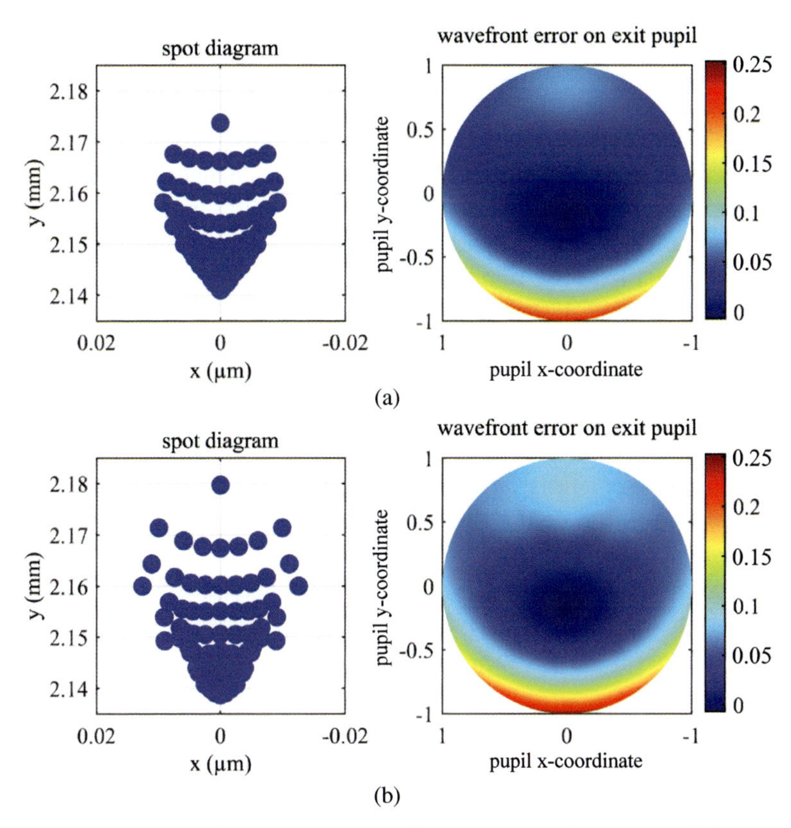

Figure 6.54 Spot diagrams and wavefront error on the exit pupil for an incidence angles of 2° for (a) the Si–Ge doublet and (b) the GRIN replacement.

6.5 Dispersion Correction in qTO-Enabled GRIN Lenses

Sections 6.3.1 and 6.3.2 showcased a variety of wide-angle lenses, while Section 6.3.4 demonstrated a new method of ARC design for achieving wideband performance in flattened GRIN lenses. By using qTO to enable these designs, they can be implemented with all-dielectric materials and without the need for anisotropic metamaterials, which rely on resonance phenomena to achieve their desired material parameters. While these designs have different types of broadband performance, they are both subject to the effects of underlying material dispersion, *i.e.*, materials with a wavelengthor frequency-dependent index of refraction. While dispersion in RF dielectric materials is often negligible, dispersive effects in optical and IR materials are more significant. Moreover, the performance of focusing optics is extremely sensitive to the changes in refractive index caused by dispersion. Often diffraction-limited focusing performance is required across the entire optical spectrum and over a specified FOV. While traditional wide-angle lenses inspired the qTO-enabled designs in Sections 6.3.1 and 6.3.2, qTO itself can also serve as the motivation for achieving chromatically corrected GRIN lenses. However, it is important to first understand some of the classical achromatic theory and designs.

To understand the impact dispersion has on the focusing performance of lenses, first consider a commonly used lens material like silicon (Si). A depiction of the dispersion behavior of Si [10] is presented in Fig. 6.55.

As can be seen, Si has more significant dispersion over the MWIR region than at the LWIR wavelengths. Now recall the equation for a thin lens of refractive index n in air [17]:

$$\frac{1}{f} = (n-1)\left[\frac{1}{R_f} - \frac{1}{R_b}\right], \tag{6.38}$$

where R_f and R_b are the front- and back-surface radii of curvature, respectively, and f is the focal length of the lens. As previously discussed, the lens index is wavelength dependent (*i.e.*, $n \rightarrow n(\lambda)$), and therefore, the lens focal length is also wavelength dependent (*i.e.*, $f \rightarrow f(\lambda)$) ultimately leading to a drift of the lens focus from the desired focal plane location. This behavior is commonly called chromatic aberration, and a depiction of the phenomenon is provided in Fig. 6.56.

Here the blue, green, and red rays correspond to the short, middle, and long wavelengths over the desired range. For a target focal plane location coincident with the middle wavelength focus, chromatic aberration will lead to a smearing of colors on the image plane. Lenses made of more dispersive materials will see more chromatic aberration than if they were made of less dispersive materials. The traditional metric used to quantify the amount of dispersion in a material is the Abbe number [23], which is defined as:

$$V = \frac{n_c - 1}{n_s - n_l}, \tag{6.39}$$

where c, s, and l subscripts on the refractive index n refer to the center, short, and long wavelengths, respectively. An additional measure of dispersion is δn, which is proportional to the slope of the material's dispersion curve:

$$\delta n = n_s - n_l \tag{6.40}$$

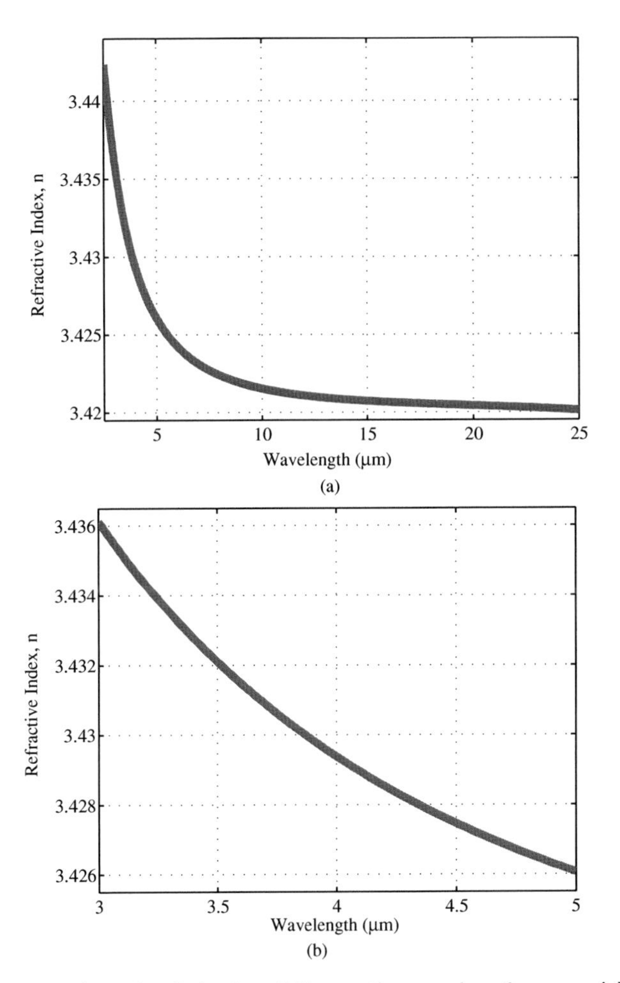

Figure 6.55 Dispersion behavior of Si over the wavelength ranges (a) LWIR 1–25 μm and (b) the MWIR range 3–5 μm.

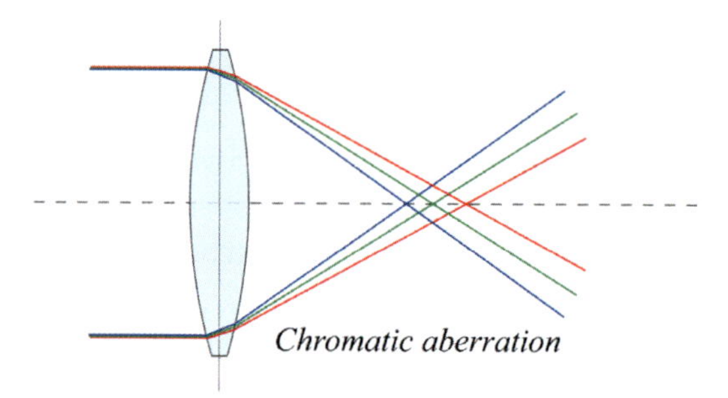

Figure 6.56 Chromatic aberration of homogeneous singlet. Image by Dr Bob from Wikimedia Commons. Used under CC-BY-SA 3.0. https://upload.wikimedia.org/wikipedia/en/a/aa/Chromatic_aberration_lens_diagram.svg.

This measure is called the effective material dispersion slope, and it is important when discussing color-correcting GRINs. Since the Abbe number is inversely proportional to the dispersion slope, a small Abbe number means high dispersion, while a large Abbe number means low dispersion. Consequently, the lens focal drift δf is inversely proportional to the Abbe number, as shown in Fig. 6.57.

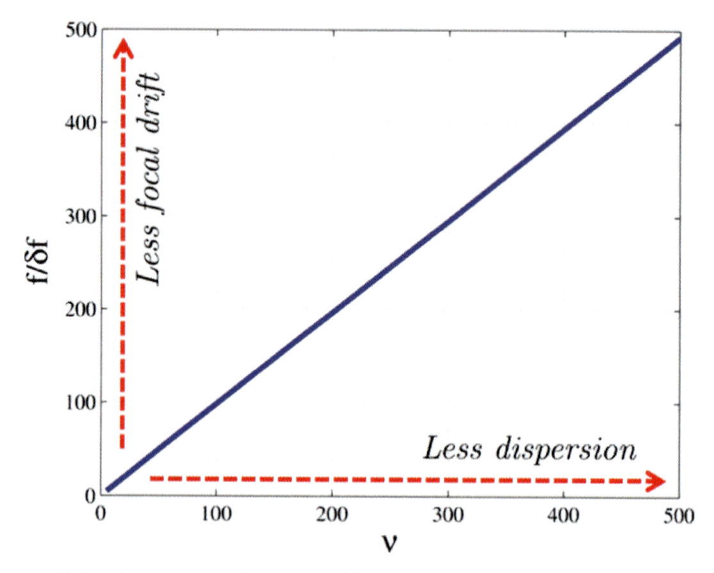

Figure 6.57 Lens focal drift versus Abbe number.

Because of this, there is a long history of homogeneous optical designs consisting of two to three lenses, which correct for chromatic aberration [9]. One of these traditional approaches is the so-called "achromatic doublet," which is a compound element made up of two lenses possessing different dispersion properties [23]. The requisite individual lens powers are the only unknowns in the design and can be found by solving the following system of equations:

$$\phi_1 + \phi_2 = \phi_{\text{Total}}$$

$$\frac{\phi_1}{\nu_1} + \frac{\phi_2}{\nu_2} = 0 \tag{6.41}$$

The total doublet power is simply the sum of the two lens powers, while the focal drift ($\delta f = f/\nu$) of the first lens cancels out that of the second, and vice versa. One of the earliest examples of an achromatic doublet is the Littrow doublet, in which one element is a concave negative-power element made of a flint glass (high refractive index and low Abbe number) and the other element is a convex positive-power element made of crown glass (low refractive index and large Abbe number). An example is shown in Fig. 6.58. While the achromatic doublet is a powerful design, for a given total lens power, the only degree of freedom (DOF) available is the choice of lens materials. Since there are no known optical materials with negative Abbe numbers ($\nu < 0$), when using homogeneous lenses, it is absolutely required to have more than one lens for color correction. Moreover, one of the lenses has to be diverging, which causes there to be excess power in the positive lens. This results in lenses that have large curvatures. However, the more different the dispersion properties are of the two materials, the more these curvatures can be relaxed. Furthermore, the performance of achromatic lenses can be increased by introducing more elements. A common three-element design is the apochromat lens, which can achieve even larger degrees of color correction [48]. These are traditionally triplets with spacing between each lens, as shown in Fig. 6.59.

While these lenses show significant improvement in color correction over a traditional singlet, there are strict tolerances on the spacings between the lenses, and these devices will experience significant reflections at each interface due to impedance mismatches. With a GRIN singlet, the spatial distribution of the refractive index introduces an additional DOF into the design process, which gives GRIN lenses their potential to correct for material dispersion [50,

51]. Power is provided by both refraction from the surface and from the GRIN distribution itself. In a similar way, the gradient itself possesses an effective Abbe number that can be radically different from the underlying materials that make up the GRIN and thus provide an avenue for correcting wavelength-dependent aberrations. As discussed in Section 6.4, by constructing a color-corrected GRIN lens, achromatic doublet pairs can be replaced by a single inhomogeneous lens, which could greatly reduce system complexity and cost. While achromatic GRIN lenses can be constructed with a variety of geometries and GRIN profiles, their solutions are extremely dependent on the underlying material properties, their dispersive behaviors in particular. An index distribution generated with qTO is typically more complex than simple radial- or axial- only GRIN lenses, but exploring the geometrical-optics solutions to these examples can form a theoretical foundation for color correction with GRIN lenses.

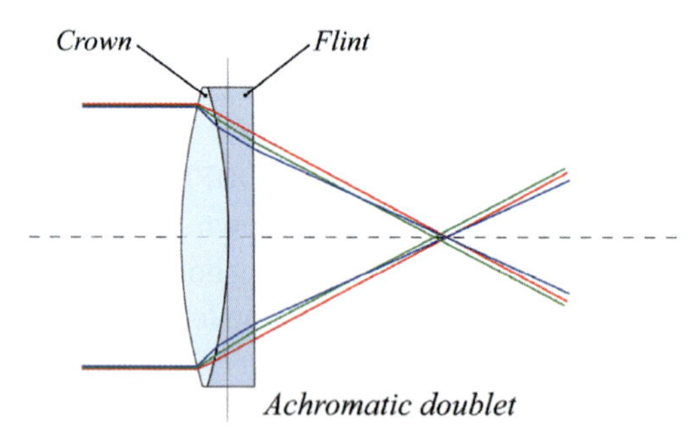

Figure 6.58 Achromatic doublet. Image by Dr Bob from Wikimedia Commons. Used under CC-BY-SA 3.0. https://upload.wikimedia.org/wikipedia/commons/1/15/Lens6b-en.svg.

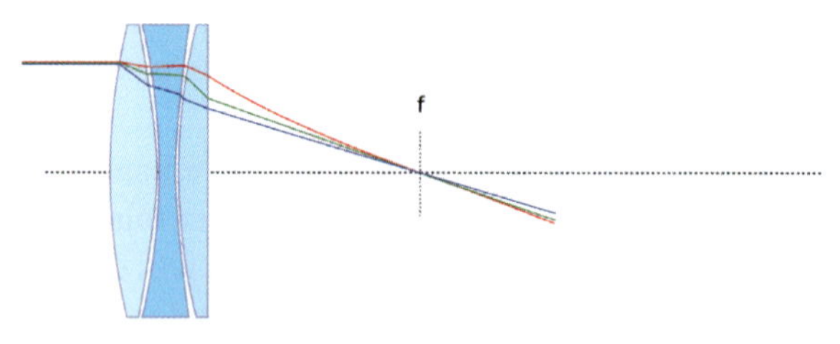

Figure 6.59 Apochromatic triplet. Image by Dr Bob from Wikimedia Commons. Used under CC-BY-SA 3.0. https://upload.wikimedia.org/wikipedia/commons/1/15/Lens6b-en.svg.

6.5.1 Geometrical-Optics Inspired Solution

Analytical expressions that describe the behavior of radial, axial, and spherical GRIN profiles have been extensively studied [4, 3, 50]. These expressions use paraxial approximations but provide insight into the physics of GRIN lenses. This section will describe the theory behind color correction for simple index profiles. In addition, the geometrical parameters to provide color correction for a given material will be provided and trade-offs between performance and complexity will be discussed.

6.5.1.1 Radial GRIN

The simplest GRIN profile is one which possess a quadratic variation in index along the radial direction. Consider such a radial GRIN whose index distribution is given by

$$n(r,z) = n_0 + \alpha_r \left(\frac{r}{D/2} \right)^2 \tag{6.42}$$

where α_r controls concavity and strength of the GRIN profile and D is the diameter of the lens. Since this index distribution has no axial dependence, a single value n_S can be used to describe the effective index on the front and back lens surfaces. The mean index can be used for simplicity and occurs at a radial position $\sqrt{1/3}D/2$. The mean index is then assigned to n_S. It follows then that the surfaces have an Abbe number defined in the usual way

$$V_S = \frac{n_{S_2} - 1}{n_{S_1} - n_{S_3}} \tag{6.43}$$

while the Abbe number of the GRIN is defined as [8]

$$V_G = \frac{\Delta n_2}{\Delta n_1 - \Delta n_3} \tag{6.44}$$

where the subscripts 1, 2, 3 refer to the index values at the short, center, and long wavelengths, respectively. By combining the powers and dispersions of the surfaces and GRIN, one can form a new aggregate lens system and can prescribe it to possess a desired effective power and dispersion. The achromatic system of equations for this lens is given as

$$\phi_S + \phi_G = \phi_T$$

$$\frac{\phi_S}{v_S} + \frac{\phi_G}{v_G} = \delta\phi_T \qquad (6.45)$$

which are similar to those of the traditional achromatic doublet (Eq. 6.41). In this convention, ϕ_T and $\delta\phi_T$ are the target (*i.e.*, total) system power and short-long wavelength focal drift, respectively, while ϕ_S and ϕ_G are the surface and GRIN powers at the central wavelength, respectively. Recall that the focal drift is inversely proportional to the target Abbe number $\delta\phi_T = \phi_T/v_T$. Solving this linear system of equations in the achromatic limit (*i.e.*, $\delta\phi_T \to 0$) yields solutions for the surface and GRIN powers:

$$\phi_S = \left(\frac{v_S}{v_S - v_G}\right)\phi_T, \quad \phi_G = \left(\frac{v_G}{v_G - v_S}\right)\phi_T \qquad (6.46)$$

The requisite surface power can then be found by

$$\phi_S = (n_{S_2} - 1)\left(\frac{1}{R_f} - \frac{1}{R_b}\right) = \left(\frac{v_S}{v_S - v_G}\right)\phi_T \qquad (6.47)$$

Next, by using the lens maker's equation, expressions for the individual surface radii of curvature can be found:

$$R_f = \left[\frac{1}{R_b} + \left(\frac{v_S}{v_S - v_G}\right)\frac{\phi_T}{n_{S_2} - 1}\right]^{-1}$$

$$R_b = \left[\frac{1}{R_f} - \left(\frac{v_S}{v_S - v_G}\right)\frac{\phi_T}{n_{S_2} - 1}\right]^{-1} \qquad (6.48)$$

Likewise, by substituting in an expression for the GRIN power [49], the lens thickness can be determined.

$$\phi_G = -2\alpha_r \frac{T}{(D/2)^2} = \left(\frac{v_G}{v_S - v_G}\right)\phi_T \qquad (6.49)$$

which results in a thickness T given by

$$T = \frac{1}{2|\alpha_r|}\left(\frac{v_G}{v_S - v_G}\right)\left(\frac{D}{2}\right)^2 \phi_T \qquad (6.50)$$

Note that while the thickness is explicitly defined, there is some flexibility in values for the surface radii since one can be specified and the other solved for. Geometries such as meniscus, biconvex, plano-convex, and convex-plano are all accessible in an achromatic radial GRIN. An axial GRIN, however, can offer a different range of solutions.

6.5.1.2 Axial GRIN

Consider, now, an axial GRIN lens whose index profile is given by

$$n(r, z) = n_0 + \alpha_z \left(\frac{z}{T}\right) \quad (6.51)$$

where n_0 is the background index and T is the lens thickness. An interesting property of the purely axial GRIN to note is that in the paraxial limit, the GRIN possesses no power itself (*i.e.*, $\phi_G \rightarrow 0$). This behavior is quite different from the radial GRIN. Also differing from the radial GRIN is that there are distinct index values and, therefore, Abbe numbers on the front and back lens surfaces. The achromatic system of equations for the axial GRIN lens is then

$$\phi_f + \phi_b = \phi_T$$

$$\frac{\phi_f}{v_f} + \frac{\phi_b}{v_b} = \delta\phi_T \quad (6.52)$$

where $\phi_{f,b}$ and $v_{f,b}$ are the front- and back-surface powers and Abbe numbers, respectively. Again, in the achromatic limit ($\delta\phi_T \rightarrow 0$), the solution is of the form

$$\phi_f = \left(\frac{v_f}{v_f - v_b}\right)\phi_T, \quad \phi_b = \left(\frac{v_b}{v_b - v_f}\right)\phi_T \quad (6.53)$$

Next, the front- and back-surface radii of curvature can be solved directly.

$$R_f = \left(\frac{n_{f_2} - 1}{\phi_T}\right)\left(\frac{v_f - v_b}{v_f}\right), \quad R_b = \left(\frac{1 - n_{b_2}}{\phi_T}\right)\left(\frac{v_b - v_f}{v_b}\right) \quad (6.54)$$

Unlike in the radial regime, there are no explicit requirements on the lens thickness for axial GRINS. However, both surface curvatures are precisely prescribed. Recall that the opposite was true for the purely radial regime. On the other hand, a radial-axial profile can combine the features of these different GRIN types to expose a wider range of solutions and geometrical trade-offs.

6.5.1.3 Radial-axial GRIN

A GRIN profile that combines radial and axial profiles is, to first order, an approximation to the spherical GRIN [11], which has seen significant interest recently. The contributions of these components are approximately orthogonal and are simply summed together as

$$n(r, z) = n_0 + \alpha_r \left(\frac{r}{D/2}\right)^2 + \alpha_z \left(\frac{z}{T}\right) \tag{6.55}$$

with the condition that $|\alpha_r| + |\alpha_z| = \Delta n$. Front- and back-surface indices are evaluated at axial distances $z = 0$, T, respectively, and at a radial location $\sqrt{1/3}\,D/2$ while their corresponding Abbe numbers are calculated in the usual way. With this GRIN choice, there are unique values for the front- and back-surface index and Abbe numbers while offering non-zero power from the radial portion of the GRIN, which is exactly the combination of properties that was desired. By specifying, in advance, the lens thickness and radial/axial ratio ($|\alpha_r|/|\alpha_z|$), the GRIN power (ϕ_G) and Abbe number (ν_G) are then both constants and can be moved to the right-hand side of the standard system of equations. These equations are then written as

$$\begin{gathered}\phi_f + \phi_b = \phi_T - \phi_G \\ \frac{\phi_f}{\nu_f} + \frac{\phi_b}{\nu_b} = \delta\phi_T - \frac{\phi_G}{\nu_G}\end{gathered} \tag{6.56}$$

which, in the achromatic limit ($\delta\phi_T \to 0$), possess a solution of the form

$$\begin{aligned}\phi_f &= \left(\frac{\nu_f}{\nu_f - \nu_b}\right)(\phi_T - \phi_G) + \left(\frac{\nu_f \nu_b}{\nu_f - \nu_b}\right)\frac{\phi_G}{\nu_G} \\ \phi_b &= \left(\frac{\nu_b}{\nu_b - \nu_f}\right)(\phi_T - \phi_G) + \left(\frac{\nu_f \nu_b}{\nu_b - \nu_f}\right)\frac{\phi_G}{\nu_G}\end{aligned} \tag{6.57}$$

Next, solving for expressions of the surface radii yields

$$R_f = \left(\frac{n_{f_2} - 1}{\phi_T + (v_b / v_G - 1)\phi_G}\right)\left(\frac{v_f - v_b}{v_f}\right)$$

$$R_b = \left(\frac{1 - n_{b_2}}{\phi_T + (v_f / v_G - 1)\phi_G}\right)\left(\frac{v_b - v_f}{v_b}\right) \tag{6.58}$$

which in the zero GRIN-power limit ($\phi_G \to 0$) reduce to the expressions for the axial GRIN (1.54). Since the GRIN power ϕ_G is linearly dependent on the lens thickness (1.49), the front- and back-surface powers are also thickness dependent. By moving away from the purely radial and axial limits, there exists a wider solution and trade-off space, which is the desired result of the radial-axial GRIN analysis.

6.5.1.4 Geometrical trade-offs

A common material pairing for achromatic singlets in the IR regime, Si and Ge are ideal candidates for analysis due to their desirable properties [5]. When looking over the mid-IR (MIR) regime (3–5 μm), Si and Ge possess refractive index values of 3.4253 and 4.0249 at a wavelength of 4 μm and Abbe numbers of 236.5 and 107.5, respectively, thus facilitating a range of potential designs. The index distribution of the mixture is governed by volume-filling-fraction mixing rules of the form $n_{\mathrm{Mix}} = fn_H + (1 - f)n_L$, where n_H and n_L refer to the high- and low-index materials, respectively. With the analytical expressions developed in Section 6.5.1.3, one can analyze the system for potential solutions. A summary of this is given in Fig. 6.60, where the sum of the absolute values of the surface radii (Fig. 6.60a) and inverse sum of the surface curvatures (Fig. 6.60b) relative to the lens diameter are plotted as a function of the relative thickness and ratio of radial/axial contribution to the GRIN and shown on a $\log_{10}$ scale to highlight the dynamic range of values. Ratios of 1,0 mean the GRIN is purely radial or purely axial, respectively, while positive and negative ratios indicate that the GRIN is converging ($\phi_G > 0$) or diverging ($\phi_G < 0$) in accordance with (1.55). All cases utilize the full Δn range of the material system and are assigned a focal length corresponding to f/2.5 to stress the validity of these expressions.

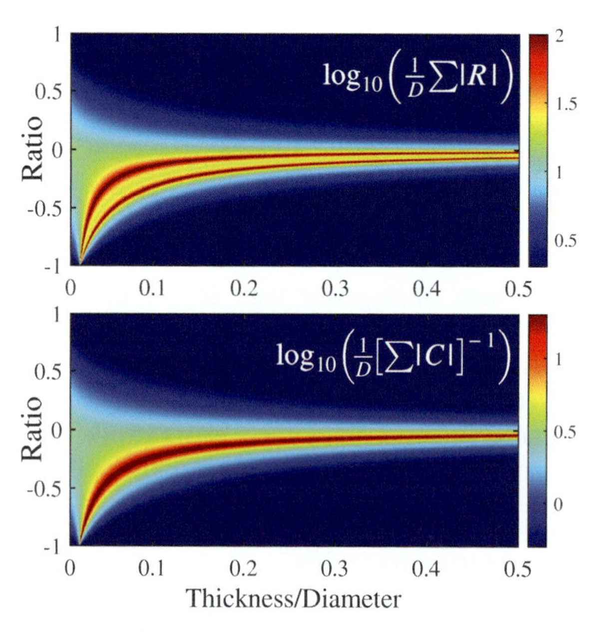

Figure 6.60 Lens surface curvature as a function of the GRIN ratio and lens thickness for a *f*/2.5 Si–Ge achromatic radial-axial singlet: (top) total surface radii and (bottom) total surface curvature.

As can be seen in Fig. 6.60, the trade-offs between thickness, surface curvature, and GRIN type (*i.e.*, radial, axial, or radial-axial) are easy to distinguish. Furthermore, the majority of potential solutions exist in the negative ratio half-space. In Fig. 6.60a, values of 2 (dark red) and 0.3 (dark blue), on the $\log_{10}$ scale, correspond to solutions with total surface radii that are 100 and 2 times the lens diameter, respectively. These solutions exist in the negative half-space due to the necessity to balance out the positive dispersion of the Si–Ge mixture (ν_G = 33.5), while a GRIN system with a negative Abbe number ($\nu_G < 0$) would have solutions existing in the positive half-space. The two dark red "branches" correspond to half-plano geometries where the surfaces transit from positive to negative curvatures (or vice versa) with the upper and lower branches belonging to the front and back surfaces, respectively. Meanwhile, Fig. 6.60b shows the effective "flatness" of the lens where the flattest solutions exist in-between the two branches seen in Fig. 6.60a. An interesting feature in the solution-space is the purely radial crossing point at $T/D \approx 0.02$ (*i.e.*, a 50:1 aspect ratio), which corresponds to the requisite thickness (1.50) of radial GRIN lenses. In general, moving away from the purely radial solution and into the radial-axial (*i.e.*, spherical) regime results in an increase in lens thickness in order to maintain a solution without radical curvature. While this potentially limits the range of geometries achievable, not all lens thicknesses nor GRIN profiles may be realizable from a manufacturing standpoint. The purely axial solution is also accessible for all lens thicknesses and is realizable due to the large background indices of Si and Ge.

While the geometrical-optics analysis of achromatic GRIN singlets was simple for the radial, axial, and radial-axial (*i.e.*, ≈ spherical) regimes, analysis for higher-order profiles such as those generated by qTO would be much more involved. However, the takeaway is clear: the dispersive properties of the surfaces and GRIN profile need to balance out in order to yield the desired achromatic performance. Moreover, this outcome can also be seen when examining the problem from a qTO-inspired perspective.

6.5.2 Transformation-Optics Inspired Solution

As discussed in Section 6.4, qTO cannot explicitly be used to develop a color-corrected lens in the optical regime. However, it can be used as a thought experiment to inspire a methodology for generating a color-correcting GRIN lens. When coupled with the geometrical-optics solutions developed above, a powerful classically inspired set of design rules can be developed.

Consider a homogeneous lens that introduces zero chromatic aberration to an incident ray bundle. In order for the lens to behave like an achromat, it would have to either be made of a non-dispersive material ($\nu = \infty$) or have a wavelength-dependent geometry. The former condition is the trivial solution, while the latter would require an adaptive wavelength-dependent geometry. Obviously, such a design is physically intractable when the lens needs to simultaneously operate over a wide wavelength range. However, as has been discussed, qTO provides the framework for getting one geometry to "behave" like another, at least in an electromagnetic sense. With qTO, an achromatic lens can be achieved then by transforming the requisite lens geometry at each wavelength to a single desired shape, which is then realized through a resulting wavelength-dependent inhomogeneous (*i.e.*, GRIN) refractive index distribution.

The required curvatures for such a homogeneous lens are shown in Fig. 6.61a at three different wavelengths: short, center, and long. If these proposed geometries are considered as existing in the “virtual space,” qTO can be applied to transform them to a single geometry that exists in the “physical space.” In practice, the geometries at the short and long wavelengths are transformed back to the geometry at the center wavelength and the desired behaviors at these wavelengths are embedded in a spatially varying refractive index distribution as seen in Fig. 6.61b. If we assume a target geometry of a homogeneous biconvex at the center wavelength and a background material with normal dispersion (*i.e.*, $dn/d\lambda < 0$), then the GRIN needs to be slightly diverging at the short wavelength to counteract the excess power of the surface attributed to the higher base material index of refraction. Similarly, a slightly converging profile at the long wavelength is necessary to increase the optical surface power to maintain the desired total power over all three wavelengths. This correction is achieved with a finite refractive material index range over which there is also a finite amount of material dispersion. The dispersion curves for the min, mid, and max refractive index values in the lens are shown in Fig. 6.62 and compared to the ideal non-dispersive case. These dispersion curves have Abbe numbers of $\nu = 30$, $\nu = 50$, $\nu = 100$, and $\nu = \infty$, respectively. Note that this is only a hypothetical material system that has its min and max index dispersion lines all pass through the same value at the center wavelength, a phenomenon not possible with conventional materials. However, this thought experiment captures the physics required for color correction with GRIN media: A spatially varying Abbe number profile is required to compensate for the underlying dispersions of the background material and lens surfaces.

To illustrate the potential of a qTO-inspired achromatic GRIN singlet, consider the flattened biconvex lens shown in Fig. 6.63. For this example, a fictitious GRIN material is considered with various assumptions for its underlying dispersion properties. The two cases considered here are that of a homogeneous Abbe number distribution and that of a material with parallel dispersion lines (*i.e.*, δn = constant). The former case is the condition for a homogeneous lens and the latter guarantees that the magnitude of the gradient Δn is unchanging with wavelength. Depictions of these two cases are given in Fig. 6.64 and ray trace results are summarized in Table 6.1.

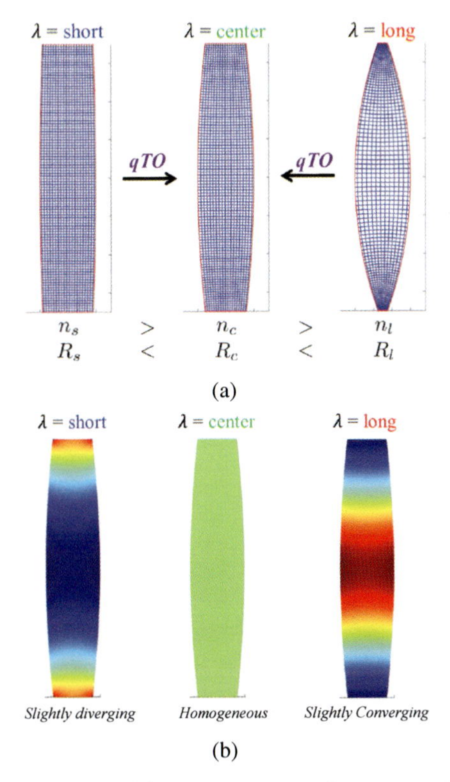

Figure 6.61 Depiction of (a) geometric transformation via qTO and (b) resulting color-correcting GRIN profiles.

Table 6.1 Chromatic performance of two dispersion cases

Dispersion case	RMS spot diameter	$f/\delta f$
Constant Abbe	85 μm	50
Constant δn	1 μm	10,000

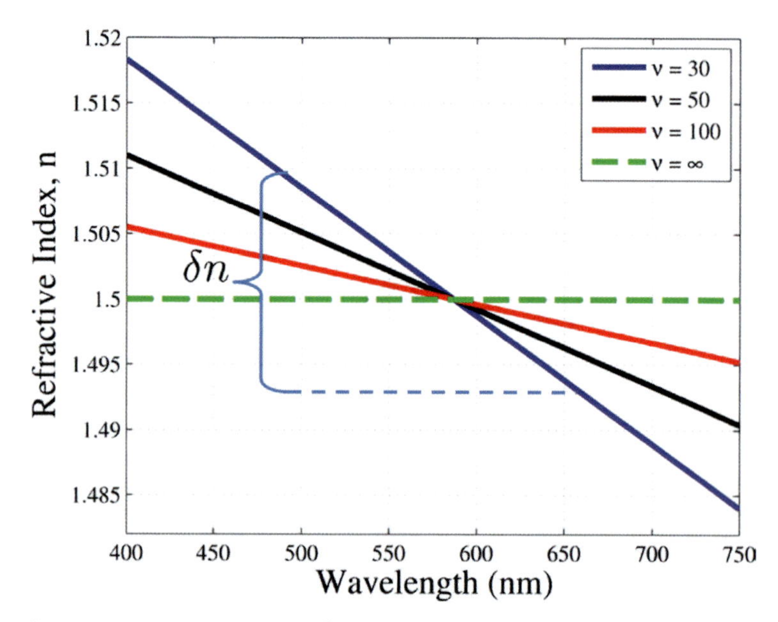

Figure 6.62 Dispersion curves for materials of different Abbe numbers.

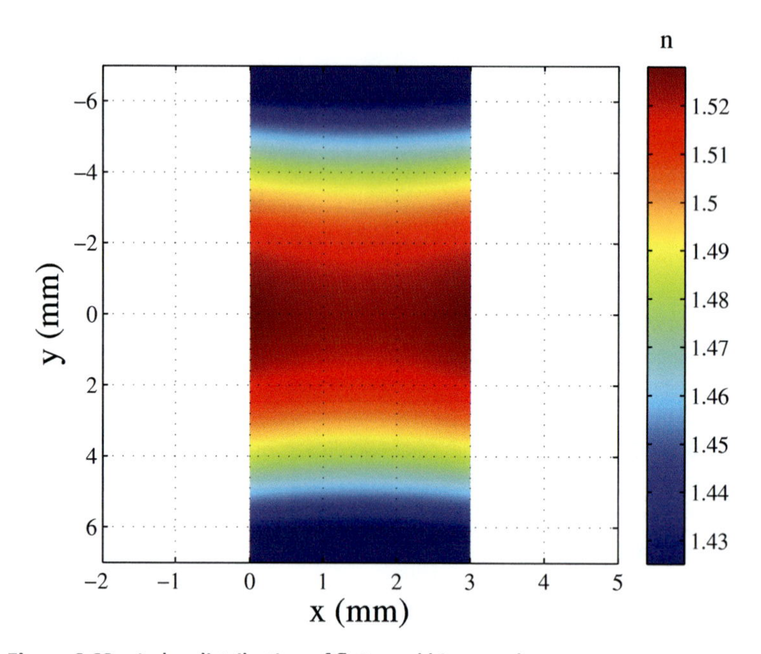

Figure 6.63 Index distribution of flattened biconvex lens.

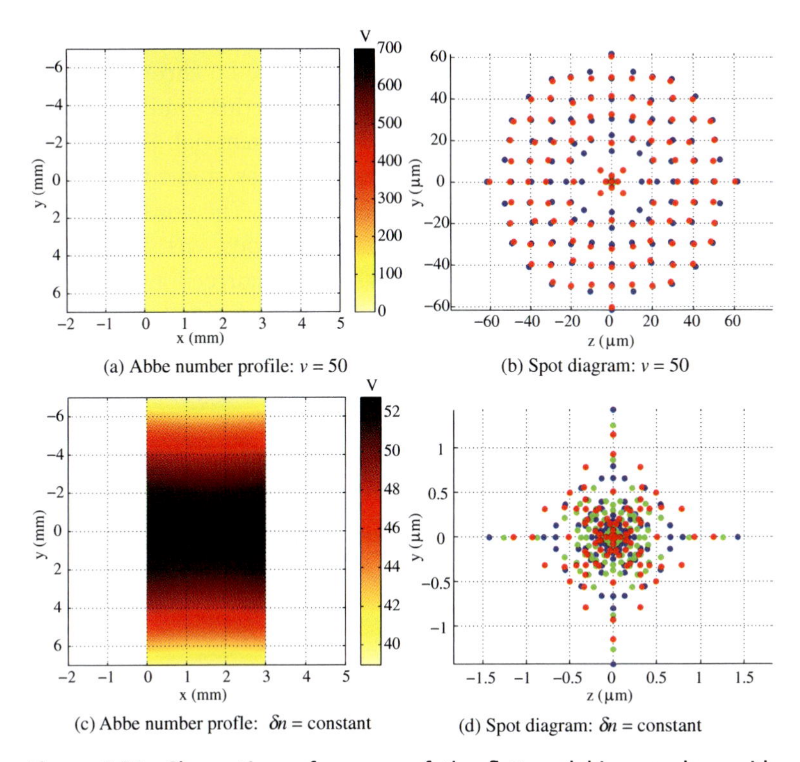

(a) Abbe number profile: $v = 50$

(b) Spot diagram: $v = 50$

(c) Abbe number profle: δn = constant

(d) Spot diagram: δn = constant

Figure 6.64 Chromatic performance of the flattened biconvex lens with dispersion assumptions of (a, b) constant Abbe number and (c, d) parallel dispersion lines.

The constant Abbe number case had a value of $v = 50$, while the constant δn case possesses a spatially varying Abbe number with a median value $v = 50$. The lens was chosen to have $f/5$ performance and illuminated with the typical F, d, and C (486.13 nm, 587.56 nm, 656.27 nm) spectral lines. Figure 6.64b shows that the constant Abbe number case is diffraction limited at the center wavelength (the green dots in the spot diagram) but shows significant chromatic aberration at the other two wavelengths leading to focal drift of $f/\delta f = 50$, which is what the equivalent homogeneous lens would possess. Meanwhile, the spot diagram of the constant δn case (see Fig. 6.64d) shows that all three wavelengths are diffraction limited, and thus chromatic aberration has been nearly eliminated in the lens. In fact, the lens achieves a $f/\delta f$ value of 10,000 due to its ideal dispersion behavior. Because $\Delta n(\lambda)$ is constant, the GRIN power is unchanging with wavelength and there will be no chromatic focal drift in the flat lens. What is interesting is that this condition exists regardless of the Abbe number of the base material. Figure 6.65 shows the focal drift behavior of the two cases studied here as a function of the base materials Abbe number. The constant Abbe number case behaves exactly as a homogeneous lens does, while the δn = constant case achieves a level of color correction more than two orders of magnitude higher even as the base material's Abbe number approaches zero (*i.e.*, infinitely dispersive).

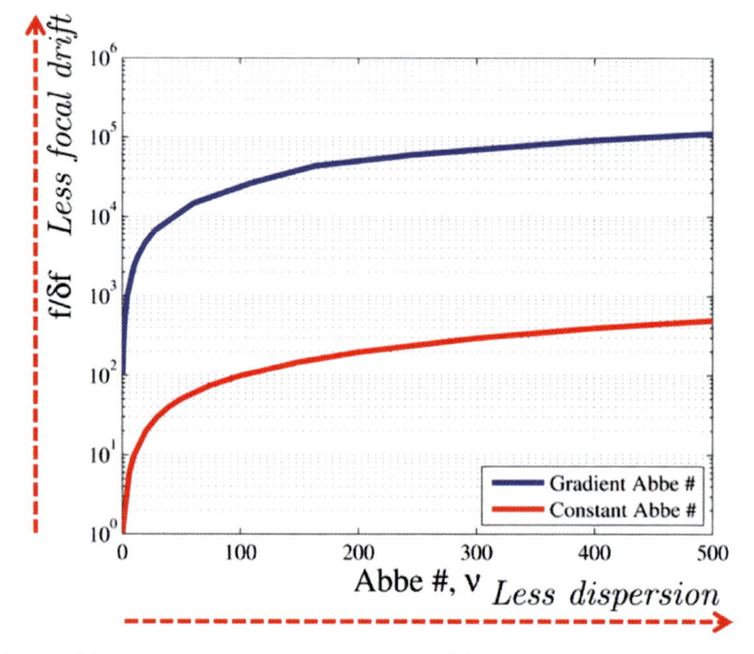

Figure 6.65 Dispersion curves for materials of different Abbe numbers.

While the color-correcting condition for a flat GRIN lens is to have parallel dispersion lines and thus a constant GRIN power with wavelength, this behavior will change as power is introduced in the lens surfaces. The total system power should be conserved, while the dispersion of the gradient will have to oppose that introduced by the surfaces. While the mathematics governing color correction was laid out in Eq. 6.41, it is not immediately clear what impact this has on the material system or geometry. To illustrate this effect, the ideal dispersion behaviors necessary to correct chromatic aberration were found for bisymmetric lenses of varying surface curvatures using CMA-ES [15, 18]. The results for lenses with front surface radii curvature of ±50 mm are summarized in Fig. 6.66.

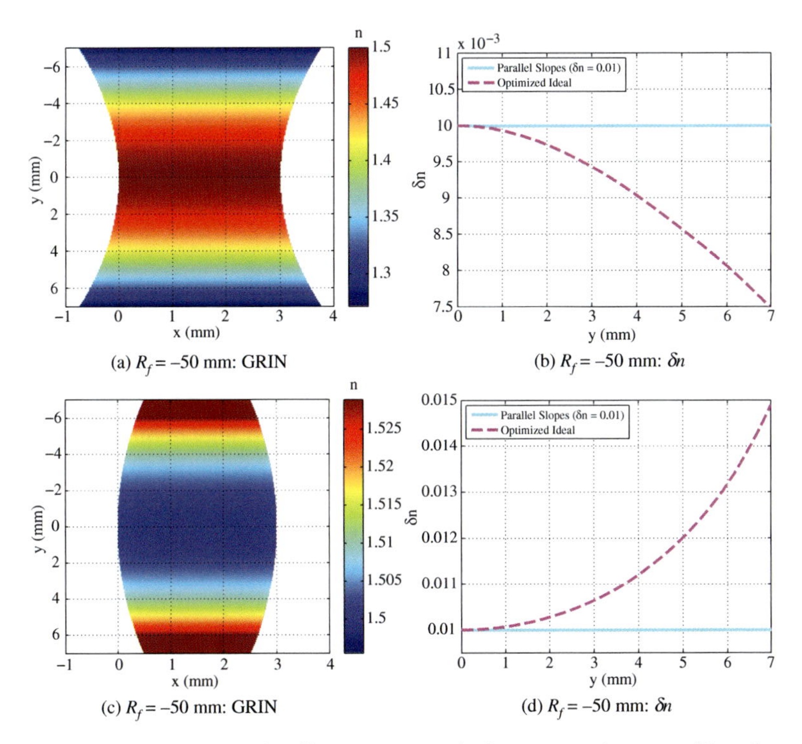

Figure 6.66 GRIN and effective material dispersion slope profiles for bisymmetric *f*/5 lenses with radii of curvature (a,c) *R* = −50 mm and (b, d) *R* = 50 mm.

For the biconcave lens, a converging GRIN profile is needed to maintain the desired *f*/5 focal length and radially varying δn value that decreases with radial position from the constant value of the parallel slopes case. Meanwhile, the highly curved biconvex lens required a diverging profile to counteract the positive surface powers and a δn that is spatially varying, but is increasing with radial position. Of course the flat lens condition exists between these two extremes; therefore, there will be a trend as the curvature goes from positive to negative and vice versa. This trend is captured in Fig. 6.67.

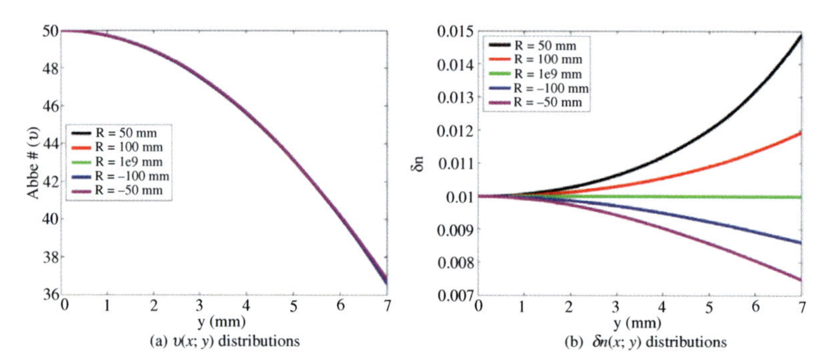

Figure 6.67 The optimal Abbe number and effective material dispersion slope profiles for several bisymmetric *f*/5 radial GRIN lenses.

As previously discussed, the δn trend does indeed sweep from positive to negative while passing through the constant case for an infinitely flat lens geometry. Moreover, as $\delta n(x, y)$ and the index value $n(x, y)$ are functions of the radial position, so is the Abbe number $v(x, y)$. What is interesting is that each lens geometry was found to require the same distribution $v(x, y)$ to maintain their color-correction behavior. The changes in index distribution are balanced by the variations in the dispersion slope to yield the unvarying Abbe number profiles. Similar behaviors are seen for lenses of different *f*/#: all geometries will require the same profile. While only focusing lenses were investigated in this section, these results can be extended to other qTO-enabled devices outside the optical regime. Since chromatic aberration can be corrected with materials that are highly dispersive in nature, this provides a potential path forward for extending the bandwidth of devices that incorporate resonant metamaterials.

This TO-inspired discussion gives insight into the dispersive properties of these optical systems. The previous section gave approximate analytical design equations for color correction, while this section examined the phenomenon through the perspective of qTO, showcasing the potential of using qTO to better understand the physics of a given problem.

6.6 Conclusion

Due to the invariance of Maxwell's equations in different coordinate systems, the electromagnetic behavior of a device with a complicated geometry can be recreated in a different device whose geometry is simpler through an appropriate mapping of inhomogeneous anisotropic material properties. These materials can be physically realized through anisotropic metamaterials, which are typically lossy and resonant. Through a series of approximations and assumptions, however, a numerical algorithm (qTO) can be developed that yields inhomogeneous isotropic metamaterials for any general geometrical transformation. These materials can be realized through much simpler, all-dielectric designs, enabling broadband performance. This chapter discussed broadband lens designs using qTO. Classical lens designs such as the Luneburg lens and the Fresnel lens can be modified to fit within a more convenient form factor, while still maintaining broadband and wide-FOV performance. To minimize reflections due to discontinuities of material properties at lens boundaries, qTO can be employed to design high-performance broadband ARCs. In addition to these direct applications of qTO, the approach can also be used for a thought experiment to arrive at an analytical method of color correction in GRIN lenses. Finally, an alternative method to qTO called wavefront matching was introduced. This method alleviates some of the limitations of qTO by enabling the explicit prescription of the desired material index range and subsequently correcting for chromatic dispersion. Unfortunately, the method requires a sophisticated numerical optimizer to be effective, while qTO only requires the numerical solution of Laplace's equation, mathematically a much simpler problem. There is a long, storied history in the field of lens design, and with it comes a vast library of classic designs that can be mined to develop a qTO-derived inhomogeneous metamaterial lens.

Acknowledgments

Some of the work reported in this chapter was supported in part by DARPA/AFRL under contract number FA8650-12-C-7225.

References

[1] Aghanejad, I., Abiri, H., Yahaghi, A. and Ramezani, R. (2012). A high-gain lens antenna based on transformation optics, in *Antennas and Propagation Conference (LAPC), 2012 Loughborough (IEEE)*, pp. 1–4.

[2] Balanis, C. A. (1989). *Advanced Engineering Electromagnetics* (Wiley), ISBN 9780471621942.

[3] Bociort, F. (1994). *Imaging Properties of Gradient-Index Lenses* (Köster).

[4] Bociort, F. (1996). Chromatic paraxial aberration coefficients for radial gradient-index lenses, *JOSA A*, **13**, pp. 1277–1284.

[5] Brocker, D. E., Campbell, S. D. and Werner, D. H. (2015). Color-correcting gradient-index infrared singlet based on silicon and germanium mixing, *Antennas and Propagation & USNC/URSI National Radio Science Meeting, 2015 IEEE International Symposium on*, pp. 788–789.

[6] Chang, Z., Zhou, X., Hu, J. and Hu, G. (2010). Design method for quasi-isotropic transformation materials based on inverse Laplaces equation with sliding boundaries, *Optics Express*, **18**, 6, pp. 6089–6096.

[7] Chen, M., Chang, H.-C., Chang, A. S., Lin, S.-Y., Xi, J.-Q. and Schubert, E. (2007). Design of optical path for wide-angle gradient-index antireflection coatings, *Applied Optics*, **46**, 26, pp. 6533–6538.

[8] Corsetti, J. A., McCarthy, P. and Moore, D. T. (2013). Color correction in the infrared using gradient-index materials, *Optical Engineering*, **52**, 11, pp. 112109/1–6.

[9] Daumas, M. (1972). *Scientific Instruments of the Seventeenth and Eighteenth Centuries and Their Makers*, (Praeger Publishers).

[10] Edwards, D. F. and Ochoa, E. (1980). Infrared refractive index of silicon, *Applied Optics*, **19**, 24, pp. 4130–4131.

[11] Flynn, R. A., Fleet, E. F., Beadie, G. and Shirk, J. S. (2013). Achromatic GRIN singlet lens design. *Optics Express*, **21**, 4, pp. 4970–4978, doi:10.1364/OE.21.004970, URL http://www.ncbi.nlm.nih.gov/pubmed/23482029.

[12] Forberich, K., Dennler, G., Scharber, M. C., Hingerl, K., Fromherz, T. and Brabec, C. J. (2008). Performance improvement of organic solar cells with moth eye anti-reflection coating, *Thin Solid Films*, **516**, 20, pp. 7167–7170.

[13] Fuchs, B., Lafond, O., Rondineau, S. and Himdi, M. (2006). Design and characterization of half Maxwell fish-eye lens antennas in millimeter waves, *IEEE Transactions on Microwave Theory and Techniques*, **54**, 6, pp. 2292–2300.

[14] Gibson, D. J., Bayya, S. S., Sanghera, J. S., Nguyen, V., Scribner, D., Maksimovic, V., Gill, J., Yi, A. Y., Deegan, J. and Unger, B. (2014). Layered chalcogenide glass structures for IR lenses, Proc. SPIE, **9070**, pp. 90702I/1–5.

[15] Gregory, M. D., Bayraktar, Z. and Werner, D. H. (2011). Fast optimization of electromagnetic design problems using the covariance matrix adaptation evolutionary strategy, *IEEE Transactions on Antennas and Propagation*, **59**, 4, pp. 1275–1285.

[16] Greisukh, G. I. and Bobrov, S. T. (1997). *Optics of Diffractive and Gradient Index Elements and Systems*, (SPIE Publications).

[17] Greivenkamp, J. E. (2004). *Field Guide to Geometrical Optics*, Vol. 1, (SPIE Press Bellingham, Washington).

[18] Hansen, N. and Ostermeier, A. (1996). Adapting arbitrary normal mutation distributions in evolution strategies: The covariance matrix adaptation, *Proceedings of the 1996 IEEE International Conference on Evolutionary Computation*, **96**, pp. 312–317.

[19] Hemenger, R. P., Garner, L. F. and Ooi, C. S. (1995). Change with age of the refractive index gradient of the human ocular lens, *Investigative Ophthalmology & Visual Science*, **36**, 3, pp. 703–707.

[20] Hristov, H. D. (1996). The multi-dielectric Fresnel zone-plate antenna-A new candidate for dbs reception, *IEEE Transactions on Antennas and Propagation*, **1**, pp. 746–749.

[21] Hunt, J., Kundtz, N., Landy, N., Nguyen, V., Perram, T., Starr, A. and Smith, D. R. (2011). Broadband wide angle lens implemented with dielectric metamaterials, *Sensors*, **11**, 8, pp. 7982–7991.

[22] Jones, C. E., Atchison, D. A., Meder, R. and Pope, J. M. (2005). Refractive index distribution and optical properties of the isolated human lens measured using magnetic resonance imaging (MRI), *Vision Research*, **45**, 18, pp. 2352–2366.

[23] Kidger, M. J. (2001). *Fundamental Optical Design*, (SPIE).

[24] King, H. and Wong, J. (1972). An experimental study of a balun-fed open-sleeve dipole in front of a metallic reflector, *IEEE Transactions on Antennas and Propagation*, **20**, 2, pp. 201–204.

[25] Kundtz, N., Roberts, D., Allen, J., Cummer, S. and Smith, D. (2008). Optical source transformations, *Optics Express*, **16**, 26, pp. 21215–21222.

[26] Kundtz, N. and Smith, D. R. (2010). Extreme-angle broadband metamaterial lens, *Nature Materials*, **9**, 2, pp. 129–132.

[27] Kwon, D.-H. (2014). Transformation electromagnetics and optics, *Forum for Electromagnetic Research Methods and Application Technologies*, **1**.

[28] Kwon, D.-H. and Werner, D. H. (2008). Transformation optical designs for wave collimators, flat lenses and right-angle bends, *New Journal of Physics*, **10**, 11, p. 115023.

[29] Kwon, D.-H. and Werner, D. H. (2010). Transformation electromagnetics: An overview of the theory and applications, *Antennas and Propagation Magazine, IEEE*, **52**, 1, pp. 24–46.

[30] Lai, Y., Ng, J., Chen, H., Han, D., Xiao, J., Zhang, Z.-Q. and Chan, C. (2009). Illusion optics: The optical transformation of an object into another object, *Physical Review Letters*, **102**, 25, pp. 253902/1–4.

[31] Li, J. and Pendry, J. (2008). Hiding under the carpet: A new strategy for cloaking, *Physical Review Letters*, **101**, 20, pp. 203901/1–4.

[32] Li, Y., Zhang, J., Zhu, S., Dong, H., Wang, Z., Sun, Z., Guo, J. and Yang, B. (2009). Bioinspired silicon hollow-tip arrays for high performance broadband anti-reflective and water-repellent coatings, *J. Mater. Chem.*, **19**, 13, pp. 1806–1810.

[33] Liu, R., Ji, C., Mock, J., Chin, J., Cui, T. and Smith, D. (2009). Broadband ground-plane cloak, *Science*, **323**, 5912, pp. 366–369.

[34] Liu, Y. and Zhang, X. (2011). Metamaterials: A new frontier of science and technology, *Chemical Society Reviews*, **40**, 5, pp. 2494–2507.

[35] Loewen, E. G. and Popov, E. (1997). *Diffraction Gratings and Applications*, (CRC Press).

[36] Lu, B. (2014). Transformation optics methodology review and its application to antenna lens designs, Ph.D. thesis, Pennsylvania State University.

[37] Lüneberg, R. K. (1944). *Mathematical Theory of Optics*, (Brown University).

[38] Luo, Y., Zhang, J., Ran, L., Chen, H. and Kong, J. A. (2008). Controlling the emission of electromagnetic source, *PIERS*, **4**, pp. 795–800.

[39] Marchand, E. W. (1978). *Gradient Index Optics*, (Academic Press).

[40] Maxwell, J. C. (1854). Problems, *Cambidge and Dublin Math. J.*, **8**, pp. 188–195.

[41] Morgan, K. L., Brocker, D. E., Campbell, S. D., Werner, D. H. and Werner, P. L. (2015). Transformation-optics-inspired anti-reflective coating design for gradient index lenses, *Optics Letters*, **40**, 11, pp. 2521–2524.

[42] Olver, P. J. (2015). Complex analysis and conformal mapping, *University of Minnesota Lecture Notes*, http://www.math.mn.edu/~olver/ln_/cml.pdf.

[43] Pierscionek, B. K. and Chan, D. C. (1989). Refractive index gradient of human lenses, *Optometry & Vision Science*, **66**, 12, pp. 822–829.

[44] Popa, B.-I., Allen, J. and Cummer, S. A. (2009). Conformal array design with transformation electromagnetics, *Applied Physics Letters*, **94**, 24, pp. 244102/1–4.

[45] Rahm, M., Cummer, S. A., Schurig, D., Pendry, J. B. and Smith, D. R. (2008). Optical design of reflectionless complex media by finite embedded coordinate transformations, *Physical Review Letters*, **100**, 6, pp. 1–5.

[46] Rahm, M., Roberts, D., Pendry, J. and Smith, D. (2008). Transformation-optical design of adaptive beam bends and beam expanders, *Optics Express*, **16**, 15, pp. 11555–11567.

[47] Raut, H. K., Ganesh, V. A., Nair, A. S. and Ramakrishna, S. (2011). Anti-reflective coatings: A critical, in-depth review, *Energy & Environmental Science*, **4**, 10, pp. 3779–3804.

[48] Ren, D. and Allington-Smith, J. R. (1999). Apochromatic lenses for near-infrared astronomical instruments, *Optical Engineering*, **38**, 3, pp. 537–542.

[49] Saleh, B. and Teich, M. (2007). *Fundamentals of Photonics*, Wiley Series in Pure and Applied Optics (Wiley), ISBN 9780471358329, URL https://books.google.com/books?id=Ve8eAQAAIAAJ.

[50] Sands, P. (1971). Inhomogeneous lenses, V. chromatic paraxial aberrations of lenses with axial or cylindrical index distributions, *JOSA*, **61**, 11, pp. 1495–1500.

[51] Sands, P. J. (1971). Inhomogeneous lenses, II. Chromatic paraxial aberrations, *J. Opt. Soc. Am.*, **61**, 6, pp. 777–783, doi:10.1364/JOSA.61.000777, URL http://www.osapublishing.org/abstract.cfm?URI=josa-61-6-777.

[52] Schurig, D. (2008). An aberration-free lens with zero f-number, *New Journal of Physics*, **10**, 11, pp. 115034/1–11.

[53] Smith, D. R., Schultz, S., Markoš, P. and Soukoulis, C. M. (2002). Determination of effective permittivity and permeability of metamaterials from reflection and transmission coefficients, *Physical Review B*, **65**, 19, pp. 195104/1–5.

[54] Smith, D. R., Urzhumov, Y., Kundtz, N. B. and Landy, N. I. (2010). Enhancing imaging systems using transformation optics, *Optics Express*, **18**, 20, pp. 21238–21251.

[55] Southwell, W. H. (1983). Gradient-index antireflection coatings, *Optics Letters*, **8**, 11, pp. 584–586.

[56] Tichit, P. H., Burokur, S. N., Germain, D. and Lustrac, A. D. (2011). Coordinate-transformation-based ultra-directive emission, *Electronics Letters*, **47**, 10, pp. 1–2.

[57] Tichit, P. H., Burokur, S. N. and Lustrac, A. D. (2009). Ultra directive antenna via transformation optics, *Journal of Applied Physics*, **105**, pp. 104912/1–3.

[58] Tsai, Y.-J., Larouche, S., Tyler, T., Lipworth, G., Jokerst, N. M. and Smith, D. R. (2011). Design and fabrication of a metamaterial gradient index diffraction grating at infrared wavelengths, *Optics Express*, **19**, 24, pp. 24411–24423.

[59] Vasic, B., Isic, G., Gajic, R. and Hingerl, K. (2010). Controlling electromagnetic fields with graded photonic crystals in metamaterial regime, *Optics Express*, **18**, 19, pp. 20321–20333.

[60] Wang, X., Wu, Q., Turpin, J. P. and Werner, D. H. (2013). Rigorous analysis of axisymmetric transformation optics lenses embedded in layered media illuminated by obliquely incident plane waves, *Radio Science*, **48**, 3, pp. 232–247.

[61] Werner, D. H. and Kwon, D.-H. (2014). *Transformation Electromagnetics and Metamaterials: Fundamental Principles and Applications*, (Springer).

[62] Wood, R. W. (1905). *Physical Optics*, (Macmillan, New York).

[63] Wu, Q., Jiang, Z. H., Quevedo-Teruel, O., Turpin, J. P., Tang, W., Hao, Y. and Werner, D. H. (2013). Transformation optics inspired multibeam lens antennas for broadband directive radiation, *IEEE Transactions on Antennas and Propagation*, **61**, 12, pp. 5910–5922.

[64] Yang, R., Tang, W. and Hao, Y. (2011). A broadband zone plate lens from transformation optics, *Optics Express*, **19**, 13, pp. 12348–12355.

Chapter 7

Broadband Chirality in Twisted Metamaterials

Amir Nader Askarpour,[a] Yang Zhao,[b] and Andrea Alù[c]
[a]*Department of Electrical Engineering, Amirkabir University of Technology, 424 Hafez Avenue, Tehran, Iran*
[b]*Department of Materials Science and Engineering, Stanford University, 496 Lomita Mall, Durand 159, Stanford, CA 94305, USA*
[c]*Department of Electrical and Computer Engineering, The University of Texas at Austin, 1 University Station C0803, Austin, TX 78712, USA*
alu@mail.utexas.edu

By iteratively applying translation and rotation operations on a metasurface, we produce a distinct class of structures known as twisted metamaterials. Wave propagation through these metamaterials exhibits new phenomena, which are not available in conventional periodic metamaterials, such as a broadband chiral response. Here we review and discuss electromagnetic wave propagation in these structures. A generalized Floquet analysis is developed and applied to obtain rigorous modal solutions for twisted metamaterials. We discuss this analytical technique, which can be applied to lossy as well as to lossless twisted metamaterials.

Broadband Metamaterials in Electromagnetics: Technology and Applications
Edited by Douglas H. Werner
Copyright © 2017 Pan Stanford Publishing Pte. Ltd.
ISBN 978-981-4745-68-0 (Hardcover), 978-1-315-36443-8 (eBook)
www.panstanford.com

Then we discuss the wave polarization properties and potential applications of twisted metamaterials for broadband polarizer designs.

7.1 Introduction

Since Erasmus Bartholinus discovered the phenomenon of double refraction in Iceland spar (calcite) in 1669, scientists have been fascinated by light propagation through complex media. Christian Huygens discovered that each of the two refracted rays in the calcite crystal can be extinguished by placing another crystal in the ray path, and then rotating the second crystal in the right direction by an appropriate amount. This discovery was one of the first hard proofs of the existence of the polarization of light (or as Newton put it, *sides of light*). After almost a hundred years, Thomas Young realized that the polarization arises from the transverse wave nature of light. From then onward, research on light polarization and its applications can be traced without any considerable gap. Another breakthrough in manipulating the polarization of light came unexpectedly in the late 19th century by the discovery of the cholesteric liquid crystals, and other types of liquid crystals were developed during the 20th century. These crystals have the potential of manipulating the polarization of light over a certain bandwidth, and this potential has been exploited in numerous liquid crystal–based devices. In recent years, with the advent of artificial materials, also known as *metamaterials* [1–27], new opportunities have arisen to manipulate electromagnetic waves, especially in the context of light polarization.

Many biological species [29–32] are also capable of detecting the polarization of light with their visual system. Although the human eye lacks this ability, we have made up for this deficiency by controlling and detecting of the polarization state of light through technical instruments. The detection of circular polarization over a wide bandwidth is more challenging compared to the detection and control over linear polarization. Nevertheless, there seems to be no fundamental reason forbidding us from achieving large bandwidths for the devices that manipulate circular polarization of electromagnetic waves. As we discuss in the following, *twisted metamaterials* hold the key to wideband manipulation of circularly polarized waves in a planarized, easy-to-integrate geometry.

Sub-wavelength engineered inclusions arrayed in a periodic lattice and designed to collectively support exotic electromagnetic properties are called metamaterials. In conventional metamaterial designs, the inclusions are often identical and placed with the same orientation in the array. The response of metamaterials in the frequency bands of interest can be made drastically different from each constituent or from naturally occurring materials exploiting collective resonances. Although technological advances have enabled the fabrication of metamaterials with exotic properties, it is also very important to delve into the theoretical foundations of the interaction between electromagnetic waves and these novel structures. By formulating the problem analytically, it is often possible to see new frontiers that one can exploit to achieve even larger degrees of freedom in the design and fabrication of engineered metamaterials.

Three-dimensional periodic metamaterials can be seen as one-dimensional arrays of *metasurfaces*. In this sense, metasurfaces are the two-dimensional counterpart to metamaterials [18–20]. These planarized structures can also exhibit extraordinary electromagnetic properties, such as negative index of refraction [21], sub-diffraction imaging [22], nanocircuitry [23–25], and cloaking [26, 27]. Clearly, their fabrication is much easier than the one of three-dimensional metamaterials, and their ultimate thin structures make them more desirable for many applications [28].

An extension to the idea of periodic metamaterials is obtained by introducing a sequential rotation between adjacent inclusions. The simplest way to apply this rotation is to consider a three-dimensional twisted metamaterial as a stack of two-dimensional metasurfaces. There are two approaches to introduce rotation in this structure. The first one consists in rotating the inclusion in each unit cell, keeping the lattice geometry unaltered. In this approach, each metasurface is a perfectly periodic two-dimensional array and all metasurfaces have the same lattice vectors. However, the inclusions in the metasurfaces are rotated with respect to each other. The result of this operation is a structure that reminds us of cholesteric liquid crystals in their geometry. The second approach is to build up a three-dimensional metamaterial from a stack of metasurfaces, while rotating each metasurface with respect to its adjacent neighbors. This will result in the rotation of the lattice vectors in each metasurface relative to its neighbors. However, in this approach the relative orientation of the inclusion with respect to the lattice vectors remains the same throughout the structure. It has been shown that the macroscopic response stemming from these two procedures can be very similar to each other, due to the sub-wavelength size of the inclusions and of the distance between neighboring ones [85]. Although these metamaterials are similar to cholesteric liquid crystals in their lattice geometry, we can design and fabricate them to achieve stronger light–matter interactions at the unit cell level with top-down techniques.

The idea of introducing rotation in an otherwise periodic structure is not by any means new. It has been used in developing a range of newly designed artificial structures, with applications to negative refractive index and partial focusing [33–39], perfect lenses [40], broadband polarizers [41], optically active media [42–48], polarization rotators [49], enhanced reflectors [50], unidirectional optical waveguides [51], photonic crystals [52, 53], chromatic aberration correctors [54], giant chirality [55], and super-chiral light [56, 57]. *Twisted metamaterials* have also been used in high-power microwave sources [58–60] to enhance the interaction between electromagnetic waves and an electron beam. The same concept can be used to obtain circular dichroism in optical waveguides [61, 62].

Introducing the rotation between adjacent metasurfaces breaks the translational symmetry of the structure; therefore, the methods used to find modal solutions to periodic arrays [63–66] cannot be applied without modification. Different techniques, including geometrical optics [67, 68], coupled mode theory [69, 70], Riccati's equation modeling [71], finite element methods [72], Green's function methods [73, 74], and the transfer matrix methods [75–84] have been used in the literature to analyze twisted structures ranging from one-dimensional twisted chain of nanoparticles to three-dimensional models of liquid crystals.

As we will discuss in the following, a twisted array of plasmonic metasurfaces could provide us with the necessary tools in order to control the state of polarization over a relatively large bandwidth. In the following sections, we first discuss a generalized Floquet analysis to model twisted arrays of metasurfaces. We discuss the underlying physical phenomena governing the interaction of waves and twisted metamaterials and provide insight into different properties of wave propagation inside these structures. We validate the accuracy of our approach and its limitations by comparing the results from our analytical model with full-wave numerical simulations based on finite-integration techniques [87]. Then we apply this analytical model to explore the possibilities of achieving large bandwidths using simple inclusions in a twisted metamaterial.

7.2 Modal Solution to Twisted Metamaterials

A twisted metamaterial is an array of particles in which a sequential translation and rotation about the twist axis produces a structure indistinguishable from the original structure. A slice of the twisted metamaterial is shown in Fig. 7.1. In this figure, each layer constitutes a metasurface parallel to the x–y plane and the metamaterial is made of an infinite number of these metasurfaces stacked on top of each other with a constant distance d between the neighboring layers. The twist axis of the presented twisted metamaterial coincides with the z-axis. Therefore, a translation in the z-direction and a rotation about this axis brings the structure back to itself. We index the metasurfaces with integer numbers n, with $n = 0$ corresponding to the metasurface at $z = 0$. The first metasurface ($n = 1$) residing directly above the $n = 0$ metasurface is rotated by a constant angle θ about the twist axis. As we move to larger values of n, the amount of rotation increases linearly with the index, *i.e.*, the nth metasurface residing at $z = nd$ is rotated by $n\theta$ radians about the twist angle. It is worth mentioning that in general this structure is not a periodic structure. For a countable set of twist angles θ, we can find a *supercell* along the twist axis that repeats itself throughout the structure, but for other values of the twist angles, no exact repetition can be found along the twist axis.

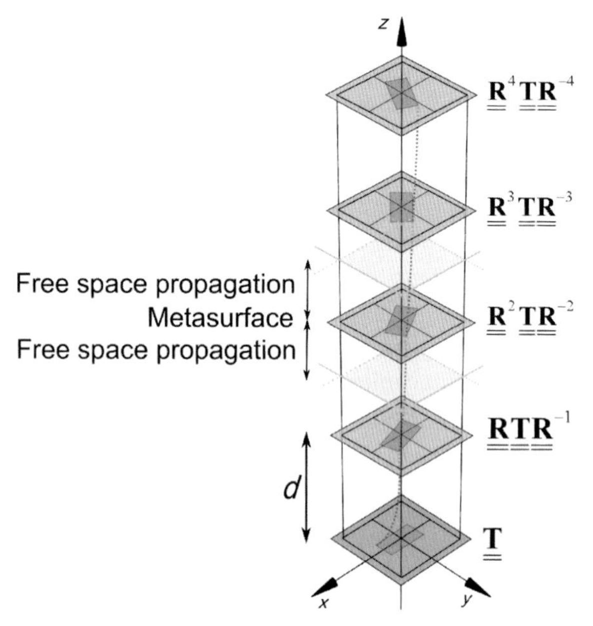

Figure 7.1 Schematic geometry of a twisted metamaterial. Each layer is rotated by a constant angle compared to its immediate neighbor. The transfer matrix of each twisted unit cell can be obtained by suitably rotating the transfer matrix of the first unit cell. A twisted unit cell consists of a propagation length d in free-space and an ultrathin metasurface in the middle. Adapted from [86].

A *unit cell* of the structure consists of a free-space part with the length $d/2$, a metasurface with a homogenized surface admittance tensor followed by another free-space block with the length $d/2$. We define the transfer matrix of a metasurface sandwiched between two $d/2$ free-space blocks in the presence of transverse electromagnetic waves as

$$\begin{bmatrix} E_{x,2} \\ E_{y,2} \\ \eta_0 H_{x,2} \\ \eta_0 H_{y,2} \end{bmatrix} = \underline{\underline{\mathbf{T}}} \begin{bmatrix} E_{x,1} \\ E_{y,1} \\ \eta_0 H_{x,1} \\ \eta_0 H_{y,1} \end{bmatrix} \tag{7.1}$$

where subscripts 1 and 2 correspond to the input and output sides of the unit cell. Both electric (E) and magnetic (H) fields in the above equation are measured on two planes located at the same distance $d/2$ before and after the metasurface. From the above definition, it is clear that the transfer matrix $\underline{\underline{\mathbf{T}}}$ is a 4×4 matrix. The elements of the transfer matrix can be related to the surface admittance of a homogenized metasurface [85], the free-space propagation between the metasurface and the boundaries of the twisted unit cell in the z-direction. The expression of these elements are listed in the Appendix. Various properties and constraints on this transfer matrix are derived in Ref. [86].

7.2.1 Construction of the Eigenvalue Problem

If the transfer matrix of the 0th metasurface is $\underline{\underline{\mathbf{T}}}$, then the transfer matrix of the nth metasurface can be easily obtained by applying the rotation operation n times. We show the rotation matrix by $\underline{\underline{\mathbf{R}}}$, so the transfer matrix of the nth metasurface becomes $\underline{\underline{\mathbf{R}}}^n\,\underline{\underline{\mathbf{T}}}\,\underline{\underline{\mathbf{R}}}^{-n}$. The rotation matrix is also 4×4 and is defined as follows.

$$\underline{\underline{\mathbf{R}}} = \begin{bmatrix} \cos\theta & -\sin\theta & 0 & 0 \\ \sin\theta & \cos\theta & 0 & 0 \\ 0 & 0 & \cos\theta & -\sin\theta \\ 0 & 0 & \sin\theta & \cos\theta \end{bmatrix} \tag{7.2}$$

It can be easily shown that $\underline{\underline{\mathbf{R}}}^n$ is equivalent to the above matrix with θ replaced by $n\theta$. Using these definitions, the fields that enter the $(n+1)$th twisted unit cell are related to the field at the entrance of the nth unit cell as

$$\mathbf{f}_{n+1} = \underline{\underline{\mathbf{R}}}^n \underline{\underline{\mathbf{T}}}\,\underline{\underline{\mathbf{R}}}^{-n} \mathbf{f}_n . \tag{7.3}$$

In the above equation, $\mathbf{f}$ is the 4×1 vector containing the transverse field elements, similar to the left-hand side of Eq. (7.1).

The fields in the structure are assumed to have the following form

$$\mathbf{f}_{n+1} = \underline{\underline{\mathbf{R}}}\, e^{i\beta d}\, \mathbf{f}_n , \tag{7.4}$$

which guarantees that the field at the input of any twisted unit cell is a rotated and phase-shifted copy of the fields at the input of the previous cell. In the above equation, β is the complex wavenumber in the twisted array. From Eqs. (7.3) and (7.4), the following modal equation is obtained:

$$\left(\underline{\underline{\mathbf{R}}}^n \underline{\underline{\mathbf{T}}}\,\underline{\underline{\mathbf{R}}}^{-n} - \underline{\underline{\mathbf{R}}}\, e^{i\beta d} \right) \mathbf{f}_n = 0 \tag{7.5}$$

From the above equation, it is possible to obtain the propagation constant and the modal solution to the transverse field vectors. Let β be a function of n, e.g., $\beta(n)$. Suppose that we have found $\beta(n)$ for a specific n. If we solve Eq. (7.3) for $\mathbf{f}_n$ and replace it in Eq. (7.5), after some straightforward algebraic manipulations, we obtain the following equation:

$$\left(\underline{\underline{\mathbf{R}}}^{n+1}\underline{\underline{\mathbf{T}}}\underline{\underline{\mathbf{R}}}^{-(n+1)} - \underline{\underline{\mathbf{R}}}\mathrm{e}^{i\beta(n)d}\right)\mathbf{f}_{n+1} = 0 \tag{7.6}$$

This means that if the complex propagation constant β satisfies the modal equation for a specific n, it also satisfies the modal equation for all integer values of n. In other words, the complex propagation constant is constant for all the layers of the twisted metamaterial. Therefore, we only need to consider the $n = 0$ case (Eq. (7.7)) and solve for the propagation constant:

$$\left(\underline{\underline{\mathbf{T}}} - \underline{\underline{\mathbf{R}}}\mathrm{e}^{i\beta d}\right)\mathbf{f}_0 = 0 \tag{7.7}$$

Finding the numerical solution to Eq. (7.7) is straightforward. We only need to obtain the eigenvalues of the $\underline{\underline{\mathbf{T}}}\,\underline{\underline{\mathbf{R}}}^{-1}$ matrix. Since this matrix is a 4×4 matrix, it has four eigenvalues in general. Two of the eigenvalues correspond to waves propagating in the positive z-direction, and two of them correspond to waves propagating in the negative z-direction. It has been shown that in a time-reversal symmetric case, the propagation constants form a pair: If β is a solution to Eq. (7.7), then $-\beta^*$ also satisfies this equation, with a different field vector [86].

7.2.2 A Twisted Metamaterial with Perfectly Conducting Inclusions

To understand the behavior of the twisted metamaterials, we examine a metamaterial made of a simple inclusion. Here we consider a cuboid (rod) structure with dimensions 50 nm × 300 nm × 50 nm made of perfect electric conductor (PEC) patterned in a square lattice with periods of 350 nm in the transverse plane. The surface admittance of this metasurface is obtained using a full-wave numerical simulation and is shown in Fig. 7.2. When the longest axis of the PEC rod is along the y-axis, the dominant term in the admittance tensor is Y_{yy}. The off-diagonal terms of the admittance tensor are negligible.

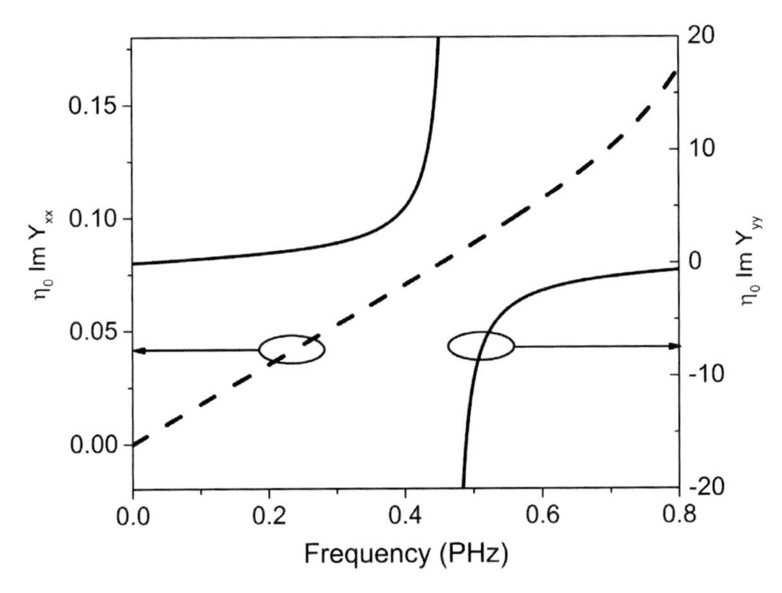

Figure 7.2 Normalized imaginary part of the admittance of a metasurface with PEC rod inclusions. The dimensions of the rod are 50 nm × 300 nm × 50 nm. The lattice constant is 350 nm and the long side of the rods is oriented in the y direction. Y_{xx} (dashed line) is much smaller than Y_{yy} (solid line) and Y_{xy} and Y_{yx} terms are negligible and are not shown here.

The dispersion diagram of this metamaterial without any twist (*i.e.*, $\theta = 0$) is shown in Fig. 7.3. The distance between the adjacent metasurfaces is $d = 100$ nm. The dispersion diagram shows four distinct modes. Two of these modes are propagating in the negative z-direction, and two of them are propagating in the positive z-direction. Figure 7.3 shows that modes 2 and 3 have linear dispersion diagrams, and they experience zero loss when propagating in the structure. The electric field in this pair of modes is polarized normal to the long axis of the rods. The other pair of modes, with electric field polarized parallel to long axis of the rods, interacts more strongly with the structure. Because of this strong interaction, there is a frequency band ($0.8 < kd < 1.4$) in which this pair of modes does not propagate. In this frequency band, modes 1 and 4 are evanescent and modes 2 and 3 are propagating. Therefore, this structure can act as a linear polarizer in this frequency band, associated with the resonance of the inclusions.

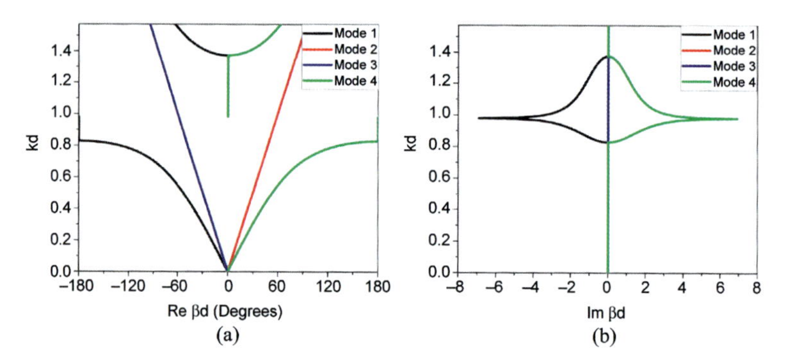

Figure 7.3 The dispersion diagram of an array of nano-rods without twist. (a) Real part and (b) imaginary parts of the propagation constant.

Now we introduce a twist in the array. Figures 7.4 and 7.5 depict the results with increasing twist angle in 10° steps. The real and imaginary parts of the complex propagation constant β are shown in Figs. 7.4 and 7.5, respectively. Examining Fig. 7.4 reveals that at low frequencies ($kd \approx 0$), the propagation constant is purely real and βd is equal to the twist angle. This can be explained by the fact that at low frequencies, the electromagnetic wave and the metasurfaces interact very weakly. This weak interaction can be modeled by assuming that the transfer matrix becomes an identity matrix at DC frequency. In this way, the modal Eq. (7.7) is simplified to

$$\underline{\underline{\mathbf{R}}}\mathbf{f}_0 = \mathrm{e}^{-i\beta d}\,\mathbf{f}_0\,. \tag{7.8}$$

It is easy to show that the eigenvalues of the rotation matrix $\underline{\underline{\mathbf{R}}}$ are $\exp(i\theta)$ and $\exp(-i\theta)$. Therefore, it becomes inevitable that at DC frequency, the propagation constant of different modes should satisfy $\beta d = \pm\theta$. The fields corresponding to the eigenvectors of the rotation matrix are circularly polarized waves. However, since these modes are degenerate at DC frequency, we can combine them to obtain any desired polarization.

If we approximate the transfer matrix $\underline{\underline{\mathbf{T}}}$ with its Taylor series for small values of kd, it is possible to obtain an approximation for the propagation constant at these frequencies. Therefore, the slope of the kd–βd diagram at small values of kd can be found. Interestingly, the slope of this linear portion of the dispersion diagram is independent of the rotation angle.

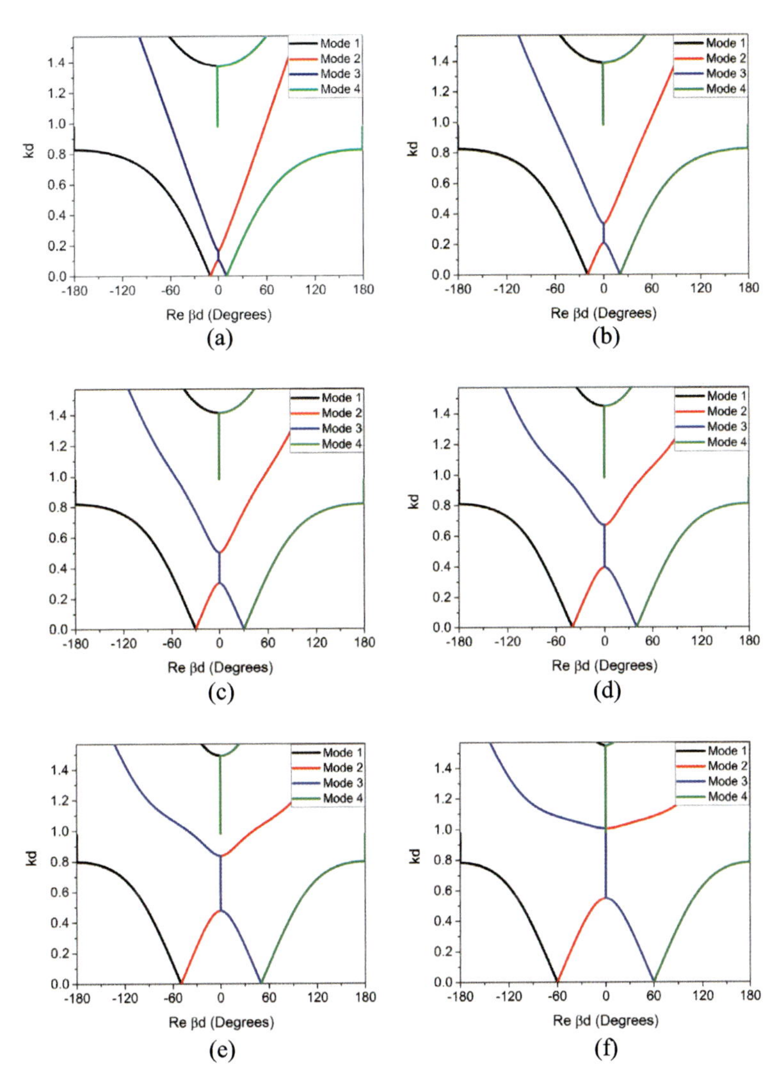

Figure 7.4 Real part of the propagation constant for a twist angle equal to (a) 10, (b) 20, (c) 30, (d) 40, (e) 50, (f) 60 degrees.

7.2.3 Effect of the Twist Angle on the Stopband

Another important feature of these diagrams is the dependency of the bandwidth of the stopbands on the twist angle. As we introduce

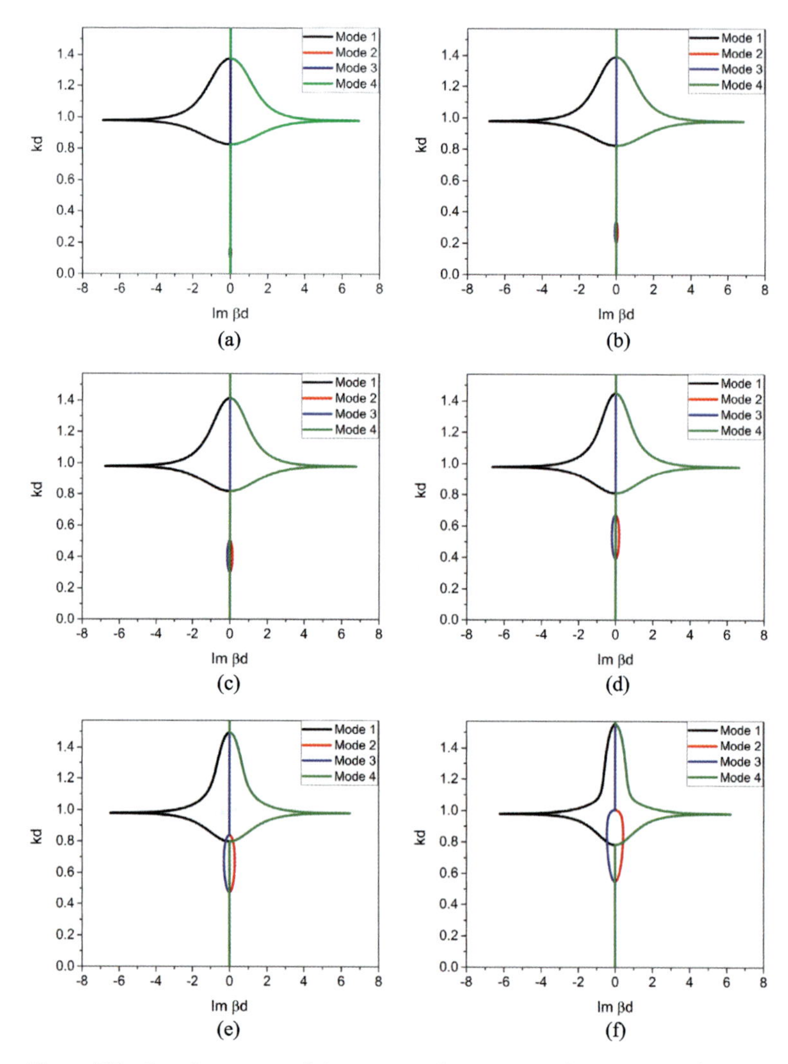

Figure 7.5 Imaginary part of the propagation constant for a twist angle equal to (a) 10, (b) 20, (c) 30, (d) 40, (e) 50, (f) 60 degrees.

a nonzero twist angle, the bandwidth in which modes 1 and 4 are evanescent starts to increase. As the twist angle increases between 0 and 90°, the lowest frequency of the stopband at about $kd = 0.8$ decreases, while the highest frequency of the stopband increases. As a result, the bandwidth increases from 258 THz at zero twist angle to 366 THz at 60° twist. For larger twist angles, the appearance of complex modes complicates the propagation phenomena. This will be discussed later in the chapter. Furthermore, a second stopband for modes 2 and 3 emerges for nonzero twist angles. Both the lowest and the highest frequencies of this second stopband increase as the twist angle increases. Nevertheless, the bandwidth of this stopband also increases by increasing the twist angle. At 60° twist angle, the second bandwidth reaches 217 THz.

Figure 7.6 depicts the aforementioned stopbands as a function of the twist angles. The yellow and cyan regions corresponds to the stopband for modes 1 and 4, and modes 2 and 3, respectively. The two stopbands change differently with the twist angle. For twist angles larger than ~50°, the two stopbands merge, where the structure starts to exhibit a complete stopband with no propagating modes and thus can act as a stopband filter. For example, with a 60° twist angle, the twisted metamaterial passes modes 1 and 4 from 263 THz to 375 THz, while stopping modes 2 and 3. Below 263 THz, all modes propagate, and from 375 THz to 480 THz, there are no propagating modes in the structure.

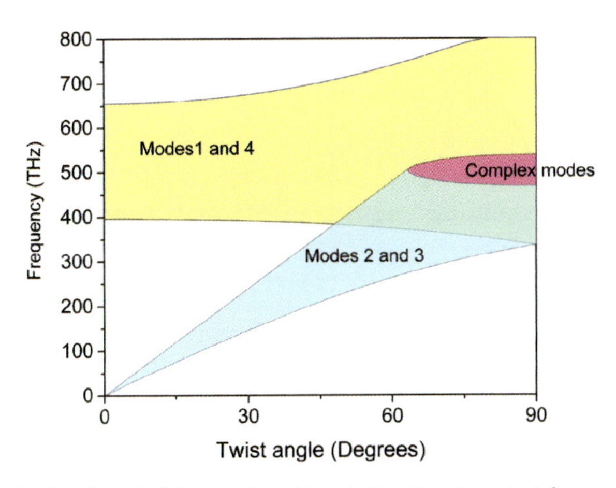

Figure 7.6 Stopband of the modes of the twisted metamaterial.

When the twist angles are larger than ~63°, complex modes appear in the dispersion diagram of the twisted metamaterials. The propagation constants of complex modes have nonzero real and imaginary parts simultaneously. The propagation constants form a complex conjugate pair and their negative counterparts (*i.e.*, β, β^*, $-\beta$, $-\beta^*$). The dispersion diagrams for 70°, 80°, and 90° twist angles are shown in Fig. 7.7. The complex modes are clearly seen at these twist angles. Interestingly, the bandwidth of the complex modes also increases as the twist angle increases. Although these complex modes cannot be excited directly by plane waves in free space, they may affect the evanescent spectrum and the local density of states in the proximity of a slab of twisted metamaterials.

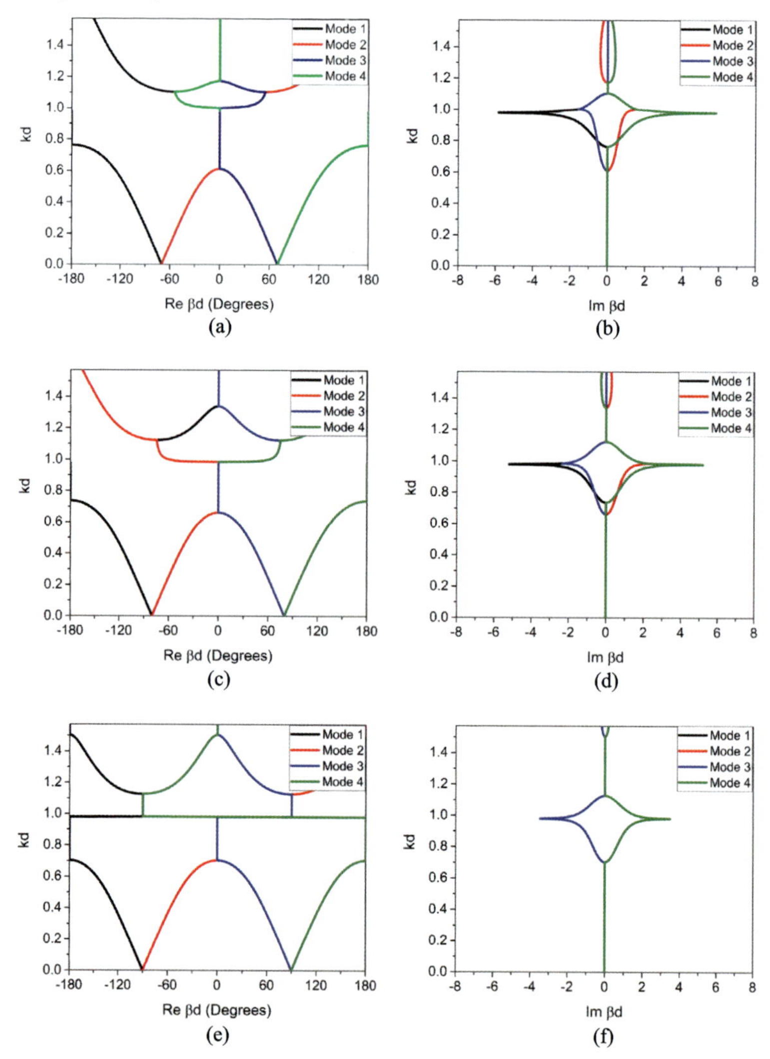

Figure 7.7 (a), (c), (e) Real and (b), (d), (f) imaginary parts of the propagation constant of the twisted metamaterial with (a), (b) 70, (c), (d) 80 and (e), (f) 90 degrees twist angles.

The propagation constants found from the above analysis are different in nature from the propagation constant in a homogenized medium or a periodic array without rotation. In addition to a phase shift due to the propagation constant that is defined in Eq. (7.4), the fields also experience a rotation. Nevertheless, it is possible in some cases, discussed in the next section, to find a simple relationship between the propagation constant of the waves inside the twisted metamaterials and the propagation constant in a periodic structure.

7.3 Supercell and Periodic Structures

As discussed earlier, the geometry of a twisted metamaterial with an arbitrary twist angle is not necessarily periodic in the direction of the twist axis. However, there is a discrete infinite number of cases for which we can find a periodicity along the twist direction. A trivial case is when the twist angle is an integer multiple of 360°. In these cases, we can define a *supercell* in the structure that captures the periodicity of the twisted metamaterial.

The twisted metamaterial is a periodic structure, if and only if the orientation of the inclusion in the zeroth unit cell is exactly the same as the Nth twisted unit cell. For example, for an inclusion without any symmetry in shape, a twist angle of 60° creates a periodic structure, where each unit cell in the periodic structure (a supercell) comprises six twisted unit cells. If the inclusion possesses some symmetry, the number of twisted unit cell in a supercell can be reduced. Because of the mirror symmetry of the PEC rods used in the above calculations, in a twisted metamaterial with a 60° twist angle, a supercell contains only three twisted unit cells.

If we take a twisted metamaterial with a supercell containing N twisted unit cells as a general example, by using Eq. (7.4), the field vector at the output of the supercell ($\mathbf{f}_N$) is related to that at the input side ($\mathbf{f}_0$) as

$$\mathbf{f}_N = \underline{\underline{\mathbf{R}}}^N e^{iN\beta d} \mathbf{f}_0 \tag{7.9}$$

We expect $\mathbf{f}_N$ to be exactly the same as $\mathbf{f}_0$ with a phase shift. If the propagation constant in the periodic structure is β', the field at the output of the supercell is

$$\mathbf{f}_N = e^{i\beta' d'} \mathbf{f}_0, \tag{7.10}$$

where d' is the length of the supercell, which is equal to Nd. By combining Eqs. (7.9) and (7.10) and using the eigenvalues of the rotation matrix $\underline{\underline{\mathbf{R}}}$ (*i.e.*, $\exp(\pm i\theta)$), the propagation constant in the periodic structure can be found in terms of the propagation constant in the twisted metamaterial.

$$\beta' d' = N\beta d \pm N\theta \tag{7.11}$$

For example, in the twisted metamaterial comprising PEC rod inclusions with a twist angle of 60° where $N = 3$, the propagation constant in the periodic structure is $\beta' d' = 3\beta d \pm \pi$. Since a phase shift outside $-\pi$ to π range in a periodic structure is equivalent to a phase shift in that range (by adding/subtracting an integer multiple of 2π), the above equation results in stretching the dispersion diagram of the twisted metamaterial and folding it back to fit in the aforementioned range. This stretch and fold procedure results in an apparently more complicated dispersion diagram for the periodic structure. However, all features of this complicated dispersion diagram can be explained by the theory developed here.

7.3.1 Comparison with Full-Wave Simulations

Equation (7.11) describes the relation between the propagation constant of the twisted metamaterial and that of a periodic structure, which enables us to compare them with full-wave numerical simulations. Let us consider a twisted metamaterial with the PEC rod inclusions and $d = 100$ nm as an example. The twist angle is 60°. A periodic structure with a supercell composed of three rods has been numerically studied using CST to obtain its dispersion diagram. The results of this full-wave calculation are compared with the results of the above stretch and fold procedure in Fig. 7.8. For simplicity, we only plot the real part of the propagation constant, since our full-wave simulations cannot compute evanescent modes. It is seen that the analytical solutions agree very well with the full-wave calculations, except that the numerical simulation was also unable to obtain the complex modes present in the structure. Note that unlike the dispersion diagram of a twisted metamaterial, the dispersion relation of the periodic structure starts at zero, as expected.

The dispersion diagrams of Fig. 7.8a start at zero frequency with two modes (modes 2 and 4) propagating in the forward direction. As the propagation constant of the second mode crosses π/d' (*i.e.*, $\beta' d' = 180°$), it becomes evanescent since this corresponds to $\beta d = 0$. Mode 4 enters from the other side of the dispersion diagram at $-180°$ after crossing $\beta' d' = 180°$, which is not shown here. Similarly, at this frequency ($kd \approx 0.6$), mode 1 reaches $\beta' d' = -180°$ and enters from the $+180°$ side of the dispersion diagram. Starting at $kd \approx 0.8$, mode 1 becomes evanescent and we see a stopband up to $kd \approx 1.0$. The full-wave calculations have a similar behavior, although the exact numbers are slightly different. The small bandgaps at $kd \approx 0.65$, 1.15, and 1.4 are due to the fact that we have neglected the thickness of the metasurfaces in our analysis.

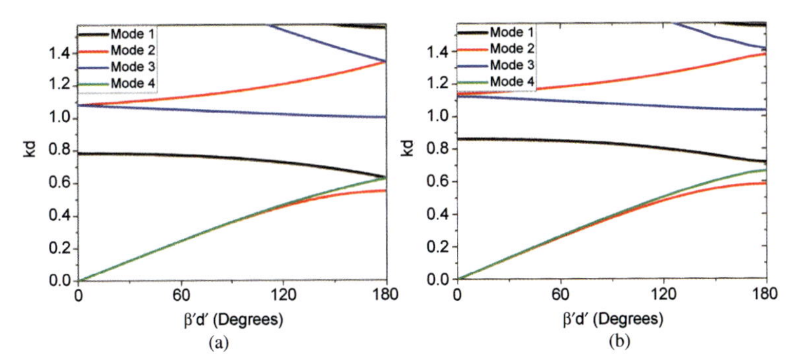

Figure 7.8 The dispersion diagram of a twisted metamaterial with super cell periodicity consisting of three twisted unit cells in the direction of propagation. Dispersion diagram obtained from (a) analytical results, and (b) numerical simulations.

At higher frequencies ($kd > 1$), more discrepancies between full-wave calculations and our analytical solution are caused by the fact that the homogenized surface impedance can no longer describe the exact behavior of the metasurfaces. This is expected since at higher frequencies, the transverse periodicity of the metasurfaces is getting comparable to the free-space wavelength. In addition, at higher frequencies, the difference between rotating the entire metasurfaces and rotating individual inclusions becomes more pronounced; the former is considered in our analytical solution, while the latter is considered in our full-wave numerical simulations [85].

7.4 Polarization

Up to this point, we have only considered the propagation constant of the modes in a twisted metamaterial. In order to make use of this class of materials in new polarizer designs, we need to understand the polarization of the modes propagating in the twisted arrays. The polarization of the modes can be found by solving for the eigenvectors of Eq. (7.7). In addition, decomposing the fields into right- and left-handed circularly polarized (RCP and LCP) plane waves provides us several other advantages. For example, notions of group velocity and Poynting vector can be easily extended for waves propagating in a twisted metamaterial.

An arbitrary forward propagating transverse electric field vector is written as a superposition of RCP and LCP as follows:

$$\mathbf{E}_n = a_{\mathrm{RF}}\mathbf{E}_{\mathrm{RF}} + a_{\mathrm{LF}}\mathbf{E}_{\mathrm{LF}} \tag{7.12}$$

In the above equation, subscripts R and L represent RCP and LCP, and F stands for forward propagating waves. The electric field vectors are defined as $\mathbf{E}_{\mathrm{RF}} = (\hat{x} + i\hat{y})$ and $\mathbf{E}_{\mathrm{LF}} = (\hat{x} - i\hat{y})$ for LCP and RCP waves. Substituting these fields into Eq. (7.4) results in

$$\mathbf{E}_{n+1} = a_{\mathrm{RF}}\, \mathrm{e}^{i(\beta d - \theta)}\, \mathbf{E}_{\mathrm{RF}} + a_{\mathrm{LF}} e^{i\theta}\, \mathrm{e}^{i(\beta d + \theta)}\, \mathbf{E}_{\mathrm{LF}}\,. \tag{7.13}$$

Equation (7.13) indicates that an RCP plane wave travels with a wave vector equal to $\beta - \theta/d$ parallel to the twist axis of the metamaterial. Likewise, an LCP plane wave has a wave vector equal to $\beta + \theta/d$. Backward waves can be obtained in a similar approach by defining the appropriate circularly polarized field vectors and assuming a $-\beta d$ propagation constant. Since each individual term in Eq. (7.12) (*i.e.*, RCP and LCP parts) is not necessarily a mode of the structure, it is not possible to define a single phase velocity for the propagating modes in the structure. This is due to the fact that some parts of a single mode travel with $\beta - \theta/d$ and some parts of it travel with $\beta + \theta/d$ phase constants. In addition, it is not possible in general to excite a purely right- or left-handed wave in the structure. However, as the waves propagate along the twist direction, both of them are required to satisfy the symmetry demands of the structure. Nevertheless, since the twist angle θ is constant, it is possible to define the group velocity as $v_g = \partial\omega/\partial\beta$ for propagating modes. A similar result can be obtained for transverse magnetic fields in the structure.

The polarization state of each of the four modes present in the twisted metamaterial can be understood by examining the ratio of a_{RF} and a_{LF} coefficients in Eq. (7.12). The absolute value of this ratio shows the relative amount of energy present in the RCP and LCP parts of each mode. Figure 7.9 shows this ratio (in logarithm scale) for the twisted metamaterial discussed in the previous sections with a twist angle of 60°.

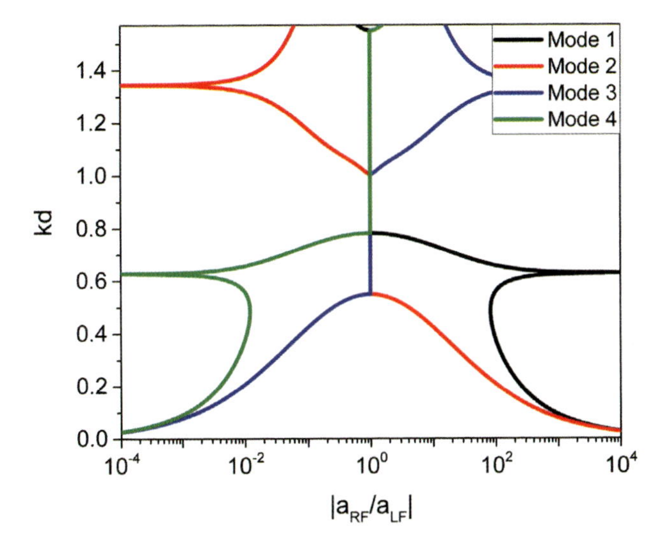

Figure 7.9 The polarization state of propagating modes in a twisted metamaterial.

At very low frequencies, where there is no strong interaction between the twisted metamaterial and the electromagnetic waves, the modes are strongly circularly polarized, which corresponds to the propagating modes in the free space. As the frequency increases, the modes become elliptical; however, they retain their senses of rotation before they become evanescent. In order to assign absolute sense of rotation to the modes, we need to know the direction of propagation of these modes. This can be achieved by examining the group velocity of the modes. Since the group velocity is related to the slope of the dispersion diagram, we can conclude that modes 2 and 4 propagate in the positive z-direction and modes 1 and 3 propagate in the negative z-direction. Therefore, at low frequencies, before the stopbands, modes 2 and 3 are right handed and modes 1 and 4 are left handed. This result is of the utmost importance for the design of circular polarizers. While modes 2 and 3 reach their stopband at $kd \approx 0.55$, modes 1 and 4 continue to propagate in the twisted metamaterial up to $kd \approx 0.78$. Moreover, within this frequency range ($0.55 < kd < 0.73$), these modes provide an almost pure LCP propagation ($|a_{RF}/a_{LF}| < 0.1$).

A similar situation exists after the stopband. Modes 2 and 3 are propagating, and modes 1 and 4 are evanescent. In this case, mode 2 propagates in the positive z-direction and mode 3 propagates in the opposite direction. These two modes have a left-handed polarization. This part of the frequency from $kd \approx 1.0$ up to $kd \approx 1.55$ can also be exploited to achieve wideband circular polarizers. In the next section, we aim at optimizing the structure to maximize the bandwidth for circular polarizers.

7.5 Broadband Polarizer Design

An important aspect of the twisted metamaterials is their ability to support one pair of propagating modes with a strong sense of handedness. This can be used to design circular polarizers with large bandwidths. To increase the bandwidth, we have studied effects of the twist angle θ and the distance between the metasurfaces d. These parametric studies can be performed in a much shorter time compared to full-wave numerical simulations.

In the following, we calculate the usable bandwidth, which is the portion of the frequency band where modes 2 and 3 are evanescent and modes 1 and 4 have a strong sense of handedness (*i.e.*, $|a_{RF}/a_{LF}| < 0.1$ or $|a_{RF}/a_{LF}| > 10$). This bandwidth is plotted as a color map in Fig. 7.10. As we can see, the best results are obtained for a twist angle of about 40° and a 70 nm separation between metasurfaces. However, as the separation d decreases, the accuracy of this approach decays. This is due to the fact that the effects of coupling between metasurfaces are neglected in our analytical approach. Moreover, the metasurfaces have a finite thickness, and the lower limit of the separation has to be larger than this thickness. However, if we only consider the usable bandwidth or fractional bandwidth, a twist angle between 40° to 50° is optimal for PEC rods with the aforementioned dimensions.

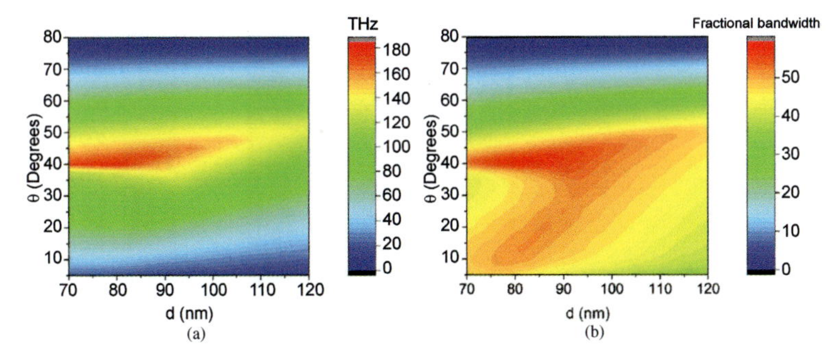

Figure 7.10 Usable (a) bandwidth and (b) fractional bandwidth for a polarizer in a twisted metamaterial versus the twist angle and the separation between the metasurfaces.

In practice, polarizers cannot have infinite layers of rotated metasurfaces. To realize a polarizer with a reasonable thickness and a small number of layers, the attenuation constant of evanescent modes becomes important. The average attenuation constant in the stopband for the evanescent modes is plotted against the twist angle θ and the separation d in Fig. 7.11. It is seen that for the optimal values of the parameters θ and d, where the bandwidth is maximized, the attenuation constant is rather small. Therefore, a polarizer that uses a twisted metamaterial with such θ and d needs to be several layers thick to suppress the undesired polarization effectively. For example, a slab of twisted metamaterial with $\theta = 40°$ and $d = 90$ nm, which exhibits an optimal fractional bandwidth, has an attenuation constant equal to 2 μm^{-1}. Therefore, to attenuate the undesired polarization to 0.1 of its value at the entrance, we need 13 layers of the metasurfaces. A large number of layers is less desired from a practical point of view. Hence, it is advisable to move to larger twist angles with a relatively smaller bandwidth but a larger attenuation constant. In this way, we can obtain thin polarizers at the expense of the bandwidth.

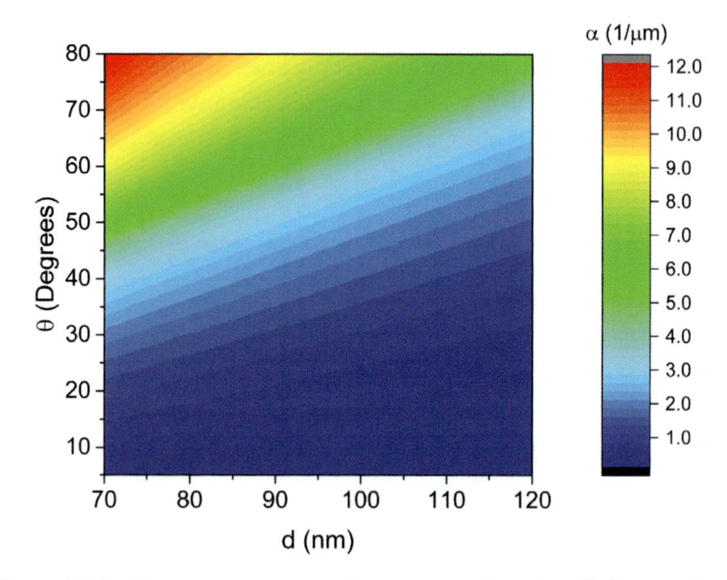

Figure 7.11 The average attenuation constant (α = Im β) for a twisted metamaterial as a function of the twist angle and the separation between metasurfaces.

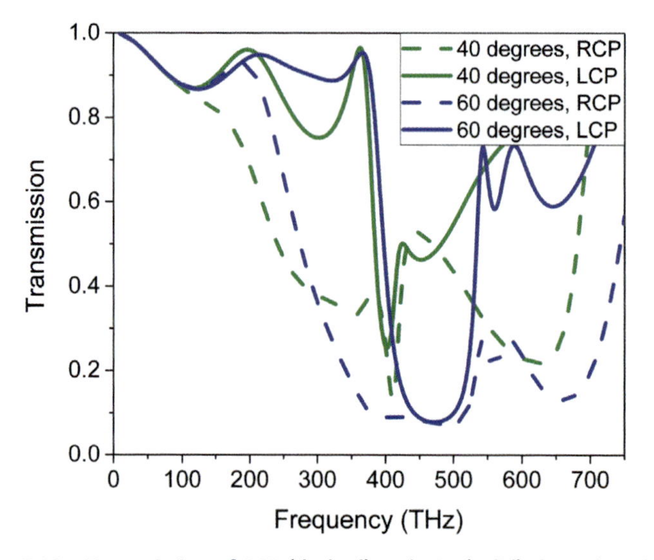

Figure 7.12 Transmission of RCP (dashed) and LCP (solid) through a slab of twisted metamaterial with twist angles of 40 (green) and 60 (blue) degrees.

To examine these issues, we have simulated a slab of twisted metamaterial using full-wave simulation. The slab is made of five metasurfaces with a separation of 100 nm between them. The transmission coefficients for RCP and LCP incident waves are plotted against frequency for 40° and 60° twist angles in Fig. 7.12. The usable bandwidth for 40° twisted metamaterial is 143 THz starting from 176 THz and ending at 319 THz. For 60° twisted metamaterial, the bandwidth starts at 263 THz and ends at 374 THz, constituting a total 111 THz bandwidth. It is clear that 40° twisted structure has a larger bandwidth at a lower center frequency. However, as discussed earlier, the attenuation constant for this metamaterial is small and the undesired polarization can pass through the slab to some extent. The 60° twisted slab has a smaller bandwidth but attenuates the undesired polarization to a greater extent.

7.6 Conclusion

In this chapter, we have modeled physical mechanisms of wave propagation in a twisted metamaterial, focusing on its broadband chiral response. Twisted metamaterials are composed of stacks of metasurfaces with each metasurface sequentially rotated about its normal with respect to its adjacent neighbors. We have shown that this type of structures support two pairs of modes. Each pair can be either propagating or evanescent, depending on the frequency and the configuration of the twisted metamaterial. We have examined the modal solution to a twisted metamaterial for various twist angles and validated the results using full-wave numerical simulations. We have also examined the polarization states of these modes and have shown that in some important cases, they have a strong sense of rotation.

Based on these analyses, we have exploited the twisted metamaterial to design wideband circular polarizers. In order to maximize the bandwidth of operation of these polarizers, we can swiftly perform a parametric analysis of the twisted metamaterials, thanks to the analytical model described in this chapter, which provides us with numerically efficient procedures to obtain their modal solutions. By controlling the twist angle of these metamaterials, we are able to optimize the bandwidth of the polarizers for an ultrathin planarized nanophotonic device that can selectively control circular polarizations.

Appendix

The elements of the matrix $\underline{\underline{\mathbf{T}}}$ is presented for the reader's reference as a function of free-space wavenumber k, the separation between metasurfaces d, and the elements of the admittance matrix of the metasurface Y.

$$T_{11} = \cos kd + \frac{j\eta_0}{2} Y_{xx} \sin kd$$

$$T_{12} = \frac{j\eta_0}{2} Y_{xy} \sin kd$$

$$T_{13} = -\eta_0 Y_{xy} \left[\sin \frac{kd}{2} \right]^2$$

$$T_{14} = \sin \frac{kd}{2} \left[-2j \cos \frac{kd}{2} + \eta_0 Y_{xx} \sin \frac{kd}{2} \right]$$

$$T_{21} = \frac{j\eta_0}{2} Y_{yx} \sin kd$$

$$T_{22} = \cos kd + \frac{j\eta_0}{2} Y_{yy} \sin kd$$

$$T_{23} = -\sin \frac{kd}{2} \left[-2j \cos \frac{kd}{2} + \eta_0 Y_{yy} \sin \frac{kd}{2} \right]$$

$$T_{24} = \eta_0 Y_{yx} \left[\sin \frac{kd}{2} \right]^2$$

$$T_{31} = \eta_0 Y_{yx} \left[\cos \frac{kd}{2} \right]^2$$

$$T_{32} = \cos \frac{kd}{2} \left(\eta_0 Y_{yy} \cos \frac{kd}{2} + 2j \sin \frac{kd}{2} \right)$$

$$T_{33} = \cos kd + \frac{j\eta_0}{2} Y_{yy} \sin kd$$

$$T_{34} = -\frac{j\eta_0}{2} Y_{yx} \sin kd$$

$$T_{41} = -\cos \frac{kd}{2} \left[\eta_0 Y_{xx} \cos \frac{kd}{2} + 2j \sin \frac{kd}{2} \right]$$

$$T_{42} = -\eta_0 Y_{xy} \left[\cos\frac{kd}{2} \right]^2$$

$$T_{43} = -\frac{j\eta_0}{2} Y_{xy} \sin kd$$

$$T_{44} = \cos kd + \frac{j\eta_0}{2} Y_{xx} \sin kd$$

Acknowledgments

This work was supported by the Air Force Office of Scientific Research, the Simons Foundation and the Welch Foundation with grant no. F-1802.

References

1. Shelby, R.A., Smith, D.R., and Schultz, S. (2001). Experimental verification of a negative index of refraction, *Science*, **292**, pp. 77–79.
2. Smith, D.R., Pendry, J.B., and Wiltshire, M.C.K. (2004). Metamaterials and negative refractive index, *Science*, **305**, pp. 788–792.
3. Alù, A., and Engheta, N. (2005). Achieving transparency with plasmonic and metamaterial coatings, *Phys. Rev. E*, **72**, pp. 016623.
4. Zhang, S., Fan, W.J., Panoiu, N.C., Malloy, K.J., Osgood, R.M., and Brueck, S.R.J. (2005). Experimental demonstration of near-infrared negative-index metamaterials, *Phys. Rev. Lett.*, **95**, pp. 137404.
5. Dolling, G., Enkrich, C., Wegener, M., Soukoulis, C.M., and Linden, S. (2006). Simultaneous negative phase and group velocity of light in a metamaterial, *Science*, **312**, pp. 892–894.
6. Leonhardt, U. (2006). Optical conformal mapping, *Science*, **312**, pp. 1777–1780.
7. Schurig, D., Mock, J.J., Justice, B.J., Cummer, S.A., Pendry, J.B., Starr, A.F., and Smith, D.R. (2006). Metamaterial electromagnetic cloak at microwave frequencies, *Science*, **314**, pp. 977–980.
8. Engheta, N. (2007). Circuits with light at nanoscales: Optical nanocircuits inspired by metamaterials, *Science*, **317**, pp. 1698–1702.
9. Lezec, H.J., Dionne, J.A., and Atwater, H.A. (2007). Negative refraction at visible frequencies, *Science*, **316**, pp. 430–432.

10. Rockstuhl, C., Lederer, F., Etrich, C., Pertsch, T., and Scharf, T. (2007). Design of an artificial three-dimensional composite metamaterial with magnetic resonances in the visible range of the electromagnetic spectrum, *Phys. Rev. Lett.*, **99**, pp. 017401.

11. Shalaev, V.M. (2007). Optical negative-index metamaterials, *Nat. Photon.*, **1**, pp. 41–48.

12. Soukoulis, C.M., Linden, S., and Wegener, M. (2007). Negative refractive index at optical wavelengths, *Science*, **315**, pp. 47–49.

13. Tsakmakidis, K.L., Boardman, A.D., and Hess, O. (2007). "Trapped rainbow" storage of light in metamaterials, *Nature*, **450**, pp. 397–401.

14. Landy, N.I., Sajuyigbe, S., Mock, J.J., Smith, D.R., and Padilla, W.J. (2008). Perfect metamaterial absorber, *Phys. Rev. Lett.*, **100**, pp. 207402.

15. Papasimakis, N., Fedotov, V.A., Zheludev, N.I., and Prosvirnin, S.L. (2008). Metamaterial analog of electromagnetically induced transparency, *Phys. Rev. Lett.*, **101**, pp. 253903.

16. Valentine, J., Zhang, S., Zentgraf, T., Ulin-Avila, E., Genov, D.A., Bartal, G., and Zhang, X. (2008). Three-dimensional optical metamaterial with a negative refractive index, *Nature*, **455**, pp. 376–379.

17. Plum, E., Zhou, J., Dong, J., Fedotov, V.A., Koschny, T., Soukoulis, C.M., and Zheludev, N.I. (2009). Metamaterial with negative index due to chirality, *Phys. Rev. B*, **79**, pp. 035407.

18. Kuester, E.F., Mohamed, M.A., Piket-May, M., and Holloway, C.L. (2003). Averaged transition conditions for electromagnetic fields at a metafilm, *IEEE Trans. Antennas Propag.*, **51**, pp. 2641–2651.

19. Holloway, C.L., Mohamed, M.A., Kuester, E.F., and Dienstfrey, A. (2005). Reflection and transmission properties of a metafilm: With an application to a controllable surface composed of resonant particles, *IEEE Trans. Electromagn. Compat.*, **47**, pp. 853–865.

20. Gordon, J.A., Holloway, C.L., and Dienstfrey, A. (2009). A physical explanation of angle-independent reflection and transmission properties of metafilms/metasurfaces, *IEEE Antennas Wireless Propag. Lett.*, **8**, pp. 1127–1130.

21. Beruete, M., Navarro-Cia, M., Falcone, F., Campillo, I., and Sorolla, M. (2010). Single negative birefringence in stacked spoof plasmon metasurfaces by prism experiment, *Opt. Lett.*, **35**, pp. 643–645.

22. Chen, P.Y., and Alù, A. (2011). Subwavelength imaging using phase-conjugating nonlinear nanoantenna arrays, *Nano Lett.*, **11**, pp. 5514–5518.

23. Alù, A., and Engheta, N. (2008). Tuning the scattering response of optical nanoantennas with nanocircuit loads, *Nat. Photon.*, **2**, pp. 307–310.
24. Alù, A., and Engheta, N. (2008). Input impedance, nanocircuit loading, and radiation tuning of optical nanoantennas, *Phys. Rev. Lett.*, **101**, pp. 043901.
25. Sun, Y., Edwards, B., Alù, A., and Engheta, N. (2012). Experimental realization of optical lumped nanocircuits at infrared wavelengths, *Nat. Mater.*, **11**, pp. 208–212.
26. Alù, A. (2009). Mantle cloak: Invisibility induced by a surface, *Phys. Rev. B*, **80**, pp. 245115.
27. Burokur, S.N., Daniel, J.P., Ratajczak, P., and de Lustrac, A. (2010). Tunable bilayered metasurface for frequency reconfigurable directive emissions, *Appl. Phys. Lett.*, **97**, pp. 064101.
28. Zhao, Y., Liu, X., and Alù, A. (2014). Recent advances on optical metasurfaces, *J. Opt.*, **16**, p. 123001.
29. Chiou, T.H., Kleinlogel, S., Cronin, T., Caldwell, R., Loeffler, B., Siddiqi, A., Goldizen, A., and Marshall, J. (2008). Circular polarization vision in a stomatopod crustacean, *Curr. Biol.*, **18**, pp. 429–434.
30. Roberts, N.W., Chiou, T.H., Marshall, N.J., and Cronin, T.W. (2009). A biological quarter-wave retarder with excellent achromaticity in the visible wavelength region, *Nat. Photon.*, **3**, pp. 641–644.
31. Schwind, R. (1991). Polarization vision in water insects and insects living on a moist substrate, *J. Comp. Physiol. A*, **169**, pp. 531–540.
32. Shashar, N., Rutledge, P.S., and Cronin, T.W. (1996). Polarization vision in cuttlefish: A concealed communication channel, *J. Exp. Biol.*, **199**, pp. 2077–2084.
33. Morgado, T.A., Marcos, J.S., Maslovski, S.I., and Silveirinha, M.G. (2012). Negative refraction and partial focusing with a crossed wire mesh: Physical insights and experimental verification, *Appl. Phys. Lett.*, **101**, 021104.
34. Brazhe, R.A., and Meftakhutdinov, R.M. (2007). Negative optical refraction in crystals with strong birefringence, *Tech. Phys.*, **52**, 793.
35. Huang, H., and Hung, Y. (2013). Numerical analysis and the effective parameter retrieval of helical metamaterials, *Photonic and Phononic Properties of Engineered Nanostructures III*, **8632**, 1.
36. Nemirovsky, J., Rechtsman, M.C., and Segev, M. (2012). Negative radiation pressure and negative effective refractive index via dielectric birefringence, *Opt. Express*, **20**, 8907.

37. Demetriadou, A., and Pendry, J.B. (2009). Extreme chirality in Swiss roll metamaterials, *J. Phys.: Condens. Matter*, **21**, 376003.
38. Plum, E., Zhou, J., Dong, J., Fedotov, V., Koschny, T., Soukoulis, C., and Zheludev, N. (2009). Metamaterial with negative index due to chirality, *Phys. Rev. B*, **79**, 35407.
39. Pendry, J.B. (2004). A chiral route to negative refraction, *Science*, **306**, 1353.
40. Zharov, A.A., Zharova, N.A., Noskov, R.E., Shadrivov, I. V., and Kivshar, Y.S. (2005). Birefringent left-handed metamaterials and perfect lenses for vectorial fields, *New J. Phys.*, **7**, 220.
41. Zhao, Y., Belkin, M.A., and Alù, A. (2012). Twisted optical metamaterials for planarized ultrathin broadband circular polarizers, *Nat. Commun.*, **3**, 870.
42. Wu, L., Yang, Z., Cheng, Y., Lu, Z., Zhang, P., Zhao, M., Gong, R., Yuan, X., Zheng, Y., and Duan, J. (2013). Electromagnetic manifestation of chirality in layer-by-layer chiral metamaterials, *Opt. Express*, **21**, 5239.
43. Christofi, A., Stefanou, N., Gantzounis, G., and Papanikolaou, N. (2012). Giant optical activity of helical architectures of plasmonic nanorods, *J. Phys. Chem. C*, **116**, 16674.
44. Decker, M., Zhao, R., Soukoulis, C.M., Linden, S., and Wegener, M. (2010). Twisted split-ring-resonator photonic metamaterial with huge optical activity, *Opt. Lett.*, **35**, 1593.
45. Decker, M., Ruther, M., Kriegler, C.E., Zhou, J., Soukoulis, C.M., Linden, S., and Wegener, M. (2009). Strong optical activity from twisted-cross photonic metamaterials, *Opt. Lett.*, **34**, 2501.
46. Liu, H., Genov, D., Wu, D., Liu, Y., Liu, Z., Sun, C., Zhu, S., and Zhang, X. (2007). Magnetic plasmon hybridization and optical activity at optical frequencies in metallic nanostructures, *Phys. Rev. B*, **76**, 073101.
47. Bai, B., Svirko, Y., Turunen, J., and Vallius, T. (2007). Optical activity in planar chiral metamaterials: Theoretical study, *Phys. Rev. A*, **76**, 023811.
48. Andryieuski, A., Menzel, C., Rockstuhl, C., Malureanu, R., Lederer, F., and Lavrinenko, A. (2010). Homogenization of resonant chiral metamaterials, *Phys. Rev. B*, **82**, 235107.
49. Rogacheva, A.V., Fedotov, V.A., Schwanecke, A.S., and Zheludev, N.I. (2006). Giant gyrotropy due to electromagnetic-field coupling in a bilayered chiral structure, *Phys. Rev. Lett.*, **97**, 177401.
50. Mitov, M., and Dessaud, N. (2006). Going beyond the reflectance limit of cholesteric liquid crystals, *Nat. Mater.*, **5**, 361.

51. Hadad, Y., and Steinberg, B.Z. (2012). One way optical waveguides for matched non-reciprocal nanoantennas with dynamic beam scanning functionality, *Opt. Express*, **21**, A77.

52. Ozaki, M., Matsuhisa, Y., Yoshida, H., Ozaki, R., and Fujii, A. (2007). Photonic crystals based on chiral liquid crystal, *Phys. Status Solidi (a)*, **204**, 3777.

53. Gevorgyan, A., and Harutyunyan, M. (2007). Chiral photonic crystals with an anisotropic defect layer, *Phys. Rev. E*, **76**, 031701.

54. Costa, J., and Silveirinha, M. (2012). Achromatic lens based on a nanowire material with anomalous dispersion, *Opt. Express*, 20, 13915.

55. Menzel, C., Helgert, C., Rockstuhl, C., Kley, E.B., Tünnermann, A., Pertsch, T., and Lederer, F. (2010). Asymmetric transmission of linearly polarized light at optical metamaterials, *Phys. Rev. Lett.*, **104**, 253902.

56. Tang, Y., and Cohen, A.E. (2011). Enhanced enantioselectivity in excitation of chiral molecules by superchiral light, *Science*, **332**, 333.

57. Tang, Y., and Cohen, A.E. (2010). Optical chirality and its interaction with matter, *Phys. Rev. Lett.*, **104**, 163901.

58. Denisov, G., Bratman, V., Cross, A., He, W., Phelps, A., Ronald, K., Samsonov, S., and Whyte, C. (1998). Gyrotron traveling wave amplifier with a helical interaction waveguide, *Phys. Rev. Lett.*, **81**, 5680.

59. Cooke, S.J., and Denisov, G.G. (1998). Linear theory of a wide-band gyro-TWT amplifier using spiral waveguide, *IEEE Trans. Plasma Sci.*, **26**, 519.

60. Denisov, G.G., Bratman, V.L., Phelps, A., and Samsonov, S.V. (1998). Gyro-TWT with a helical operating waveguide: New possibilities to enhance efficiency and frequency bandwidth, *IEEE Trans. Plasma Sci.*, **26**, 508.

61. Shvets, G. (2006). Optical polarizer/isolator based on a rectangular waveguide with helical grooves, *Appl. Phys. Lett.*, **89**, 141127.

62. Bunch, K.J., Grow, R.W., and Baird, J.M. (1988). Backward-wave interaction using step periodic structures, *Int. J. Infrared Milli.*, **9**, 609.

63. Alù, A. (2011). First-principle homogenization theory for periodic metamaterial arrays, *Phys. Rev. B*, **83**, 081102-R.

64. Simovski, C.R. (2011). On electromagnetic characterization and homogenization of nanostructured metamaterials, *J. Opt.*, **13**, 013001.

65. Shore, R.A., and Yaghjian, A.D. (2012). Complex waves on periodic arrays of lossy and lossless permeable spheres: 1. Theory, *Radio Sci.*, **47**, RS2014.

66. Campione, S., Sinclair, M., Capolino, F. (2013). Effective medium representation and complex modes in 3D periodic metamaterials made of cubic resonators with large permittivity at mid-infrared frequencies, *Photon. Nanostruct. Fund. Appl.*, **11**, 423.
67. Ong, H.L., and Meyer, R.B. (1985). Geometrical-optics approximation for the electromagnetic fields in layered-inhomogeneous liquid-crystalline structures, *J. Opt. Soc. Am. A*, **2**, 198.
68. Ong, H.L. (1988). Origin and characteristics of the optical properties of general twisted nematic liquid-crystal displays, *J. Appl. Phys.*, **64**, 614.
69. McIntyre, P. (1978). Transmission of light through a twisted nematic liquid-crystal layer, *J. Opt. Soc. Am.*, **68**, 869.
70. McIntyre, P., and Snyder, A.W. (1978). Light propagation in twisted anisotropic media: Application to photoreceptors, *J. Opt. Soc. Am.*, **68**, 149.
71. Azzam, R.M.A., and Bashara, N.M. (1972). Simplified approach to the propagation of polarized light in anisotropic media: Application to liquid crystals, *J. Opt. Soc. Am.*, **62**, 1252.
72. Di Pasquale, F., Fernandez, F.A., Day, S.E., and Davies, J.B. (1996). Two-dimensional finite-element modeling of nematic liquid crystal devices for optical communications and displays, *IEEE J. Sel. Top. Quant. Electron.*, **2**, 128.
73. Van Orden, D., and Lomakin, V. (2011). Fundamental electromagnetic properties of twisted periodic arrays, *IEEE Trans. Antennas Propag.*, **59**, 2824.
74. Van Orden, D., Fainman, Y., and Lomakin, V. (2010). Twisted chains of resonant particles: Optical polarization control, waveguidance, and radiation, *Opt. Lett.*, **35**, 2579.
75. Berreman, D.W. (1972). Optics in stratified and anisotropic media: 4 × 4-matrix formulation, *J. Opt. Soc. Am.*, **62**, 502.
76. Berreman, D.W. (1973). Optics in smoothly varying anisotropic planar structures: Application to liquid-crystal twist cells, *J. Opt. Soc. Am.*, **63**, 1374.
77. Gagnon, R.J. (1981). Liquid-crystal twist-cell optics, *J. Opt. Soc. Am.*, **71**, 348.
78. Wöhler, H., Haas, G., Fritsch, M., and Mlynski, D.A. (1988). Faster 4 × 4 matrix method for uniaxial inhomogeneous media, *J. Opt. Soc. Am. A*, **5**, 1554.

79. Foresti, M. (1989). Plane-wave propagation in a plane-oriented nematic liquid-crystal multilayer: The case of a wave incident upon the plane containing the liquid-crystal directors, *J. Opt. Soc. Am. A*, **6**, 1254.

80. Oldano, C. (1989). Electromagnetic-wave propagation in anisotropic stratified media, *Phys. Rev. A*, **40**, 6014.

81. Georgieva, E. (1995). Reflection and refraction at the surface of an isotropic chiral medium: Eigenvalue–eigenvector solution using a 4 × 4 matrix method, *J. Opt. Soc. Am. A*, **12**, 2203.

82. Lakhtakia, A., and Weiglhofer, W.S. (1995). On light propagation in helicoidal bianisotropic mediums, *Proc. R. Soc. A*, **448**, 419.

83. Aksenova, E.V., Karetnikov, A.A., Kovshik, A.P., Kryukov, E.V., and Romanov, V.P. (2008). Light propagation in chiral media with large pitch, *J. Opt. Soc. Am. A*, **25**, 600.

84. de Vries, H. (1951). Rotatory power and other optical properties of certain liquid crystals, *Acta Crystallographica*, **4**, 219.

85. Zhao, Y., Engheta, N., and Alù, A. (2011). Homogenization of plasmonic metasurfaces modeled as transmission-line loads, *Metamaterials*, **5**, 90.

86. Askarpour, A.N., Zhao, Y., and Alù, A. (2014). Wave propagation in twisted metamaterials, *Phys. Rev. B*, **90**, 054305.

87. CST Microwave Studio™

Chapter 8

Broadband Optical Metasurfaces and Metamaterials

Jeremy A. Bossard, Zhi Hao Jiang, Xingjie Ni, and Douglas H. Werner
Department of Electrical Engineering, Penn State University, University Park, PA 16802, USA
dhw@psu.edu

8.1 Introduction

In this chapter, we investigate recent breakthroughs in the development of broadband optical metamaterials and metasurfaces that illustrate how their exotic properties can be exploited for broadband applications. In the first part of this chapter, dispersion engineering is introduced as a powerful method for exploiting the resonant properties of metamaterials over broad wavelength ranges in order to enhance practical devices. The second part of the chapter examines how the metamaterial loss can be exploited for broadband absorption in the infrared regime. A robust genetic algorithm (GA) synthesis technique is used to design super-octave and multi-octave metamaterial absorbers (MMAs) using only a single patterned metallic screen. The last part of the chapter investigates broadband

Broadband Metamaterials in Electromagnetics: Technology and Applications
Edited by Douglas H. Werner
Copyright © 2017 Pan Stanford Publishing Pte. Ltd.
ISBN 978-981-4745-68-0 (Hardcover), 978-1-315-36443-8 (eBook)
www.panstanford.com

optical metasurfaces that can control the phase and polarization of a reflected wave. Optical metasurface designs are presented along with measurement results that demonstrate broadband and wide-angle quarter-wave plate and half-wave plate functionalities. A second type of metasurface based on nanoantenna arrays that can artificially induce a phase gradient in the cross-polarized reflected or transmitted wave at an interface and, therefore, steer and focus light is also presented. Together, these metamaterial and metasurface examples illustrate the potential for nanostructured metamaterials to provide unique functionalities over broad bandwidths to facilitate practical optical devices.

8.2 Broadband Dispersion-Engineered Optical Metamaterials

8.2.1 Introduction to Dispersion Engineering

Metamaterials have been demonstrated to exhibit remarkable electromagnetic properties, including negative [81, 93], zero/low [27, 55, 102], and high [20, 82] indices of refraction. These properties, which also include independent control of the refractive index and the intrinsic impedance enable novel optical design strategies and have the potential to enhance the performance of existing devices or to introduce entirely new device functionalities. Recent examples of metamaterial devices that have emerged include artificial mirrors [16], structures with electromagnetic induced transparency [64, 105], flat wave collimating lenses [31], ultrathin absorbers [45, 58], and metamaterial emitters with customized angle- and polarization-dependent emissivity [14, 68]. The unique optical properties arise from the specific geometry and arrangement of nanoscale inclusions in the metamaterial nanostructure that are typically aligned in a periodic lattice [56].

The resonant properties resulting from the metamaterial nanostructure, including the effective refractive index and group delay, are generally strongly dependent on wavelength, which can cause signal distortion and narrow operational bandwidths [17, 32, 35, 80], limiting the widespread use of metamaterials in practical optical devices. However, broadband optical metamaterials can be realized by choosing to operate at wavelengths sufficiently far from the resonant band of the nanostructured inclusion to avoid strong dispersion. Such a strategy has been utilized in the microwave regime to design ground plane cloaks [66, 94] and Luneburg lenses [57, 69]. However, in choosing to avoid highly dispersive regions, this approach excludes negative and zero/low index values, which are among the most important and interesting regions associated with metamaterials. Another alternative strategy to avoiding resonant bands altogether is to exploit them by tailoring the dispersive properties of the metamaterial to specific device needs in order to improve existing components or to enable new optical functionalities [18, 29, 109]. This powerful dispersion engineering design technique has recently been applied at microwave frequencies to demonstrate novel broadband radiated-wave components [3, 44, 46, 61] and planar guided-wave devices [4, 36, 75].

In the following section, we show that the metamaterial dispersion of a metallodielectric fishnet structure can be controllably tailored across the negative, zero, and positive refractive index values to produce a specific broadband optical filtering function by adding deep-subwavelength inclusions into the structure. This general design strategy allows the structure to be optimized to produce an optically thin metamaterial band-pass filter with high transmission and nearly constant group delay throughout the 3.0 μm to 3.5 μm band in the mid-IR and high rejection outside of this band. The dispersion-engineered metamaterial filter is also extended to demonstrate a bifunctional metamaterial prism that both filters an incident wave and provides wavelength-dependent steering of the transmitted beam. These examples illustrate how this powerful dispersion engineering design approach overcomes the narrow bandwidth limitations of previous optical metamaterials and enables opportunities to create novel and practical metamaterial-enabled devices and components.

8.2.2 Broadband Plasmonic Metamaterial Filters with Passive Beam Steering

The unit cell configuration of the multilayer metallodielectric metamaterial considered here is shown in Fig. 8.1a. Gold (Au) is used for the metal layers, and Kapton is employed for the dielectric spacers. As our design target, we selected a filter with a passband in the mid-infrared range between 3.0 μm and 3.5 μm and stopbands on either side of the passband extending down to 2.5 μm and up to 4.0 μm. Within the passband, the effective index of refraction (or the transmission phase delay) should transition from negative through zero to positive seamlessly with a near-constant group delay. We began with a conventional fishnet nanostructure composed of a three-layer metal–dielectric–metal stack perforated with a doubly periodic array of air holes because this structure provides flexibility to adjust the effective permittivity and permeability from positive to negative values [25, 76, 93, 102]. In order to provide a more delicate manipulation over the surface plasmon polariton (SPP) modes related to the air hole waveguide and the subwavelength gap between the metal layers, two square subwavelength nano-notch cuts are made in each side of the square air holes, enabling a precise control over the effective permittivity and permeability dispersion.

A GA coupled with an efficient full-wave electromagnetic solver was used to optimize the nanostructure dimensions for the effective medium parameter dispersion required to meet the challenging multi-objective design criteria. Importantly, predefined constraints on the allowable unit cell size, air hole size, nano-notch size, and layer thicknesses were incorporated into the optimization algorithm to avoid generating structures that are impractical to fabricate. Eight-fold symmetry was enforced on the structure in the plane perpendicular to the incident wave vector to ensure a polarization-insensitive response for normally incident plane waves. Finally, experimentally measured dispersive optical properties for the constituent metal and dielectric materials were used in the full-wave solver to minimize discrepancies between the simulated and experimentally fabricated metamaterial responses. For each candidate design in the GA optimization, the complex transmission and reflection coefficients were calculated using the high frequency structure simulator (HFSS) finite-element solver. Periodic boundary conditions were assigned to the lateral walls of the simulation domain to approximate a plane wave normally incident on the structure. The effective medium parameters (ε_{eff} and μ_{eff}) within the target wavelength range were retrieved from the transmission and reflection coefficients by using an inversion algorithm [85]. These parameters were compared with the target effective medium requirements to determine the *cost* of the candidate design, which is defined by

$$\begin{cases} \text{cost } 1 = \sum_{\lambda_{n,e,p}} \left(\left| \varepsilon_{\text{eff}} - \varepsilon_{\text{tar},i} \right| + \left| \mu_{\text{eff}} - \mu_{\text{tar},i} \right| \right) + \sum_{\lambda_{\text{stop}}} \left| \frac{1}{\log(Z_{\text{eff}})} \right| \\ \text{cost } 2 = \sum_{\lambda_{\text{pass}}} \left| \tau_{\text{g}} - \tau_{\text{g,mean}} \right| \end{cases} \tag{8.1a}$$

$$\text{cost} = \text{cost } 1 + \text{cost } 2, \tag{8.1b}$$

where $\varepsilon_{\text{tar},i} = \{-1,0,1\}$ and $\mu_{\text{tar},i} = \{-1,0,1\}$ are the targeted permittivity and permeability values at λ_n, λ_e, and λ_p, respectively, while λ_{stop} refers to the wavelength range larger than λ_n and smaller than λ_p, and $\tau_{\text{g,mean}}$ is the average group delay of the sample frequency points within the passband. The group delay is calculated by

$$\tau_{\text{g}} = \frac{L\left(\text{Re}(n_{\text{eff}}) + \omega \dfrac{d(\text{Re}(n_{\text{eff}}))}{d\omega} \right)}{c_0} = \text{constant} \tag{8.2}$$

where c_0 is the speed of light in free space, L is the total thickness of the material slab, and n_{eff} is the effective index of refraction. Cost 1 minimizes the difference between the targeted and optimized values for permittivity and permeability in the passband and maximizes the impedance mismatch at the stopband frequencies to ensure high reflection. It also minimizes the absorption loss, *i.e.*, the imaginary parts of permittivity and permeability of the metamaterial. Cost 2 minimizes the group delay variation in the passband. The GA evolved solutions until it converged on a metallodielectric nanostructure with a sufficiently low overall cost value that met the target design criteria.

The simulated optical response and effective medium properties of the optimized modified fishnet nanostructure are shown in Fig. 8.2a and Fig. 8.2b, respectively. Within the 3.0 μm to 3.5 μm range, the average transmission at normal incidence is −0.9 dB, indicating that there is a good impedance match between the metamaterial and free space and low absorption loss. Notably, the maximum variation in transmitted power is less than 0.4 dB across the full passband window, which confirms that this structure provides excellent control of the effective medium parameters across a broad bandwidth.

Outside the passband, the impedance mismatch to free space results in a high reflectivity, which reduces the average transmission to less than −10 dB. The transitions between the passband and the stop-bands have steep roll-offs of ~93 dBμm^{-1} on the short-wavelength side and ~101 dBμm^{-1} on the long wavelength side. The simulated group delay is also plotted in Fig. 8.2a and exhibits a small variation from 15 fs to 27 fs across the transmission passband window from 3.0 μm to 3.5 μm. This group delay fluctuation of ~1 period within a 20% bandwidth is much lower than has been previously reported for negative index metamaterials [26, 76], which had fluctuations of ~3 periods within a 5% bandwidth. In addition, the simultaneous negative phase and group velocities can also be identified within the range from 3.65 μm to 3.70 μm, which corroborates previously reported results [76]. From these results, we expect that even better control over the effective medium parameter profiles and resulting broadband optical properties could be achieved given additional geometric design flexibility, at the expense of more challenging nanofabrication requirements.

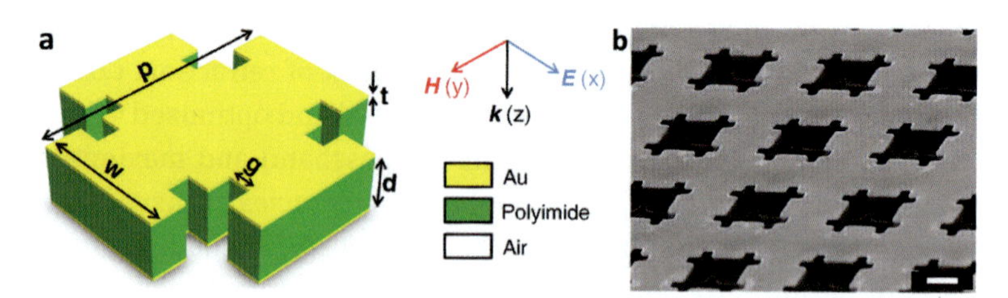

Figure 8.1 (a) The geometry and dimensions of a single unit cell of the modified fishnet nanostructure. The optimized geometry dimensions are p = 2113 nm, w = 990 nm, g = 198 nm, t = 30 nm, and d = 450 nm. (b) FESEM image of the fabricated nanostructure. Scale bar: 1 μm.

The optical properties of the modified fishnet nanostructure can be understood by examining the simulated effective permittivity, permeability, refractive index, and impedance shown in Fig. 8.2b. The effective permeability has a strong primary magnetic resonance at 3.70 μm and a weaker secondary resonance at 2.85 μm, which correspond to the transmission minima of the filter. These two magnetic resonances are strongly related to the excitation of the gap-SPP modes with transverse wave numbers expressed as $k_{spp} = iG_x + jG_y$, where $G_x = G_y = 2\pi/p$. The magnetic resonance at 3.70 μm corresponds to the $(i,j) = (1,0)$ mode, whereas the resonance at 2.85 μm denotes the $(i,j) = (1,1)$ mode [49]. The relative strengths of these two resonances give a nearly linear decreasing effective permeability that varies from $\mu_{\text{eff}} = +0.95$ to $\mu_{\text{eff}} = -0.9$ in the 3 μm to 3.5 μm range with a zero crossing at 3.25 μm. The effective permittivity varies from $\varepsilon_{\text{eff}} = +1.20$ to $\varepsilon_{\text{eff}} = -0.95$ across the same wavelength range, which is slightly different from the values of the permeability. Nevertheless, the effective impedance of this structure is well-matched to free space throughout the entire passband, leading to its low reflectance across the 3 μm to 3.5 μm band. Outside the passband, the impedance is purely imaginary, which appears to be inductive in the long wavelength range and capacitive in the short-wavelength regime. The impedance mismatch is responsible for the strong reflection observed in the stopbands. Finally, the effective refractive index, which decreases from $n_{\text{eff}} = +1.05$ at 3.0 μm to $n_{\text{eff}} = -0.92$ at 3.5 μm, has a dispersive rate of change (*i.e.*, slope) that compensates for the refractive index dispersion and produces a nearly constant group delay. The imaginary part of effective refractive index has a magnitude less than 0.15 across the entire passband, which results in low absorption loss.

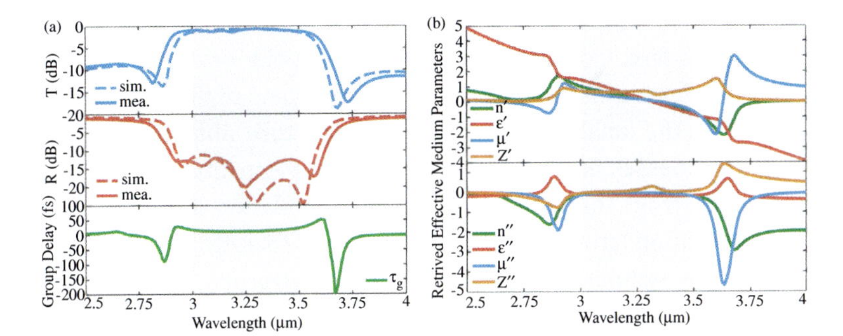

Figure 8.2 (a) Simulated and measured transmission (top) and reflection (middle) magnitudes for normally incident radiation. Simulated group delay τ_g (bottom). (b) Real (top) and imaginary (bottom) parts of the effective index of refraction n_{eff}, permittivity ε_{eff}, permeability μ_{eff}, and impedance Z_{eff} retrieved from the full-wave simulation of the metamaterial structure. The shaded region from 3.0 μm to 3.5 μm represents the passband of the metamaterial filter.

The broadband metamaterial filter was fabricated by standard electron-beam lithography (see Fig. 8.1b). In order to prevent degradation in optical properties due to substrate-induced bianisotropy [48, 103], the optically thin structure was released from the handle wafer by selective etching of a thermal oxide and then mounted on a frame for optical characterization. The measured optical properties shown in Fig. 8.2a are in strong agreement with the simulation predictions, exhibiting a high and flat in-band transmission as well as a high out-of-band reflection. The measured average insertion loss within the passband window from 2.95 μm to 3.60 μm is −1 dB. The transmission window of the fabricated structure has a maximum measured variation of only 0.5 dB over the entire passband, which confirms that the impedance match is maintained as the effective index traverses zero. The transmissions in the short and long wavelength stopband regions remain below −9.2 dB and −11 dB, respectively. This gives an average transmitted power of approximately 10% across both stopbands with an average reflectivity of −1.94 dB and an average absorption of −5.85 dB. The roll-off rates between the passband and stopbands on the short and long wavelength sides are 76 dBμm^{-1} and 91 dBμm^{-1}, respectively. The slightly increased bandwidth, decreased roll-off rates, and higher absorption as compared with those predicted by simulation are caused by a minor reduction in nano-notch size, which lowers the quality factor of the resonances within the band. The lower quality factor is attributed to the rounded corners of the nano-notch as well as the small deviations in the structure dimensions (e.g., layer thicknesses and air hole size) and the constitutive material properties from the values used in the simulation. Despite these slight deviations from the ideal performance, this strong agreement between experimental and simulated optical properties validates the design concept and shows that deep-subwavelength inclusions can be used to tailor the permittivity and permeability dispersion for broadband performance.

This design procedure can also be extended to synthesize multilayer nanostructures such as the one shown in Fig. 8.3a. This structure consists of six gold layers and five polyimide spacers, each with thicknesses of 46 nm and 76 nm, respectively. Figure 8.3b shows the simulated transmission and reflection at normal incidence for this metamaterial. A broad and flat passband can be observed ranging from 3.0 μm to 3.7 μm with an average transmission of −1.14 dB. Outside this passband, the transmission is less than −13.28 dB in the 3.8 μm to 4.5 μm and 2.5 μm to 2.9 μm stopbands. The transitions between the passband and stopbands have steep roll-offs of 142 dBμm^{-1} between 3.7 μm and 3.8 μm and 113 dBμm^{-1} between 2.9 μm and 3.0 μm. In contrast to conventional filters that rely on bulky multilayer stacks of different materials, the proposed nanostructured metamaterial is more suitable for integration into micro- and nanoscale photonic systems. It also outperforms in terms of both transmittance level and bandwidth when compared to thin frequency-selective surface-based infrared filters reported in the literature [70, 89]. The performance is well maintained for incidence angles up to 30°, due to the angle-insensitive excitation of the electric and magnetic SPP modes [47].

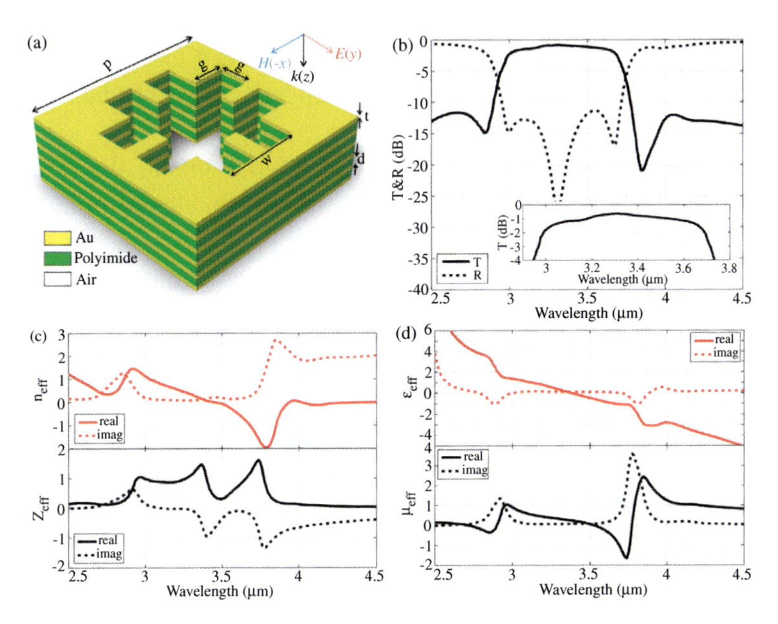

Figure 8.3 (a) Unit cell geometry of the multilayer metallodielectric metamaterial with eight square subwavelength nano-notches. The structure has six gold layers and five dielectric spacer layers. The dimensions are p = 1985 nm, w = 807 nm, g = 341 nm, t = 46 nm, and d = 76 nm. (b) Simulated transmission and reflection spectra. The inset shows a magnified view of the transmission in the passband. (c) Retrieved effective index of refraction and impedance. (d) Retrieved effective permittivity and permeability.

This multilayer metamaterial structure can be further extended to construct a broadband, bifunctional plasmonic metamaterial device by tilting the top interface to form a prism. Similar metamaterial prisms have been demonstrated previously for the RF [81] and optical [93] regimes as qualitative and quantitative evidences of negative refraction. However, these previous demonstrations were limited to only a single polarization, suffered from high reflectivity due to impedance mismatch with free space and exhibited significant losses due to the evanescent mode gap around the index zero crossing wavelength region. By contrast, the device presented here is able to simultaneously provide two functionalities: a broadband frequency-selective transmission and in-band wavelength-dependent passive beam steering, both of which are polarization independent. As Fig. 8.4a illustrates, the metamaterial prism is composed of a stair-type multilayer metamaterial structure with eight steps. The steps range in number of functional layers from four to eighteen, forming a slope angle of approximately 13°, and each step has a width of a half unit cell.

To numerically validate the metamaterial prism, the HFSS finite-element solver was employed. In the simulation domain, a portion of the metamaterial prism with a width of one unit cell in the x-direction and a length of four unit cells in the y-direction was considered. The metamaterial prism has a step every half unit cell with a slope angle of approximately 13°. To mimic a one-dimensionally infinite structure, perfect electric conducting boundary conditions were assigned to the front and back walls in the x-direction for the transverse electric (TE) polarization and perfect magnetic conducting boundary conditions were assigned for the transverse magnetic (TM) polarization. A waveguide and a wave port were employed to produce the incident waves impinging normally on the prism from the negative z-direction.

The full-wave simulated electric field distributions at different wavelengths and for both TE and TM polarizations are shown in Fig. 8.4c and Fig. 8.4d, respectively. Similar to the multilayer metamaterial filter described in Fig. 8.3, at normal incidence this prism has a passband from 3.0 μm to 3.7 μm, outside of which the incident light is strongly reflected. Figure 8.4b presents the reflection properties of the metamaterial prism for both polarizations, which shows a reflection amplitude smaller than −12 dB only within the desired beam-steering passband. The out-of-band incident wave, on the other hand, is highly rejected by the prism. By virtue of the tilted interface of the metamaterial prism and the smoothly changing index in the passband, the transmitted beam is directed at different angles upon exiting the structure, depending on the wavelength of the incident light. As can be observed from the snapshots of the electric field distribution at different wavelengths for both polarizations, outside the passband there is almost no light passing through the metamaterial prism due to the as-designed strong impedance mismatch between the metamaterial structure and free space. Within the passband, at 3.66 μm, where the effective index is close to negative unity, the beam is directed at an angle of refraction of −11°. At 3.41 μm, where the effective index is approaching zero, the beam is directed at an angle of refraction of 0° (normal to the tilted interface of the metamaterial prism), and at 3.06 μm, where the effective index is around positive unity, the beam is in the same direction as that of the incident wave. The angle of refraction as a function of wavelength is presented in Fig. 8.4e, showing a smooth transition from negative through zero to positive values, thus validating the proposed design as well as the convergence of the effective index. The curves obtained for both TE and TM polarizations agree well with each other, confirming the polarization-independent properties of this metamaterial prism. This result represents the first example of an impedance-matched metamaterial prism with a smooth negative-zero-positive index transition that works for both TE and TM polarizations. The observed transmitted wave steering is a direct consequence of the dispersive properties of the metamaterial without actively tuning the phase of each unit cell in the metamaterial prism. Thus, the passive angular deviation of the transmitted wave is wavelength dependent, similar to the behavior of leaky-wave radiators [95], although operating under a different physical mechanism.

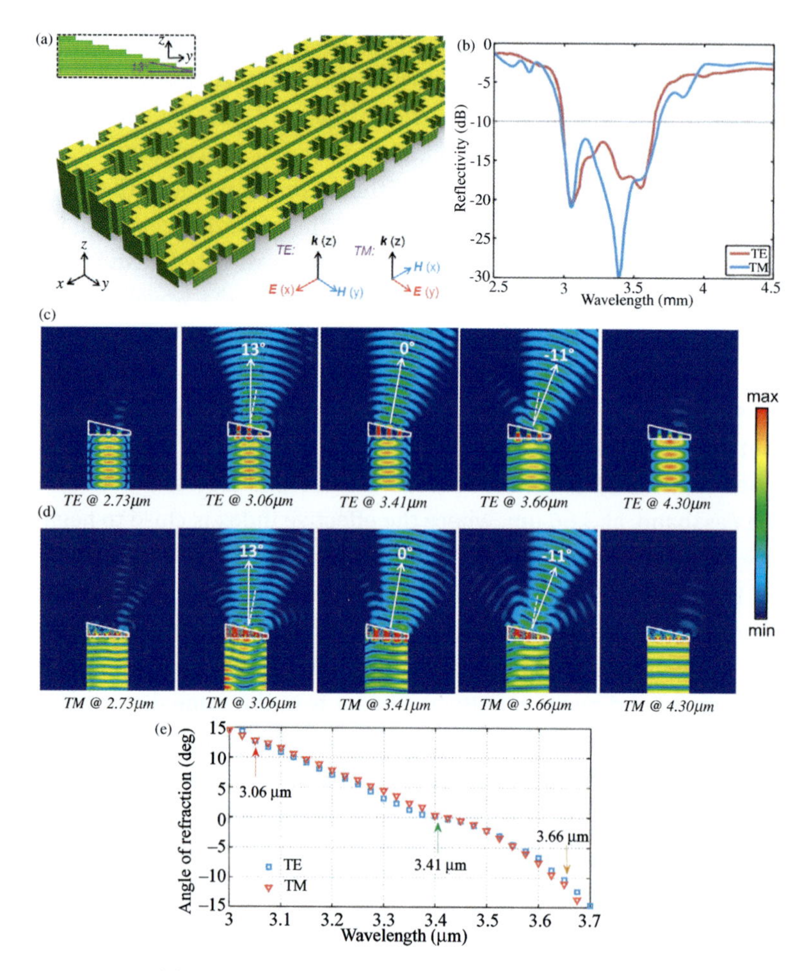

Figure 8.4 (a) Three-dimensional tilted view of the configuration of the metamaterial prism and orientation of the incident beam. The inset shows the side view of the metamaterial prism. (b) Simulated reflectivity of the actual metamaterial prism for both TE and TM polarizations. (c) Snapshots of the electric field distribution for TE polarization at different wavelengths. (d) Snapshots of the electric field distribution for TM polarization at different wavelengths. (e) Angle of refraction with respect to the top surface normal as a function of wavelength in the passband of the metamaterial prism for both TE and TM polarizations.

8.3 Broadband Metamaterial Absorbers for the Infrared

8.3.1 Introduction to Metamaterial Absorbers

A wide variety of applications exists for electromagnetic absorbers both at radiofrequency (RF) and optical regimes, ranging from spectroscopy [91, 108] and thermal imaging [23, 59] to signature control [19, 52, 58, 96]. This has motivated the development of an array of technologies for realizing absorbers with diverse functionality, including multiband [45] or broadband [7, 15, 21] performance as well as targeting specific angles [13] or polarizations [14] of light. Generally, electromagnetic absorbers rely on electric and/or magnetic losses in the constituent material properties in order to achieve energy dissipation in the form of heat. The typical approach is to employ some type of resonant structure that supports high currents or fields that amplify the absorption in the lossy material. Broadband absorption can be obtained using structures with multiple resonances, including multilayer structures with many patterned screens cascaded together [22]. By contrast, the approach described in the following sections requires only a single patterned screen to achieve broadband absorption covering one or more octaves in bandwidth in the infrared (IR), producing designs that can be fabricated without the need for performing difficult registration between patterned layers.

Electromagnetic absorbers have long been a topic of interest, with metal diffraction gratings being identified several decades ago for producing high absorption at specific wavelengths and incidence angles [43]. More recently, metallic gratings were explored for use as narrowband optical absorbers [11, 60, 92]. Similarly, another versatile absorber technology based on exploiting properties of electromagnetic bandgap (EBG) metamaterials was also first introduced for use at RF [52] and later adapted for THz and optical wavelength applications by employing micro- and nanofabrication techniques [7, 21, 38, 45, 65, 67, 104]. A more general effective medium approach involves designing structures with subwavelength inclusions that give rise to resonances in the effective permittivity or effective permeability of the medium [6, 9, 58, 90]. Near resonance, the effective parameters of such metamaterials can have large imaginary parts that produce high absorption as a wave passes through the medium. One other recent technique termed an optical black hole guides incident light to the core of the device, where an absorbing material can be used to attenuate the electromagnetic waves [71]. Among these techniques, EBG structures are particularly appealing for applications at optical wavelengths because they offer straightforward nanofabrication.

Significant progress has been made at both THz and optical wavelengths to improve the performance of metamaterial absorbers (MMA), experimentally demonstrating single- and multiband devices with near-unity absorption, polarization-independent absorption, and good performance over wide fields of view (FOV) [6, 45, 65, 67, 104]. Efforts have also been made to realize absorbers with wide bandwidths by combining resonators of different sizes [7, 21] or by optimizing the structure [5, 97] to have a wide bandwidth. Recently, researchers were able to achieve an octave of bandwidth with near-perfect absorption across the band [15].

In the following sections, we describe a technique for synthesizing EBG-based MMAs designed to have excellent (>90%) absorptivity over octave and multi-octave bandwidths and a wide FOV in the near- to mid-IR. Recent material advances such as using Pd instead of conventional Au as the metal and introducing an impedance matching superstrate into the EBG structure are incorporated into the design [15], and a GA [37] is employed to synthesize the screen geometry and structure dimensions to meet the challenging performance requirements. Two super-octave MMA designs are presented utilizing Pd and Au screens with >90% absorption over the 2 μm to 5 μm band, and two multi-octave MMA designs are synthesized with >90% absorption over the 1 μm to 5 μm band and a ±40° FOV, demonstrating the effectiveness of the proposed design technique.

8.3.2 GA Optimization of Metamaterial Absorbers

The proposed broadband EBG-based MMA structure comprises four layers, including a metallic ground layer and a perforated metallic layer sandwiched between two dielectric layers, as shown in Fig. 8.5.

Because the metallic ground completely attenuates the wave, the MMA can be placed over any arbitrary surface without affecting its performance. Hence, a supporting Si wafer is assumed in this study. Polyimide was chosen as the dielectric because of its low loss over the band of interest in the near-IR and mid-IR, and Pd was chosen as the metal because it is less conductive than commonly employed metals such as Au. The lower Pd conductivity means that it has a larger skin depth than Au, which allows for optimized screen thickness values that are reasonable to fabricate. The metamaterial is defined by a single unit cell with periodic boundary conditions in two dimensions. An electromagnetic cavity is formed between the Pd screen and ground in which large fields are excited at resonance, which causes the incident wave to be attenuated. This cavity can support multiple resonances, depending on the dimensions and screen geometry, that merge together to form a broad absorption band. Selecting complex, multi-resonant geometries is a task that is well suited for robust stochastic optimizers that can effectively solve challenging design problems with many parameters [24]. Hence, a GA was employed to evolve MMA solutions with super-octave and multi-octave absorption bandwidths.

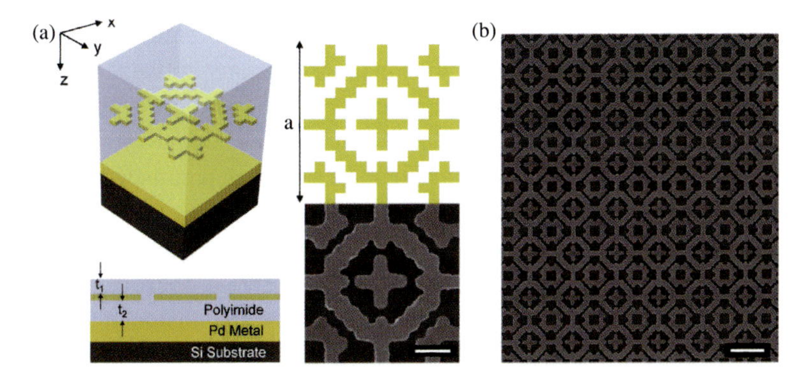

Figure 8.5 Geometry and FESEM image of a super-octave bandwidth EBG-based MMA. (a) Left: 3D schematic and cross-sectional view of the Pd-based MMA structure optimized for broadband absorption from 2 μm to 5 μm in the mid-IR. Right: Unit cell geometry and FESEM image of one unit cell. Scale bar: 200 nm. (b) Low-magnification FESEM image showing the fabricated structure. Scale bar: 600 nm.

The GA is a population-based stochastic optimizer that incorporates the principles of natural selection and survival of the fittest [37]. A population of binary strings, or chromosomes, that contain the design parameters is initially randomly generated and evaluated for performance. The chromosomes are then ranked and selected for mating to produce a new generation. Parent chromosomes selected for mating undergo a crossover operation, where the binary data is swapped between two parents after a crossover point is randomly selected. A small percentage of random mutations are also applied to the children to help the population continually explore more of the design parameter space. In order to represent the geometry in a binary string, the metallic screen is divided into a 15 × 15 binary grid, where "0" represents no metal and "1" represents metal. The other dimensions, such as layer thicknesses and the unit cell size, are also encoded as binary numbers into the chromosome. Eight-fold mirrored symmetry is also applied to the unit cell, where one-eighth of the unit cell is encoded into the chromosome, while the remaining sections of the screen geometry are retrieved by mirroring the encoded triangle. Using eight-fold symmetry was found to improve the azimuth and polarization stability of the MMA designs. Fabrication constraints were implemented in the screen generation, which remove isolated single metal pixels and also diagonal connections between metal pixels, which are difficult to accurately produce in experiment.

The performance of each chromosome is evaluated by a user-defined fitness function. In this function, the design parameters from the chromosome are decoded and used to construct the unit cell geometry, allowing its performance to be evaluated with a full-wave electromagnetic solver. The measured dispersive properties of the constituent polyimide, Pd, and/or Au films are also included in the full-wave electromagnetic simulation to ensure that the predicted performance of the MMA will agree well with experimental results. Each candidate design is simulated using a full-wave finite-element boundary integral (FEBI) technique [28], which predicts the scattering from a periodic arrangement of the unit cell geometry. The absorptivity is calculated as the difference between unity and the sum of the transmission and reflection intensities. Using the absorptivity for both TE and TM polarizations at user-specified wavelengths λ_i and incidence angles $(\theta,\phi)_i$, the cost of the design is evaluated according to

$$\text{cost} = \sum_{\lambda_i} \sum_{(\theta,\varphi)_j} (1 - A_{TE})^2 + (1 - A_{TM})^2 \tag{8.3}$$

where A_{TE} and A_{TM} are the TE- and TM-polarized absorption coefficients, respectively. The GA picks MMA geometries that have a minimized cost function, thereby evolving designs that exhibit near-unity absorption over all of the user-specified wavelengths and angles.

8.3.3 Super-Octave Metamaterial Absorbers for the Infrared

The GA optimizer was employed to identify a metallic nanostructure array that simultaneously maximizes absorption over a broad bandwidth while also giving a wide-FOV and polarization-independent response, targeting an MMA with a high minimum absorptivity of greater than 90% over an angular range of ±40° and having more than an octave of bandwidth across the 2 μm to 5 μm mid-IR range. Absorber geometries were optimized using both Au and Pd metals for the ground plane and patterned nanostructure layer in order to compare the performance of these two materials. As described later in the next section, the higher optical loss of Pd as compared to Au provides significant advantages in terms of reliable nanofabrication of the MMA structure. Polyimide was selected for the dielectric substrate and superstrate layers because of its non-dispersive optical properties over the target wavelength range. In order to accurately account for the material dispersion in the design, the measured optical constants (n and k) of the Au, Pd, and polyimide thin films were used during the design optimization.

The GA evolved all of the adjustable design parameters, including the thickness of each layer, the unit cell period, and the patterned metal feature geometry, to identify a four-layer MMA that meets the target optical performance metrics. The metal–ground plane layer thickness was fixed at 100 nm, which is several times the skin depth for both Pd and Au. By minimizing the cost function in Eq. (8.3), the GA evolves the nanostructure pattern and layer thicknesses to have near-unity absorption over all of the test wavelengths and incidence angles. For the first mid-IR broadband MMA design, 16 test wavelengths were specified in approximately equally spaced frequency intervals over the range from 2 μm to 5 μm, and the test incidence angles were chosen to be $(\theta,\varphi)_\text{i} = \{(0°,0°), (40°,0°), (40°,45°)\}$, which covers an FOV of up to ±40°. Throughout the evolutionary process, structures that do not meet predefined nanofabrication constraints were assigned a high cost and were eliminated from subsequent candidate populations [89]. This ensured that the optimized structure could be fabricated without adjustments to the screen geometry. Operating on a population of 24 chromosomes, the GA converged to the MMA structure shown in Fig. 8.5a with minimized cost in 86 generations. The optimized thicknesses are 30 nm for the patterned Pd nanostructures and t_1 = 398 nm and t_2 = 429 nm for the top and bottom polyimide layers. The unit cell period is a = 851 nm on both sides with a pixel size of 57 nm. The optimized unit cell geometry seen in Fig. 8.5a is complex, containing several identifiable features, including a cross dipole centered in the unit cell, a large loop centered in the unit cell and interconnected between unit cells, and four smaller, isolated loops centered at the corners of the unit cell. These nanostructure elements and the coupling between them support multiple electromagnetic resonances and work in consort to produce high absorption across the target broad mid-IR band.

The simulated absorptivity of the MMA is calculated from the scattering parameters according to $A = 1 - R$, where R is the reflectance. Figure 8.6a shows the simulated absorptivity for unpolarized, TE-polarized, and TM-polarized light. The normal incidence spectra in Fig. 8.6a demonstrate that the minimum spectral absorptivity for this structure is greater than 90% over the entire 1.90 μm to 5.47 μm range with a high average value of 98.8% across this band. It is evident from these spectra that the broadband absorption is due to the merging of multiple strong resonances positioned at optimized wavelengths across the band. The peak values of normal incidence absorptivity occur at 2.13 μm, 3.0 μm, 3.7 μm, and 5.05 μm. Figure 8.6a also shows two-dimensional contour plots of the predicted angular dependence of the absorptivity from normal $\theta = 0°$ up to $\theta = 55°$ off-normal incidence for all polarizations. These simulations reveal that at an incidence angle of $\theta = 55°$, more than an octave bandwidth is achieved with a minimum absorptivity of 89.5% and that an average absorptivity of 94.7% is maintained over the 2 μm to 5 μm band, which is wider than the targeted FOV used for optimization. Comparing the TE and TM responses in Fig. 8.6a, we can see that the TM absorption at the long wavelength edge of the absorption band drops off more quickly with increasing incidence angle, whereas the dip in absorptivity around 4.5 μm becomes more pronounced for TE polarization at large incidence angles. These small changes in the MMA response with increasing incidence angles are associated with slight shifts in peak position, strength, and bandwidth of the dominant resonances found at normal incidence.

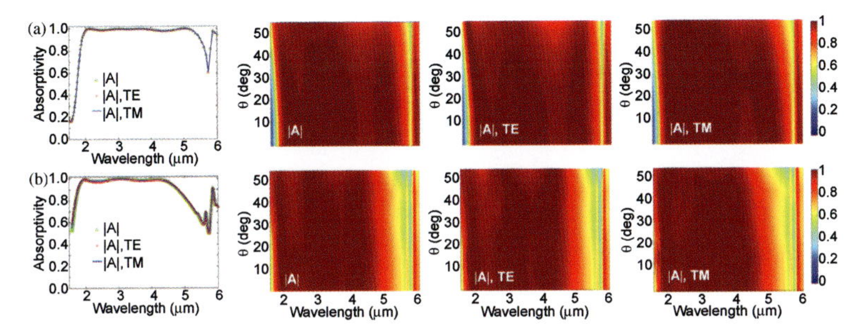

Figure 8.6 (a) Simulated and (b) measured angular dispersion of the absorption spectra for the MMA design shown in Fig. 8.5. From left to right: Normal incidence absorptivity for unpolarized, TE, and TM illumination. Contour plots of absorptivity as a function of wavelength and angle of incident from normal up to 55° incidence under unpolarized (second column), TE (third column), and TM (right column) illumination.

In order to understand the contribution of different parts of the nanostructure geometry to the broadband absorption, the electric volume currents excited within the MMA were simulated at the wavelengths with peak absorption (2.13 μm, 3.0 μm, 3.7 μm, and 5.05 μm) and at one wavelength outside the band (1.5 μm) for normally incident, TE-polarized illumination. The top views of the current distributions are shown in Fig. 8.7, revealing that outside the absorption band at 1.5 μm, the electric current in the MMA is negligible. By contrast, at wavelengths corresponding to peak in-band absorption, large electric currents and field enhancements are found on different parts of the Pd nanostructure array. The central crosses support the highest current at the shortest 2.13 μm wavelength. At the intermediate 3.00 μm and 3.70 μm wavelengths, the current distribution spreads from the central cross to include the large and small loops, and at the longest 5.05 μm wavelength, the electric current is high on all parts of nanostructure except the central cross.

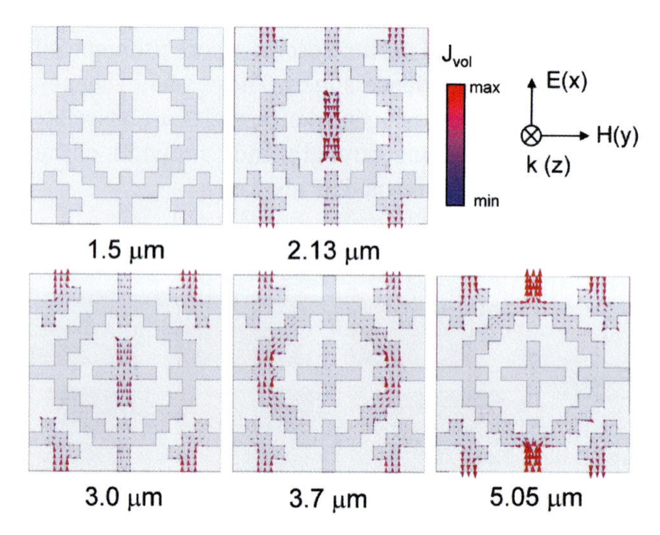

Figure 8.7 Top view of the current distributions in the nanostructured Pd screen under normal incidence illumination at the wavelengths 1.50 μm, 2.13 μm, 3.00 μm, 3.70 μm, and 5.05 μm.

Compared to other MMAs that excite magnetic resonances [45], the strong broadband absorption in this MMA comes entirely from electric resonances. The relatively thick polyimide substrate layer prevents strong coupling between the patterned nanostructure and the ground layer, inhibiting loop currents associated with magnetic responses. Cross-sectional views of the electric volume currents confirm that large currents are only present on the Pd nanostructure layer and not on the ground plane. Hence, the simulation results demonstrate that the broadband performance is achieved by exciting multiple closely spaced electric resonances on the single-layer nanostructure array.

8.3.4 Choice of Metals in Broadband Absorbers

In order to evaluate the tradeoffs of using Au versus Pd as the metal for the nanostructure array and ground layers in the four-layer broadband MMA, a second Au-based MMA was synthesized by the GA using the same cost function and performance criteria as was used for the design shown in Fig. 8.5. The optimized Au-based broadband MMA design is presented in Fig. 8.8. Similar to the Pd-based MMA simulations in Fig. 8.6, this design provides greater than 90% absorptivity over a band from 1.90 μm to 5.25 μm and a wide FOV of ±40°. The optimized layer thicknesses are t_1 = 423 nm and t_2 = 456 nm for the polyimide superstrate and substrate, respectively. The unit cell period is a = 998 nm on each side with a pixel size of 67 nm. Because Au is more conductive than Pd in the mid-IR, the optimum Au nanostructure film thickness that provides a good impedance match over such a large bandwidth is only 10 nm, which is much thinner than the 30 nm thick Pd nanostructure film. The extremely thin Au film makes the optical properties of the Au-based MMA more sensitive to small variations in metal thickness.

In order to investigate this Au-based structure experimentally, the optimized Au-based MMA was fabricated, as shown in Figs. 8.8b,c with average Au nanostructure film thicknesses that varied from 9 nm to 12 nm in increments of ~1 nm. The measured absorptivity of the four samples is compared with the simulated result in Fig. 8.8c. The fabricated Au-based MMA maintains a high absorptivity greater than 90% over a broadband between 1.95 μm and 4.80 μm. However, the structure has a lower average broadband absorptivity of 94.4% as compared to the optimized Pd-based MMA. Decreasing the Au film thickness to 9 nm greatly reduces the average absorptivity to 81.9%, while increasing the film thickness to 11 nm and 12 nm results in large fluctuations in spectral absorptivity, giving an average absorptivity of approximately 89.2% across the band. As shown in Fig. 8.8d, the absorptivity of the optimized Pd-based MMA is relativity insensitive to Pd thickness. Two samples with the optimum Pd thickness of 30 nm and two samples with an increased Pd thickness of 40 nm were fabricated and characterized, showing that increasing the Pd thickness from 30 nm to 40 nm had a negligible effect on the absorption with only a slight reduction in the long wavelength band edge from 4.81 μm to 4.48 μm. This underscores the importance of identifying and selecting constituent materials that are robust against process variations for high-performance metamaterial designs.

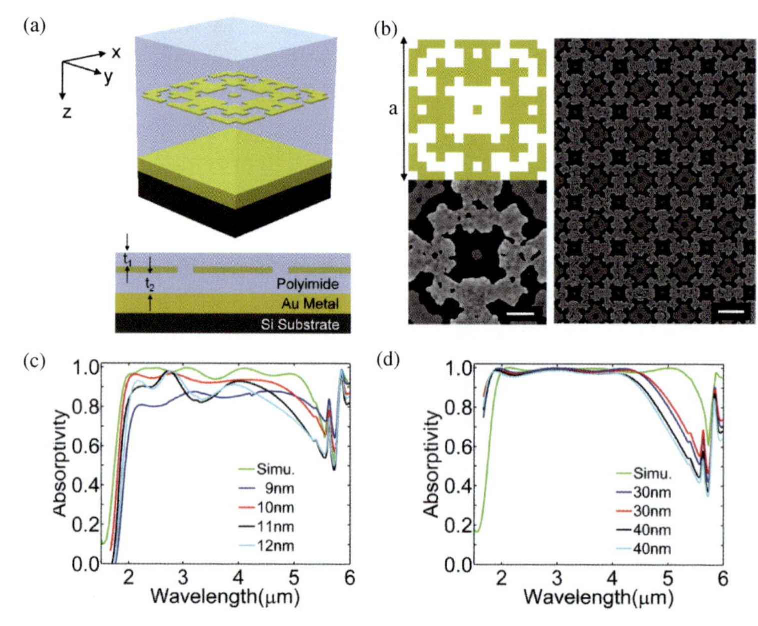

Figure 8.8 (a) Three-dimensional schematic and cross section of an Au-based broadband MMA. (b) Left: Optimized unit cell geometry and FESEM image of fabricated unit cell. Scale bar: 200 nm. Right: Low-magnification FESEM image of the fabricated sample. Scale bar: 600 nm. (c) Simulation and measurements for four Au design samples under unpolarized illumination at normal incidence. The average Au thicknesses are 9 nm, 10 nm, 11 nm, and 12 nm. The performance of these four samples varies even with only 1 nm thickness change. (d) Simulation and measurements for four Pd design samples with thicknesses of 30 nm and 40 nm under unpolarized illumination at normal incidence. Comparison of the four samples shows only a slight blue-shift at longer wavelengths for thicker Pd nanostructures.

8.3.5 Multi-Octave Metamaterial Absorbers for the Infrared

Extending the bandwidth of EBG-based MMAs to multiple octaves presents a difficult design challenge for the GA optimizer. In addition to absorbing the incident wave over the entire multi-octave bandwidth, the MMA must also maintain a good impedance match to free space over the targeted operating band to minimize reflection from the surface of the MMA. Because the optical properties of the metal are dispersive over the near-IR to mid-IR range, the optimum metallic screen thickness is changing with wavelength. Furthermore, the unit cell size is limited to be less than half of the minimum absorption wavelength, so that no higher-order diffraction modes are reflected from the surface of the MMA for all incidence angles. Hence, the GA must balance the metal layer thickness as well as find a pixelated geometry that can support electric resonances to cover the entire targeted multi-octave bandwidth. In this section, two MMA designs will be presented targeting multi-octave absorption from 1 μm to 5 μm in the near- to mid-IR ranges, which is unprecedented performance for single screen designs.

In order to meet the goal of multi-octave, near-ideal absorption, 27 approximately equally spaced wavelengths λ_i were targeted over the range from 1 μm to 5 μm, and three incidence angles $(\theta,\phi)_i$ = {(0°,0°), (40°,0°), (40°,45°)} covering an FOV of ±40° were selected for optimization. The GA optimized a population of 24 chromosomes converging on the two MMA designs shown in Fig. 8.9 after 375 generations and 250 generations, respectively. The optimized parameters for Design 1 are polyimide thicknesses PI_1 = 243 nm and PI_2 = 219 nm, a Pd screen thickness of 21 nm, and unit cell dimension p = 469 nm, while the optimized parameters for Design 2 are polyimide thicknesses PI_1 = 253 nm and PI_2 = 212 nm, a Pd screen thickness of 25 nm, and unit cell dimension p = 426 nm. The optimum parameters for these two designs share similar values, but the Pd screen geometries shown in Fig. 8.9 are complex and very different from each other. This variation in geometry indicates that there are multiple, complex screen patterns that are capable of supporting many resonances to meet the custom, broadband absorption criteria. The pixel sizes for Designs 1 and 2 are 31 nm and 28 nm, respectively, which can be accurately defined by using electron-beam lithography.

The simulated near- to mid-IR scattering responses for the synthesized designs are shown in Fig. 8.10 for unpolarized light averaged over azimuth angles. The normal and 40° off-normal plots show that the absorptivity is well above 90% for nearly the entire 1 μm to 5 μm range, indicating that the GA was able to evolve geometries that met the targeted design criteria. At normal incidence, the minimum absorptivity for Design 1 is above 90% from 1.05 μm to 5.10 μm with an average of 97.2%, and for the 40° off-normal incidence, 90% absorptivity is maintained from 1.00 μm to 4.85 μm with an average of 95.3%. Likewise, Design 2 has >90% absorptivity at normal incidence from 1.05 μm to 5.10 μm with an average of 97.3% and >90% absorptivity at 40° off-normal incidence from 1 μm to 4.85 μm with an average of 95.4%. Two-dimensional color plots of the absorptivity versus wavelength and incidence angle are shown in Fig. 8.11 for unpolarized light as well as for TE- and TM-polarized waves. These plots show that the multi-octave absorptivity is maintained well beyond the optimized FOV with the 90% absorptivity maintained to ±50° and 80% absorptivity until about ±60°. As the incidence angle approaches grazing, the absorptivity rapidly decreases. The absorptivity also falls off quickly outside of the optimized wavelength range.

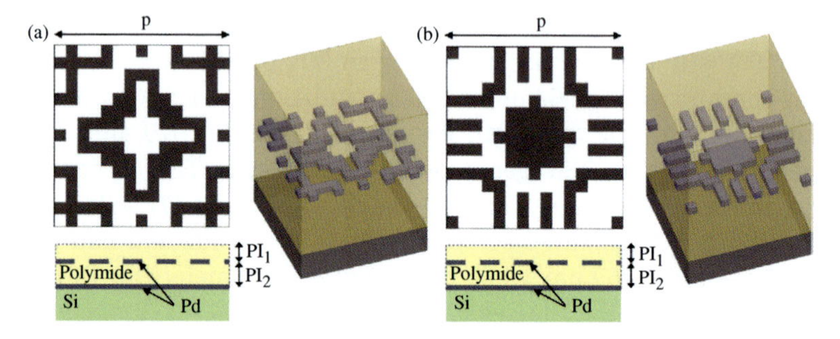

Figure 8.9 Optimized geometries for multi-octave broadband absorbers. The unit cell, consisting of a patterned Pd screen sandwiched between two polyimide layers and a Pd ground, is shown along with a 3D render. (a) Design 1 has optimized dimensions p = 469 nm, Pl_1 = 243 nm, Pl_2 = 219 nm, Pd screen thickness of 21 nm, and Pd ground thickness of 100 nm. (b) Design 2 has optimized dimensions p = 426 nm, Pl_1 = 253 nm, Pl_2 = 212 nm, Pd screen thickness of 25 nm, and Pd ground thickness of 100 nm.

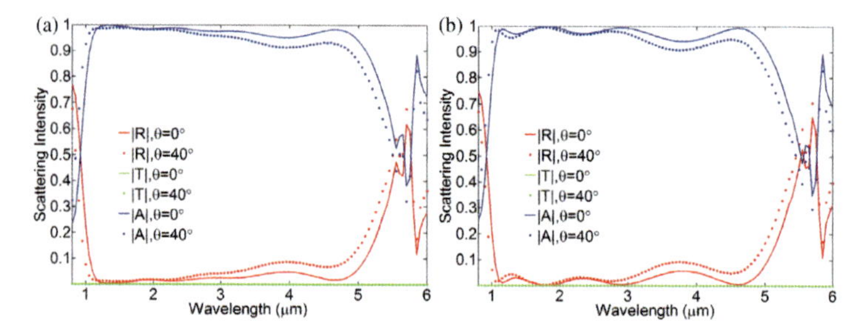

Figure 8.10 Simulated scattering intensities for (a) Design 1 and (b) Design 2 described in Fig. 8.9 at normal and θ = 40° off-normal incidence, averaged over polarization and azimuth angle.

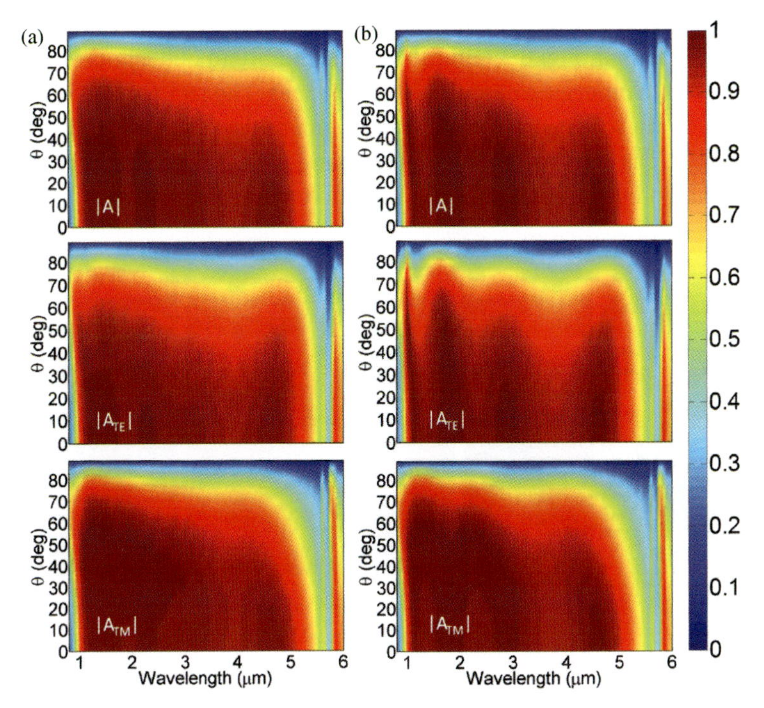

Figure 8.11 Angle-dependence of the absorptivity shown with the wavelength plotted on the horizontal axis and the incidence angle on the vertical axis for (a) Design 1 and (b) Design 2 described in Fig. 8.9. Unpolarized absorptivity is shown on the top, TE-polarized in the center, and TM-polarized at the bottom.

Similar to other MMAs studied at radiofrequency wavelengths [52], the source of the absorption is expected to be from the enhanced field surrounding the lossy metal screen. In order to demonstrate this phenomenon, relative field plots calculated 1 nm above the Pd screen under TE-polarized incidence are shown in Fig. 8.12 for four wavelengths: 0.8 μm, 1.5 μm, 4 μm, and 5 μm. The first wavelength, 0.8 μm, is below the absorption band, where the wave is almost completely reflected. At this wavelength, there is no field enhancement beyond the standing wave due to the incident wave being terminated at the Pd ground plane for both designs. By contrast, the other three wavelengths have large enhanced fields surrounding parts of the Pd screen, which result in the high absorption observed in Figs. 8.10 and 8.11. Also noteworthy are the different parts of the screen that are resonant in these three sampled wavelengths.

Multiple subtle peaks in the absorptivity spectra can be observed for both designs in Fig. 8.9, indicating that the broad absorption band is due to the merging of multiple resonances. The fact that different parts of the screen are resonant at 1.5 μm and 5 μm in Fig. 8.12 for both designs confirms that the broadband performance of the MMA is enabled by the combination of multiple resonances from a single Pd screen.

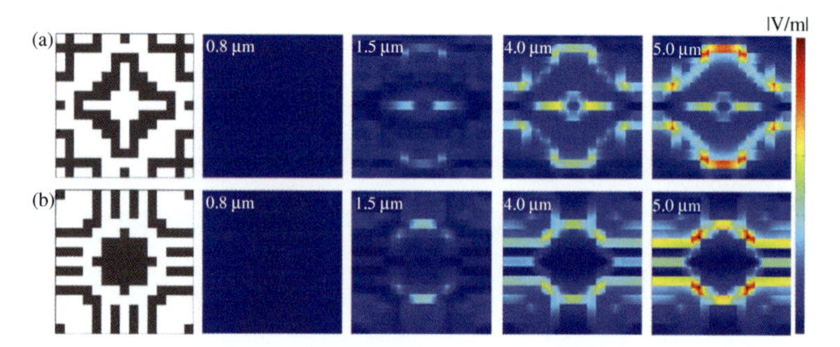

Figure 8.12 Plots of the relative E-field for (a) Design 1 and (b) Design 2 described in Fig. 8.9 at one out-of-band wavelength and three wavelengths within the absorption band.

In summary, we have shown the design synthesis of two polarization-independent, multi-octave bandwidth absorbers for the near- to mid-IR using an EBG-based structure with only a single Pd screen. These MMA designs possess greater than 90% absorptivity over an optimized FOV of ±40°. The robust GA made it possible to synthesize complex pixelized screen geometries that have multiple resonances that combine to form a broad absorption band. This allows for a broad absorption bandwidth to be accessed with a single patterned metal screen, something that would be extremely difficult to achieve by using standard design approaches such as trial and error.

8.4 Broadband Optical Metasurfaces

8.4.1 Introduction to Metasurfaces

Optical metasurfaces represent a new class of two-dimensional metamaterials comprising a single layer of metallic nanostructures that possess exceptional capabilities for manipulating light in an ultrathin, planar platform [39, 54, 101]. Metasurfaces also exhibit reduced loss and less fabrication complexity as compared to three-dimensional metamaterial devices, making them attractive for integration into practical optical systems [86]. The spectral and spatial dispersion of the metasurface optical response can also be tailored to generate a specific abrupt interfacial phase changes and cross-polarized responses on a subwavelength scale by engineering the nanostructured geometry. Because of these unique capabilities, metasurfaces have been exploited to demonstrate a variety of new physical phenomena and associated optical devices over the past few years, including anomalous reflection and refraction [40, 72, 87, 99], frequency-selective near-perfect absorption [15, 45, 65], optical wavefront manipulation [1, 73, 78], spin-hall effect of light [83, 98], spin-controlled photonics [84], polarization-dependent unidirectional SPP excitation [41, 63], and metasurface holograms [42, 74].

A potential highly desirable application for metasurfaces is optical waveplates with broadband polarization conversion and a wide FOV. Such devices could be used in systems that perform optical characterization, sensing, and communication functions [12]. Simultaneously achieving broadband and wide-angle performance is difficult using conventional multilayer stacks of birefringent materials because such structures rely on dispersive birefringence properties. By contrast, metasurfaces could provide a pathway toward broadband and wide-angle polarization conversion in an ultrathin, submicron-thick layer.

Several optically thin, metasurface-based polarization-control components have been theoretically proposed and demonstrated, including various polarizers [30, 106], near-field polarization shapers [10], and ultrathin waveplates [8, 34, 51, 53, 77, 79, 88, 100, 107]. These examples have employed homogenous arrays of weakly coupled anisotropic resonant elements, including crossed nanodipoles [30, 77] and nanoslits [8, 53, 79], L-shaped [88] or V-shaped [51] nanoantennas, and elliptical nanoholes [34]. Because the anisotropic optical responses of these elements rely on the resonance of each isolated element, where strong dispersion and impedance mismatches may exist, these structures typically suffer from a narrow FOV, limited bandwidth, and/or low efficiency.

More recently, a near-IR quarter-wave plate composed of an array of orthogonally coupled nanodipole elements was reported that achieved broadband circular-to-linear polarized light conversion. However, this example provided an average power efficiency of less than 50% [107]. In the next section, we will demonstrate that broadband and wide-angle optical waveplates can be achieved by carefully tailoring the anisotropic properties of the metasurface and the interference of light on a subwavelength scale. A super-octave bandwidth plasmonic half-wave plate as well as a quarter-wave plate are experimentally demonstrated in the near-IR regime, exhibiting polarization conversion ratios and energy efficiencies of 90% or greater.

8.4.2 Broadband Optical Metasurface-Based Waveplates

The metasurface-based half-wave plate considered here consists of a top periodic gold nanorod array and a bottom continuous gold layer, separated by a silicon dioxide spacer, as shown in the schematic in Fig. 8.13a. An *s*-polarized wave is incident on the nanostructure at an angle of ($\varphi_i = 135°$, θ_i) with its electric field oscillating in the $\hat{x}+\hat{y}$ direction. Once the incident wave makes contact with the nanostructure, the reflected wave becomes a *p*-polarized wave, with its electric field polarized in the ($\varphi_i = 135°$, $\theta = 90° - \theta_i$) direction (*i.e.*, it has a transverse component in the $-\hat{x}+\hat{y}$ direction). The multilayer nanostructure can be efficiently designed by using the interference model and the optical properties of the strongly coupled nanorod array. When the incident wave interacts with the multilayer structure, both *s*- and *p*-polarized reflected and transmitted waves are generated by the metasurface. The transmitted waves undergo multiple reflections between the metasurface and the bottom Au layer, where they interfere with one another to create the final reflected wave.

To solve the complex multi-objective design problem of achieving a half-wave plate with highly efficient polarization conversion spanning a wide spectral and angular range, a powerful covariance matrix adaptation evolutionary strategy (CMA-ES) optimization technique was employed to identify the nanostructure dimensions that meet the challenging multi-objective half-wave plate design criteria. Predefined constraints on the allowable unit cell size, nanorod length and width, as well as the thickness of metal and dielectric layers were incorporated into the optimization algorithm to avoid generating structures that are impractical to fabricate. Additionally, experimentally measured dispersive optical properties for the constituent metal and dielectric materials were used to minimize discrepancies between the theoretically predicted and experimentally characterized nanostructure response. For each design candidate, the calculated reflection coefficients for both polarizations (r_{ps}, r_{ss}) in the targeted wavelength and angular range are compared with the user-input-defined target values to determine the cost, which is expressed as:

$$\text{cost}_{\text{hwp}} = \sum_{\lambda,\theta_i} \left[\left(|r_{ps}| - 1 \right)^2 + |r_{ss}|^2 \right] \tag{8.4}$$

where λ and θ_i denote the wavelength and incident angles included in the optimization. For this work, we selected wavelengths from 640 nm to 1290 nm in steps of 50 nm and angles from 0° to 40° in steps of 10°. The CMA-ES evolved solutions until it converged on a three-layer metal–dielectric nanostructure with a sufficiently low overall cost value that achieved the desired half-wave plate optical properties. The optimized unit cell geometry that meets these design criteria is shown in Fig. 8.13b. The Au nanorods have a length of 210 nm and a width of 70 nm, which corresponds to an aspect ratio of 0.33. The periodicity in both the *x*- and *y*-directions is 252 nm, giving an inter-element gap of 42 nm in the *x*-direction, which provides strong electromagnetic coupling to the neighboring nanorods. The thickness values for the top Au nanorod array layer, the SiO_2 layer, and the bottom Au layer are 42 nm, 114 nm, and 100 nm, respectively, resulting in a structure that is only 256 nm thick. The polarization conversion ratio (PCR) of the metasurface-based half-wave plate was calculated using the semi-analytical interference model, defined as:

$$\text{PCR}_{\text{hwp}} = |r_{ps}|^2/(|r_{ps}|^2 + |r_{ss}|^2) \tag{8.5}$$

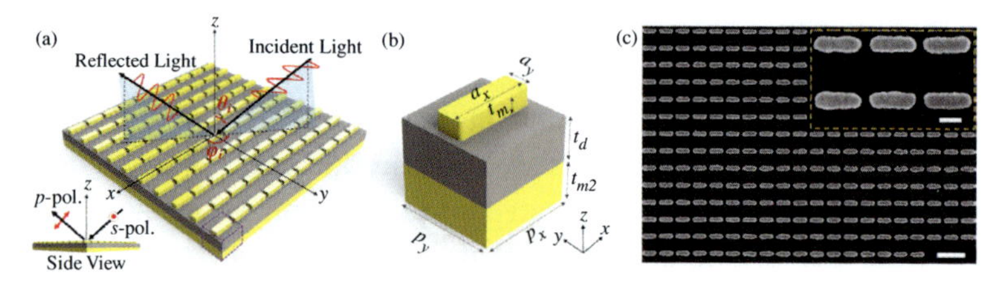

Figure 8.13 (a) Tilted three-dimensional view of the metasurface-based half-wave plate. An *s*-polarized wave incident from an angle of (θ_i, φ_i = 135°) is converted into a *p*-polarized wave upon reflection. Inset shows the side view of the nanostructure and light path. (b) Unit cell configuration of the optimized half-wave plate. The dimensions are a_x = 210, a_y = 70, p_x = 252, p_y = 252, t_m = 42, t_d = 114, and t_{m2} = 100 (all in nm). (c) Top-view FESEM image of a portion of the fabricated nanostructure showing the nanorod array. Scale bar: 400 nm. The inset shows the magnified top view of two by three unit cells. Scale bar: 100 nm.

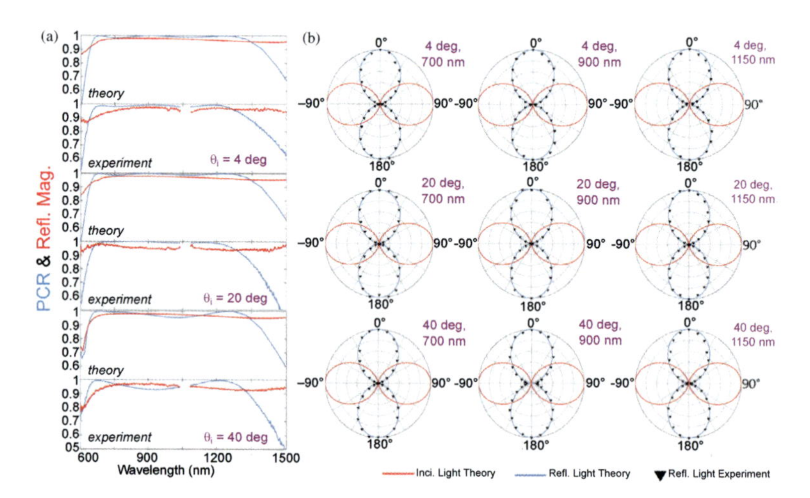

Figure 8.14 (a) Theoretically predicted and experimentally measured polarization conversion ratio (PCR) and reflection magnitude (Refl. Mag.) as a function of wavelength at different angles of incidence (4°, 20°, 40°). (b) Theoretically predicted and experimentally measured polarization state in the plane perpendicular to the wave vector at 700 nm, 900 nm, and 1150 nm for different angles of incidence (4°, 20°, 40°).

As shown by the theoretical curves plotted in Fig. 8.14a, the calculated PCR_{hwp} is greater than 94% across the targeted broad wavelength and wide angular range and remains above 90% over a wider band from 640 nm to 1400 nm. Throughout this spectral and angular range, the reflection magnitude is greater than 95%, thereby achieving a much higher power efficiency than previous plasmonic metamaterial and metasurface-based designs [33, 39, 86, 100, 107]. For the optimized two-dimensional nanorod array structure, at large incident angles, the in-band PCR_{hwp} is limited by the higher quality factors of two resonant peaks that correspond to the x- and y-directed localized SPP dipolar modes of the coupled Au nanorods as well as the increase in phase retardation in the dielectric spacer. As a result, in the spectral range bracketed by the two modes, the reflection phase difference between the x- and y-directed electric field components exceeds 180° for incident angles greater than 40°. This causes the observed decrease in the PCR_{hwp} near the middle of the band between the resonant peaks, which restricts the FOV for >90% PCR_{hwp} to approximately ±40°. The state of polarization (SOP) traces of the incident and reflected light are presented in Fig. 8.14b at several angles of incidence and wavelengths within the band. Clearly defined cosine-shaped patterns can be identified, indicating that the reflected wave possesses a high degree of linear polarization (DoLP). The SOP patterns of the incident light have their maxima along the −90°/90° directions, while the SOP patterns of the reflected light have their maxima along the 0°/180° directions. The 90° rotation over the broad spectral and wide angular range confirms that the optimized plasmonic metasurface-based half-wave plate indeed transforms a linearly polarized incident wave into a reflected wave with a cross polarization.

The optimized plasmonic metasurface-based half-wave plate was fabricated by standard electron-beam lithography followed by an Au lift-off process. The field emission scanning electron microscopy (FESEM) image of the fabricated nanostructure is shown in Fig. 8.13c. The PCR_{hwp} and reflection magnitude of the fabricated structure was characterized using a custom-built optical setup that utilizes a supercontinuum source to illuminate the structure at incident angles of 4°, 20°, and 40°. The strong peak in the power spectrum of the source at 1064 nm (the pump wavelength used to generate the supercontinuum) introduces significant measurement error; thus, data collected in the wavelength range between 1040 nm and 1080 nm are excluded from the plots.

The optical properties of the fabricated metasurface-based half-wave plate shown in Fig. 8.14a are in strong agreement with theoretical predictions. The measured PCR_{hwp} and reflection magnitude both remain above 92% over the targeted broad wavelength range from 640 nm to 1290 nm and wide FOV from 0° to 40°. Even at a larger 50° angle of incidence, the PCR_{hwp} and reflection magnitude are still 86% and 87%, respectively [50]. In comparison to the design, the long wavelength cut-off for the 90% PCR_{hwp} bandwidth at a 4° incidence angle is blue-shifted from 1400 nm to 1330 nm without a change in the short-wavelength cut-off. This small discrepancy is attributed to the wider inter-rod gap spacing in the fabricated structure compared to the optimized design dimensions, which causes a blue-shift in the x-directed localized SPP dipolar resonance without affecting the y-directed dipolar resonance as the gap increases.

The polarization conversion is experimentally verified by the measured SOP patterns shown in Fig. 8.14b. The cosine-shaped patterns confirm the high DoLP of the reflected light. In addition, the angle between the maxima of the reflected and incident light patterns is 90° throughout the entire wavelength band and the 40° FOV, indicating that the maximum light is reflected from the sample when the polarizers for the incident and reflected light are oriented orthogonally to each other. Compared to previous experimentally demonstrated plasmonic metasurface-based polarization-control devices, the nanorod array half-wave plate presented here achieves a high power efficiency over a wide spectral and angular range.

To demonstrate the versatility of the proposed platform and design approach, a metasurface-based quarter-wave plate was also optimized for the same spectral range of 640 nm to 1290 nm and wide FOV from 0 to 40°. The three-layer structure shown in Fig. 8.15a transforms a circularly polarized incident wave from the direction θ_i φ_i = 180° into a linearly polarized reflected wave in the direction θ_i φ_i = 0°, which provides an in-phase superposition of both the s- and p-polarizations. As depicted in Fig. 8.15b, the unit cell of the optimized doubly periodic array is composed of 180 nm long and 70 nm wide Au nanorods that are centered within the 240 nm × 282 nm cell. This array structure possesses a weaker degree of anisotropy compared to the half-wave plate because of the larger nanorod aspect ratio of 0.5 and wider inter-element gap of

60 nm. As plotted in Fig. 8.16a, the simulated PCR_{qwp} is greater than 90% over a wider than targeted wavelength range from 620 nm to 1500 nm for incident angles up to 50° (not shown here). The reflection magnitude is greater than 93% across the targeted spectral and angular range except for a narrow dip near 680 nm. The SOP traces presented in Fig. 8.16b show that the nanostructure transforms the incident wave with a circular SOP pattern into a reflected wave with a cosine pattern that has a maximum along the 45°/225° direction.

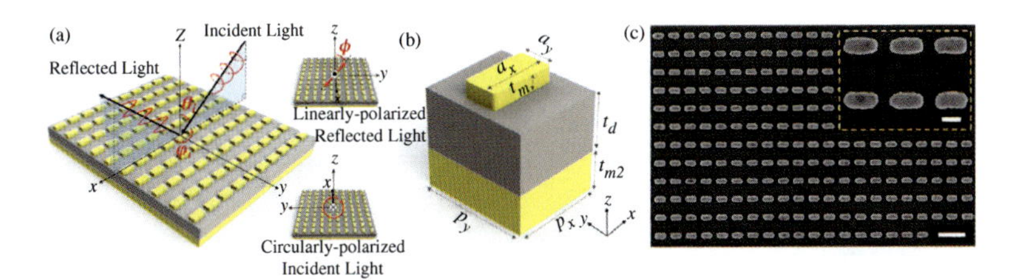

Figure 8.15 (a) Tilted three-dimensional view of a metasurface-based quarter-wave plate. A circularly polarized wave incident from an angle of θ_i φ_i = 0° is converted into a linearly polarized wave upon reflection. Inset shows the polarization of the reflected and incident waves in the plane perpendicular to the wave vector. The angle between the electric field and the plane of incidence of the reflected light is 45°. (b) Unit cell configuration of the optimized quarter-wave plate. The dimensions are a_x = 180, a_y = 90, p_x = 240, p_y = 282, t_m = 40, t_d = 150, and t_{m2} = 100 (all in nm). (c) Top-view FESEM image of a portion of the fabricated nanostructure showing the nanorod array. Scale bar: 400 nm. The inset shows the magnified top view of two by three unit cells. Scale bar: 100 nm.

The quarter-wave plate was fabricated using a similar nanofabrication process, and an FESEM image of the nanofabricated device is shown in Fig. 8.15c. The measured dimensions of the Au nanorod length, width, and minimum inter-element gap spacing are 180 ± 6 nm, 90 ± 4 nm, and 60 ± 5 nm, respectively, which accurately reflect the design dimensions. Similar to the full-wave plate, the rounded nanorod edges result in an effective gap of 68 ± 3 nm, which is wider than the design target. During the characterization process, a commercial quarter-wave plate was used to generate circularly polarized incident light. As shown in Fig. 8.16a, the measured PCR_{qwp} is greater than 91% within the targeted wavelength band and angular range, and the reflection magnitude is higher than 92%. The reduced PCR_{qwp} value in the long wavelength range can be attributed to the imperfect circular polarization of the incident wave above 1300 nm, which is outside of the operational band of the commercial quarter-wave plate. The measured SOP patterns are in strong agreement with the simulated predictions, indicating that the circularly polarized incident wave is converted to a linearly polarized reflected wave.

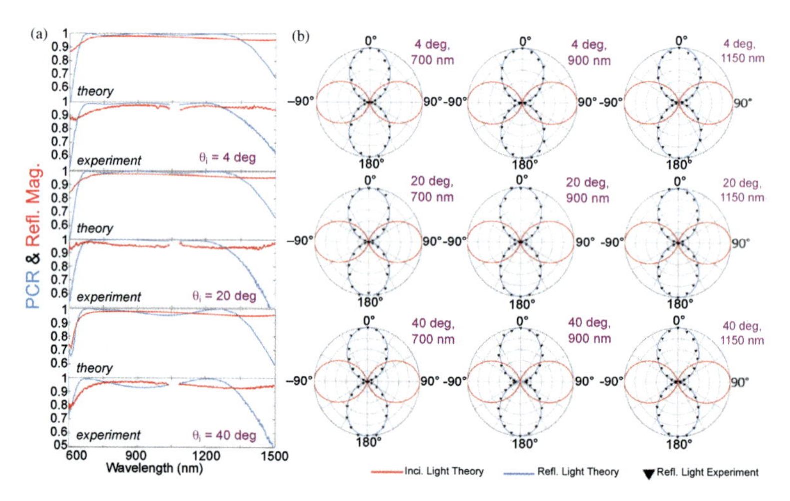

Figure 8.16 (a) Theoretically predicted and experimentally measured polarization conversion ratio (PCR) and reflection magnitude (Refl. Mag.) as a function of wavelength at different angles of incidence (4°, 20°, 40°). (b) Theoretically predicted and experimentally measured polarization state in the plane perpendicular to the wave vector at 700 nm, 900 nm, and 1150 nm for different angles of incidence (4°, 20°, 40°).

In conclusion, we have described a robust approach for realizing super-octave, highly efficient, wide-FOV plasmonic metasurface-based waveplates for the visible and near-IR. These goals were achieved by employing a global CMA-ES optimizer to tailor the spectral phase properties of a strongly coupled anisotropic nanorod array in conjunction with the interference of light between the array and the ground plane. A half-wave plate and a quarter-wave plate were validated through nanofabrication and characterization, demonstrating 90% or greater PCRs and reflection magnitudes over more than an octave bandwidth and a wide FOV. Beyond the demonstrated waveplates, the general design approach presented can be extended to nanostructured optical metasurfaces that exhibit arbitrary polarization conversion properties, as well as other unique optical functionalities.

8.4.3 Broadband Optical Light Steering with Metasurfaces

The wavefront of a light beam propagating across an interface can be modified arbitrarily by introducing abrupt phase changes at the surface of the interface. Such an artificially induced phase shift can be created by placing a metasurface comprising an array of nanoantennas at the interface. By introducing this additional phase shift, the conventional Snell's law no longer holds. A generalized version of Snell's law was proposed in Refs. [72, 99] as follows:

$$\sin(\theta_t)n_t - \sin(\theta_i)n_i = \lambda_0 \nabla\Phi/2\pi, \tag{8.6}$$

$$\sin(\theta_r) - \sin(\theta_i) = n_i^{-1}\lambda_0 \nabla\Phi/2\pi. \tag{8.7}$$

These expressions indicate that a gradient in a phase discontinuity, $\nabla\Phi$, along an interface can modify the direction of the refracted and reflected rays. Another more general interpretation of this is that $\nabla\Phi$ is essentially an additional momentum contribution introduced by breaking the symmetry at the interface. Hence, in order to conserve momentum, the light wave has to bend accordingly. Note that when there is no phase gradient $\nabla\Phi$ along the interface, Eqs. (8.6) and (8.7) reproduce the conventional Snell's law. This is conceptually different from the diffraction formula for a periodic grating since the bent light is caused by a unidirectional phase discontinuity, which is not a particular order of diffraction. Thus, by designing and engineering a phase discontinuity along an interface, one can fully control the bending of a light wave beyond what the conventional Snell's law predicts.

A plasmonic nanoantenna array, as shown in Fig. 8.17, can be used to create the phase gradient $\nabla\Phi$, where each individual nanoantenna provides a distinct phase shift. This type of V-shaped nanoantenna supports two kinds of resonant modes, a symmetric mode, and an asymmetric mode, and the superposition of those two modes allows the phase shift of the scattered light by the nanoantennas to span from 0 to 2π using eight nanoantenna configurations. Note that for this type of nanoantenna, the phase shift only happens in the cross polarization—the polarization perpendicular to the incident polarization. The eight nanoantennas shown in Fig. 8.17 form a period of a linear array in the *x*-direction and provide a phase gradient of $2\pi/\Lambda$, where Λ is the period of the array in the *x*-direction. Different phase gradients can be achieved by varying the period Λ. Full-wave simulations using the finite-element method were performed to ensure accurate designs. The simulated results for the entire nanoantenna array are shown in Fig. 8.18. The plotted electric field clearly shows that the wavefront for cross-polarized light is modified due to the phase discontinuity introduced by the nanoantenna array.

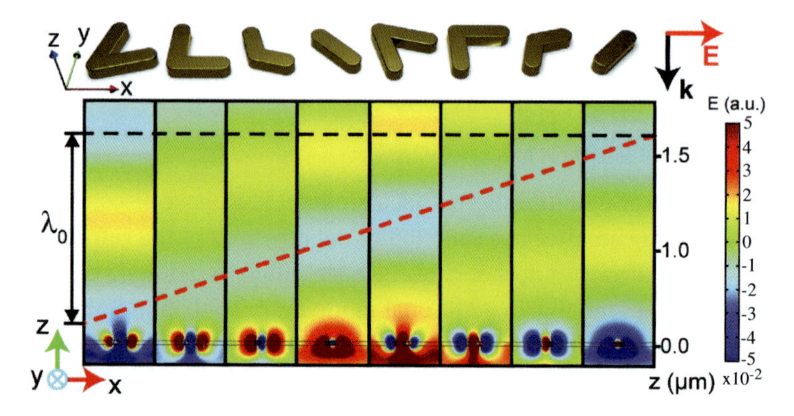

Figure 8.17 The simulated results for the individual constituent antennas in a unit cell on a silicon substrate using realistic optical properties for the materials as obtained from experiments. The false-color map indicates the cross-polarized electric field at normal incidence from the top of the sample with the incident electric field polarized in the *x*-direction. The red dashed line shows the phase difference between the antennas. The total phase shift from the eight antennas is 2π.

The metasurface consisting of that nanoantenna array was fabricated on a double-side polished silicon substrate using standard electron-beam lithography and lift-off processes. The height of the nanoantennas was about 30 nm. The metasurface was measured using the scatterometry mode in a spectroscopic ellipsometer, and the experimental measurements are shown in Fig. 8.19.

The measurement results indicate that the reflection and refraction of the light are indeed following the generalized Snell's law. In addition, the wavelength-dependent measurements showed the broadband behavior of the sample for wavelengths from 1.0 μm to 1.9 μm. Over some wavelengths, the reflection and refraction angles of the light even become "negative." In other words, they are on the same side as the incident light. This phase gradient on the surface provides an additional degree of freedom in that the light can be steered into any arbitrary direction by adding properly designed phase shifts on the surface.

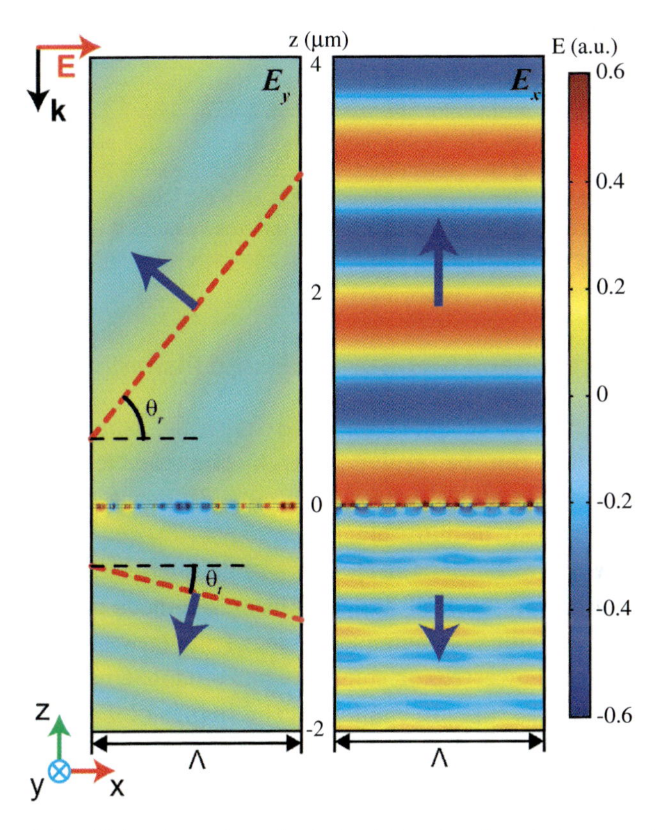

Figure 8.18 Finite-element method simulation for the eight nanoantenna array elements. This example shows a nanoantenna array with Λ = 1920 nm. Light is incident from the top with the electric field polarized in the *x*-direction. The left field map shows the cross-polarized electric field. The red dashed line shows the wavefront of the scattered light, and the blue arrow indicates the propagation direction. The right field map shows the *x*-polarized scattered electric field for the same structure, which exhibits ordinary reflection and refraction.

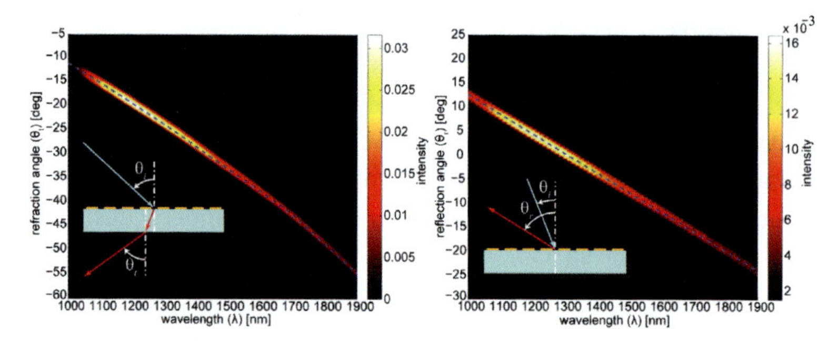

Figure 8.19 The false-color maps indicate the experimentally measured relative intensity profiles for the nanoantenna array with Λ = 1440 nm and *x*-polarized excitation. The dashed line shows the theoretical prediction of the peak position using the generalized Snell's law. (Left panel) Refraction angle versus wavelength for cross-polarized light with a 65° incidence angle. (Right panel) Reflection angle versus wavelength for cross-polarized light with a 30° incidence angle. Reprinted from Ref. 72 with permission from AAAS.

8.4.4 Broadband Metasurface-Based Planar Microlenses

The capability of changing the phase directly using resonating nanoantennas can be exploited as a powerful technique for designing planar optical lenses [1, 2, 62, 73]. Lenses made using metasurfaces can be ultrathin, planar, spherical aberration free, and even highly efficient [62]. The planar microlenses reported in [73] utilize an inverted design built on Babinet's principle, *i.e.*, instead of metallic nanoantennas, a set of similarly shaped nano-voids (Babinet-inverted, or complementary nanoantennas) milled in a thin metallic film are employed. The shapes of the nano-voids are designed by simulating each element individually using a full-wave, three-dimensional, finite-element method. Figure 8.20 shows the phase shift ranging from 0 to $7\pi/4$ in the cross-polarized light within the visible range, depending on the nano-void design.

The nano-voids were arranged in concentric rings to form the microlens geometry. The nano-voids of different shapes are distributed in such a way that the individual, discrete phase shifts created by each nano-void element cause the overall cross-polarized wave front scattered at the interface to focus at a focal distance *f*. The design of the metasurface lens is obtained through the reciprocity principle, *i.e.*, by reverse-propagating the light from a point source located at the focal point back to the lens plane. Therefore, the required relative phase shift φ for a nanoantenna located at a distance r from the center is

$$\varphi(r) = \frac{2\pi\sqrt{r^2 + f^2}}{\lambda} \tag{8.8}$$

where λ is the wavelength of the incident light in free space.

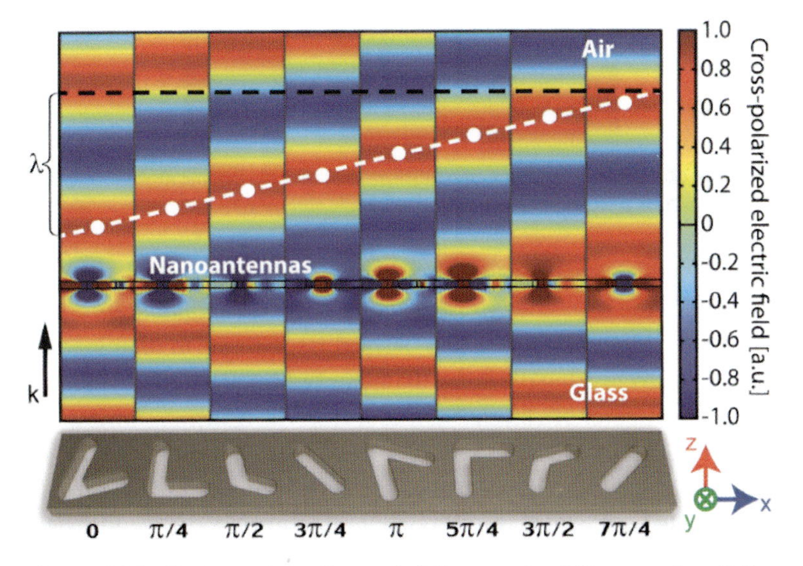

Figure 8.20 Design schematics and full-wave simulation results of the individual Babinet-inverted nanoantennas (nano-voids) at a wavelength of 676 nm. The nano-voids create discrete phase shifts from 0 to 7π/4 for cross-polarized light. The linearly polarized light enters the system from the glass substrate side of the sample. The pseudo-color field map indicates the cross-polarized light scattered from each nanoantenna, clearly revealing the discrete phase shifts.

The lens was milled in a 30 nm thick gold film using a focused ion beam. The initial metal film is deposited on a glass substrate with electron-beam vapor deposition. Figure 8.21 shows an FESEM image of the fabricated sample. Note that the focal lengths are designed for a wavelength of 676 nm, and the thickness of the samples is only about 1/22 of the operational wavelength.

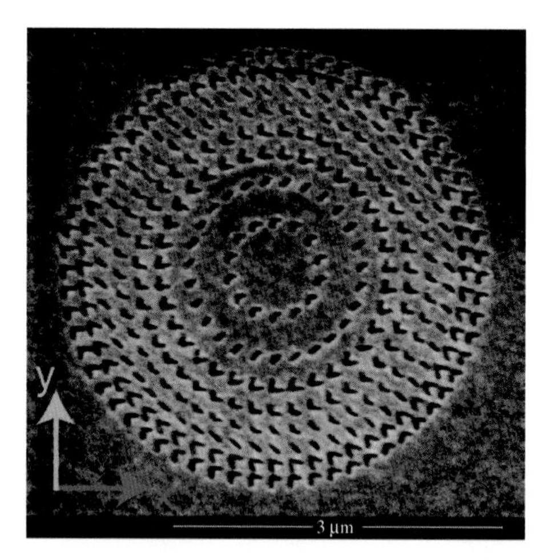

Figure 8.21 An FESEM image of a fabricated planar plasmonic metalens with a focal length of 2.5 μm at an operational wavelength of 676 nm.

The metasurface lens under test is mounted on the microscope stage with the lens side up. The sample is then illuminated from the substrate side with a linearly polarized laser. Uniform illumination is ensured by making the beam diameter orders of magnitude larger than the diameter of a given lens under test. The transmission images from the samples are then recorded by a CCD camera. A pair of perpendicular polarizers is placed in the path, one before the sample and the second before the CCD camera, to ensure that only cross-polarized light is collected in the measurement and to eliminate any possible co-polarized background light. The metasurface lenses do not produce any significant cross-polarized stray light, and for this reason, there is almost no light capable of distorting the intensity profiles obtained by the described method.

By changing the height of the stage, the intensity distribution at different distances from the surface of the metasurface lens on the transmission side can be obtained. The focal point of the objective lens and the surface of the sample are coincident at $z = 0$, and $(x, y) = (0, 0)$ is the center of the metasurface lens. Figure 8.22(a2) shows the CCD image obtained at the focal plane with 676 nm incident light. From the CCD images, three-dimensional intensity distribution profiles of the cross-polarized light are reconstructed.

Figure 8.22 shows the reconstructed light intensity distributions for cross-polarized light at three different wavelengths on the transmission side. The pseudo-color field maps obtained from simulations and measurements for each design are plotted side by side for comparison. The light is strongly focused at the expected position for each design. At their focal planes, the diameters of the focused light spot are significantly smaller than 1 μm (*i.e.*, the focal spots are at the scale of the operational wavelength).

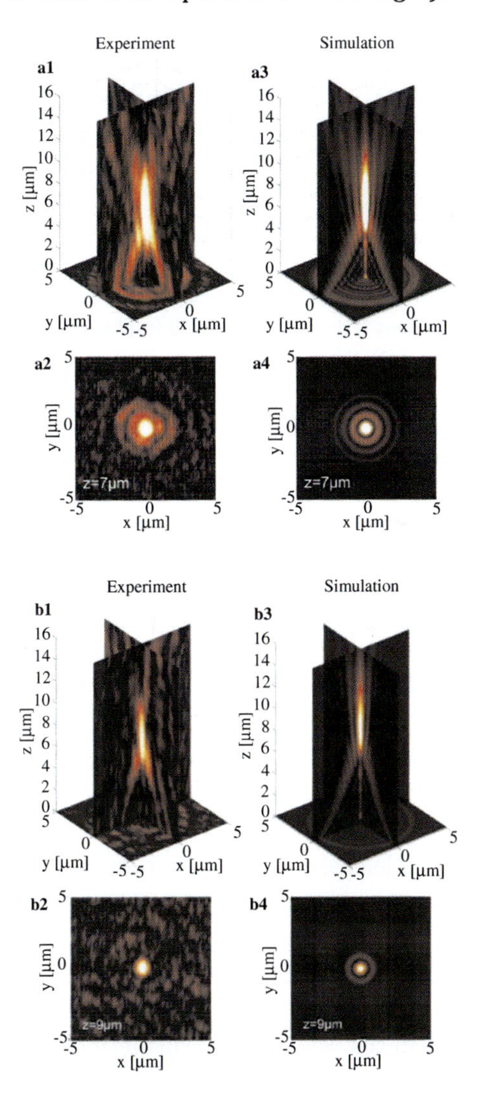

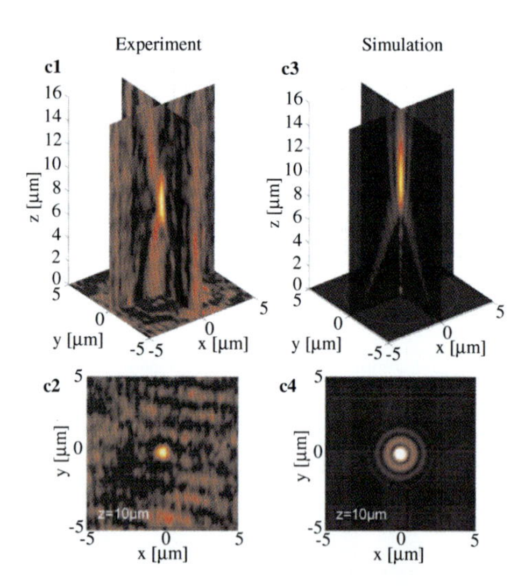

Figure 8.22 Comparison between the measured and simulated results for a metasurface lens design at incident wavelengths of (a) 676 nm, (b) 531 nm, and (c) 476 nm. The lens is designed to have a focal length of 7 μm. (a1), (a2), (b1), (b2), (c1), and (c2) Cross-polarized light intensity distributions on the transmission side reconstructed from measurements; (a3), (a4), (b3), (b4), (c3), and (c4) Simulated results under the same conditions. (a1), (a3), (b1), (b3), (c1), and (c3) Intensity distributions for two cross-sectional planes cutting through the center of the metalens; (a2), (a4), (b2), (b4), (c2), and (c4) Intensity distribution at the respective focal planes (*z*-coordinates are shown on the plots). The *x*-*y* planes in (a1), (b1), and (c1) are at $z = 0$ μm. The *x*-*y* planes in (a3), (b3), and (c3) are at $z = 0.1$ μm (avoiding the singularity at $z = 0$ μm in the simulations).

Because of the symmetry in the set of nanoantenna building blocks utilized in this design (the last four nanoantennas are identical to the first four nanoantennas upon 90° clockwise rotation), changing the linear polarization of the incident light from *x*-polarized to *y*-polarized creates just an additional π phase shift of the cross-polarized scattered light. This means that the performance of the metasurface lens should not suffer due to such a change in polarization. This prediction has been successfully verified in experiment.

In conclusion, we have demonstrated a type of metasurface lens that can be extremely small (only a few microns in size) and thin (much smaller than the wavelength). Furthermore, it is possible to design the focal length of the microlens to be on the order of the operational wavelength. The experimental and simulation results demonstrated here exhibit a focal length of f = 2.5 μm with an ultrathin planar 4 μm diameter metasurface [73].

Acknowledgment

The authors would like to thank Theresa S. Mayer and Lan Lin for their contributions on the fabrication and characterization of several of the optical metamaterial and metasurface devices that are presented in this chapter.

References

1. Aieta, F. et al. (2012). Aberration-free ultrathin flat lenses and axicons at telecom wavelengths based on plasmonic metasurfaces, *Nano Lett.*, **12**, pp. 4932–4936.
2. Aieta, F., et al. (2015). Multiwavelength achromatic metasurfaces by dispersive phase compensation, *Science*, **347**, pp. 1342–1345.
3. Alu, A., Bilotti, F., Engheta, N., and Vegni, L. (2007). Theory and simulations of conformal omnidirectional sub-wavelength metamaterial leaky-wave antenna, *IEEE Trans. Ant. Propag.*, **55**, pp. 1698–1708.
4. Antoniades, M. A., and Eleftheriades, G. V. (2005). A broadband series power divider using zero-degree metamaterial phase-shifting lines, *IEEE Microw. Wireless Compon. Lett.*, **15**, pp. 808–810.
5. Argyropoulos, C., Le, K. Q., Mattiucci, N., D'Aguanno, G., and Alù, A. (2013). Broadband absorbers and selective emitters based on plasmonic Brewster metasurfaces, *Phys. Rev. B*, **87**, 205112.
6. Avitzour, Y., Urzhumov, Y. A., and Shvets, G. (2009). Wide-angle infrared absorber based on a negative-index plasmonic metamaterial, *Phys. Rev. B*, **79**, 045131.
7. Aydin, K., Ferry, V. E., Briggs, R. M., and Atwater, H. A. (2011). Broadband polarization-independent resonant light absorption using ultrathin plasmonic super absorbers, *Nat. Comm.*, **2**, 517.
8. Baida, F. I., Boutria, M., Oussaid, R., and van Labeke, D. (2011). Enhanced-transmission metamaterials as anisotropic plates, *Phys. Rev. B*, **84**, 035107.

9. Bayraktar, Z., Gregory, M. D., Wang, X., and Werner, D. H. (2012). Matched impedance thin planar composite magneto-dielectric metasurfaces, *IEEE Trans. Antennas Propag.*, **60**, pp. 1910–1920.

10. Biagioni, P. et al. (2009). Near-field polarization shaping by a near-resonant plasmonic cross antenna, *Phys. Rev. B*, **80**, 153409.

11. Bonod, N., Tayeb, G., Maystre, D., Enoch, S., and Popov, E. (2008). Total absorption of light by lamellar metallic gratings, *Opt. Express*, **16**, 15431.

12. Born, M., and Wolf, E. (2002). *Principles of Optics*, 6th Ed. (Cambridge University Press, Cambridge, UK).

13. Bossard, J. A., and Werner, D. H. (2013). Metamaterials with angle selective emissivity in the near-infrared, *Opt. Express*, **21**, pp. 5215–5225.

14. Bossard, J. A., and Werner, D. H. (2013). Metamaterials with custom emissivity polarization in the near-infrared, *Opt. Express*, **21**, pp. 3872–3884.

15. Bossard, J. A., Lin, L., Yun, S., Liu, L., Werner, D. H., and Mayer, T. S. (2014). Near-ideal optical metamaterial absorbers with super-octave bandwidth, *ACS Nano*, **8**, pp. 1517–1524.

16. Brian, T., and Piestun, R. (2003). Total external reflection from metamaterials with ultralow refractive index, *J. Opt. Soc. Am. B*, **20**, pp. 2448–2453.

17. Cai, W. S., Chettiar, U. K., Kildishev, A. V., and Shalaev, V. M. (2007). Optical cloaking with metamaterials, *Nat. Photon.*, **1**, pp. 224–227.

18. Caloz, C., and Itoh, T. (2005). *Electromagnetic Metamaterials: Transmission Line Theory and Microwave Applications* (John Wiley-IEEE Press, USA).

19. Cheng, Q., Cui, T. J., Jiang, W. X., and Cai, B. G. (2010). An omnidirectional electromagnetic absorber made of metamaterials, *New J. Phys.*, **12**, 063006.

20. Choi, M. et al. (2011). A terahertz metamaterial with unnaturally high refractive index, *Nature*, **470**, pp. 369–373.

21. Cui, Y., Xu, J., Fung, K. H., Jin, Y., Kumar, A., and He, S. A. (2011). A thin film broadband absorber based on multi-sized nanoantennas, *Appl. Phys. Lett.*, **99**, 253101.

22. Cui, Y. et al. (2012). Ultrabroadband light absorption by a sawtooth anisotropic metamaterial slab, *Nano Lett.*, **12**, pp. 1443–1447.

23. Diem, M., Koschny, T., and Soukoulis, C. M. (2009). Wide-angle perfect absorber/thermal emitter in the terahertz regime, *Phys. Rev. B*, **79**, 033101.

24. Diest, K., ed. (2013). *Numerical Methods for Metamaterial Design* (Springer, New York, NY).

25. Dolling, G., Enkrich, C., Wegener, M., Soukoulis, C. M., and Linden, S. (2006). Simultaneous negative phase and group velocity of light in a metamaterial, *Science*, **312**, pp. 892–894.

26. Dolling, G., Wegener, M., Soukoulis, C. M., and Linden, S. (2007). Negative-index metamaterial at 780 nm wavelength, *Opt. Lett.*, **32**, pp. 53–55.

27. Edwards, B., Alù, A., Young, M. E., Silveirinha, M., and Engheta, N. (2008). Experimental verification of epsilon-near-zero metamaterial coupling and energy squeezing using a microwave waveguide, *Phys. Rev. Lett.*, **100**, 033903.

28. Eibert, T. F., Volakis, J. L., Wilton, D. R., and Jackson, D. R. (1999). Hybrid FE/BI modeling of 3-D doubly periodic structures utilizing triangular prismatic elements and an MPIE formulation accelerated by the Ewald transformation, *IEEE Trans. Antennas Propag.*, **47**, pp. 843–850.

29. Eleftheriades, G. V., and Balmain, K. G. (2005). *Negative Refraction Metamaterials: Fundamental Principles and Applications* (Wiley-IEEE Press, USA).

30. Ellenbogen, T., Seo, K., and Crozier, K. B. (2012). Chromatic plasmonic polarizers for active visible color filtering and polarimetry, *Nano Lett.*, **12**, pp. 1026–1031.

31. Enoch, S., Tayeb, G., Sabouroux, P., Guérin, N., and Vincent, P. (2002). A metamaterial for directive emission, *Phys. Rev. Lett.*, **89**, 213902.

32. Fang, N., Lee, H., Sun, C., and Zhang, X. (2005). Sub-diffraction-limited optical imaging with a silver superlens, *Science*, **308**, pp. 534–537.

33. Gansel J. K. et al. (2009). Gold helix photonic metamaterial as broadband circular polarizer, *Science*, **325**, pp. 1513–1515.

34. Gordon, R. et al. (2004). Strong polarization in the optical transmission through elliptical nanohole arrays, *Phys. Rev. Lett.*, **92**, 037401.

35. Grbic, A., and Eleftheriades, G. V. (2004). Overcoming the diffraction limit with a planar left-handed transmission-line lens, *Phys. Rev. Lett.*, **92**, 117403.

36. Gupta, S., and Caloz, C. (2009). Analog signal processing in transmission line metamaterial structures, *Radioengineering*, **18**, pp. 155–167.

37. Haupt, R. L., and Werner, D. H. (2007). *Genetic Algorithms in Electromagnetics* (Wiley, Hoboken, NJ).

38. Hibbins, A. P., Murray, W. A., Tyler, J., Wedge, S., Barnes, W. L., and Sambles, J. R. (2006). Resonant absorption of electromagnetic fields by surface plasmons buried in a multilayered plasmonic nanostructure, *Phys. Rev. B*, **74**, 073408.

39. Holloway, C. L. (2012). An overview of the theory and applications of metasurfaces: The two-dimensional equivalents of metamaterials, *IEEE Antennas Propag. Magazine*, **54**, pp. 10–35.

40. Huang, L. et al. (2012). Dispersionless phase discontinuities for controlling light propagation, *Nano Lett.*, **12**, pp. 5750–5755.

41. Huang, L. et al. (2013). Helicity dependent directional surface plasmon polariton excitation using a metasurface with interfacial phase discontinuity, *Light Sci. Appl.*, **2**, e70.

42. Huang, L. et al. (2013). Three-dimensional optical holography using a plasmonic metasurface, *Nat. Commun.*, **4**, 2808.

43. Hutley, M. C., and Maystre, D. (1976). The total absorption of light by a diffraction grating, *Opt. Commun.*, **19**, pp. 431–436.

44. Jiang, Z. H., Gregory, M. D., and Werner, D. H. (2011). Experimental demonstration of a broadband transformation optics lens for highly directive multibeam emission, *Phys. Rev. B*, **84**, 165111.

45. Jiang, Z. H., Yun, S., Toor, F., Werner, D. H., and Mayer, T. S. (2011). Conformal dual-band near-perfectly absorbing mid-infrared metamaterial coating, *ACS Nano*, **5**, pp. 4641–4647.

46. Jiang, Z. H., Wu, Q., and Werner, D. H. (2012). Demonstration of enhanced broadband unidirectional electromagnetic radiation enabled by a subwavelength profile leaky anisotropic zero-index metamaterial coating, *Phys. Rev. B*, **86**, 125131.

47. Jiang, Z. H., Lan, L., Bossard, J. A., and Werner, D. H. (2013). Bifunctional plasmonic metamaterials enabled by subwavelength nano-notches for broadband, polarization-independent enhanced optical transmission and passive beam-steering, *Opt. Express*, **21**, pp. 31492–31505.

48. Jiang, Z. H., and Werner, D. H. (2013). Compensating substrate-induced bianisotropy in optical metamaterials using ultrathin superstrate coatings, *Opt. Express*, **21**, pp. 5594–5605.

49. Jiang, Z. H. et al. (2013). Tailoring dispersion for broadband low-loss optical metamaterials using deep-subwavelength inclusions, *Sci. Rep.*, **3**, pp. 1571.

50. Jiang, Z. H. et al. (2014). Broadband and wide field-of-view plasmonic metasurface-enabled waveplates, *Sci. Rep.*, **4**, pp. 7511.

51. Kats, M. A. et al. (2012). Giant birefringence in optical antenna arrays with widely tailorable optical anisotropy, *Proc. Natl. Acad. Sci. U.S.A.*, **109**, pp. 11364–12368.

52. Kern, D. J., and Werner, D. H. (2003). A genetic algorithm approach to the design of ultra-thin electromagnetic bandgap absorbers, *Micro-Opt. Technol. Lett.*, **38**, pp. 61–64.

53. Khoo, E. H., Li, E. P., and Crozier, K. B. (2011). Plasmonic wave plate based on subwavelength nanoslits, *Opt. Lett.*, **36**, pp. 2498–2500.

54. Kildishev, A. V., Boltasseva, A., and Shalaev, V. M. (2013). Planar photonics with metasurfaces, *Science*, **339**, 1232009.

55. Kocaman, S. et al. (2011). Zero phase delay in negative-refractive-index photonic crystal superlattices, *Nat. Photon.*, **5**, pp. 499–505.

56. Koschny, T., Kafesaki, M., Economou, E. N., and Soukoulis, C. M. (2004). Effective medium theory of left-handed materials, *Phys. Rev. Lett.*, **93**, 107402.

57. Kundtz, N., and Smith, D. R. (2010). Extreme-angle broadband metamaterial lens, *Nat. Mater.*, **9**, pp. 129–132.

58. Landy, N. I., Sajuyigbe, S., Mock, J. J., Smith, D. R., and Padilla, W. J. (2008). Perfect metamaterial absorber, *Phys. Rev. Lett.*, **100**, 207402.

59. Landy, N. I., Bingham, C. M., Tyler, T., Jokerst, N., Smith, D. R., and Padilla, W. J. (2009). Design, theory, and measurement of a polarization-insensitive absorber for terahertz imaging, *Phys. Rev. B*, **79**, 125104.

60. Le Perchec, J., Quémerais, P., Barbara, A., and López-Ríos, T. (2008). Why metallic surfaces with grooves a few nanometers deep and wide may strongly absorb visible light, *Phys. Rev. Lett.*, **100**, 066408.

61. Lier, E., Werner, D. H., Scarborough, C. P., Wu, Q., and Bossard, J. A. (2011). An octave bandwidth negligible-loss radiofrequency metamaterial, *Nat. Mater.*, **10**, pp. 216–222.

62. Lin, D. et al. (2014). Dielectric gradient metasurface optical elements, *Science*, **345**, pp. 298–302.

63. Lin, J. et al. (2013). Polarization-controlled tunable directional coupling of surface plasmon polaritons, *Science*, **340**, pp. 331–334.

64. Liu, N. et al. (2009). Plasmonic analogue of electromagnetic induced transparency at the Drude damping limit, *Nat. Mater.*, **8**, pp. 758–762.

65. Liu, N., Mesh, M., Weiss, T., Hentschel, M., and Giessen, H. (2010). Infrared perfect absorber and its application as plasmonic sensor, *Nano Lett.*, **10**, pp. 2342–2348.

66. Liu, R. et al. (2009). Broadband ground-plane cloak, *Science*, **323**, pp. 366–369.
67. Liu, X., Starr, T., Starr, A. F., and Padilla, W. J. (2010). Infrared spatial and frequency selective metamaterial with near-unity absorbance, *Phys. Rev. Lett.*, **104**, 207403.
68. Liu, X. et al. (2011). Taming the blackbody with infrared metamaterials as selective thermal emitters, *Phys. Rev. Lett.*, **107**, 045901.
69. Ma, H. F., and Cui, T. J. (2010). Three-dimensional broadband and broad-angle transformation-optics lens, *Nat. Commun.*, **1**, 124.
70. Möller, K. D., Warren, J. B., Heaney, J. B., and Kotecki, C. (1996). Cross-shaped bandpass filters for the near- and mid-infrared wavelength regions, *Appl. Opt.*, **35**, pp. 6210–6215.
71. Narimanov, E. E., and Kildishev, A. V. (2009). Optical black hole: Broadband omnidirectional light absorber, *Appl. Phys. Lett.*, **95**, 041106.
72. Ni, X., Emani, N. K., Kildishev, A. V., Boltasseva, A., and Shalaev, V. M. (2012). Broadband light bending with plasmonic nanoantennas, *Science*, **335**, pp. 427.
73. Ni, X., Ishii, S., Kildishev, A. V., and Shalaev, V. M. (2013). Ultra-thin, planar, Babinet inverted plasmonic metalenses, *Light Sci. Appl.*, **2**, e27.
74. Ni, X., Kildishev, A. V., and Shalaev, V. M. (2013). Metasurface holograms for visible light, *Nat. Commun.*, **4**, 2807.
75. Okabe, H., Caloz, C., and Itoh, T. (2004). A compact enhanced-bandwidth hybrid ring using an artificial lumped-element left-handed transmission-line section, *IEEE Trans. Microw. Theory Tech.*, **52**, pp. 798–804.
76. Paul, T., Menzel, C., Rockstuhl, C., and Lederer, F. (2010). Advanced optical metamaterials, *Adv. Mater.*, **22**, pp. 2354–2357.
77. Pors, A. et al. (2011). Plasmonic metamaterial wave retarders in reflection by orthogonally oriented detuned electrical dipoles, *Opt. Lett.*, **36**, pp. 1626–1628.
78. Pors, A., Nielsen, M. G., Eriksen, R. L., and Bozhevolnyi, S. I. (2013). Broadband focusing flat mirrors based on plasmonic gradient metasurfaces, *Nano Lett.*, **13**, pp. 829–834.
79. Roberts, A., and Lin, L. (2012). Plasmonic quarter-wave plate, *Opt. Lett.*, **37**, pp. 1820–1822.
80. Schurig, D. et al. (2006). Metamaterial electromagnetic cloak at microwave frequencies, *Science*, **314**, pp. 977–980.

81. Shelby, R. A., Smith, D. R., and Schultz, S. (2001). Experimental verification of a negative index of refraction, *Science*, **292**, pp. 77–79.

82. Shin, J., Shen, J. T., and Fan, S. (2009). Three-dimensional metamaterials with an ultrahigh effective refractive index over a broad bandwidth, *Phys. Rev. Lett.*, **102**, 093093.

83. Shitrit, N., Bretner, I., Gorodetski, Y., Kleiner, V., and Hasman, E. (2011). Optical spin Hall effects in plasmonic chains, *Nano Lett.*, **11**, pp. 2038–2042.

84. Shitrit, N. et al. (2013). Spin-optical metamaterial route to spin-controlled photonics, *Science*, **340**, pp. 724–726.

85. Smith, D. R., Schultz, S., and Soukoulis, C. M. (2002). Determination of effective permittivity and permeability of metamaterials from reflection and transmission coefficients, *Phys. Rev. B*, **65**, 195104.

86. Soukoulis, C. M., and Wegener, M. (2011). Past achievements and future challenges in the development of three-dimensional photonic metamaterials, *Nat. Photon.*, **5**, pp. 523–530.

87. Sun, S. et al. (2012). High-efficiency broadband anomalous reflection by gradient metasurfaces, *Nano Lett.*, **12**, pp. 6223–6229.

88. Sung, J. et al. (2008). Nanoparticle spectroscopy: Birefringence in two-dimensional arrays of L-shaped silver nanoparticles, *J. Phys. Chem. C*, **112**, pp. 3252–3260.

89. Tang, Y., Bossard, J. A., Werner, D. H., and Mayer, T. S. (2008). Single-layer metallodielectric nanostructures as dual-band midinfrared filters, *Appl. Phys. Lett.*, **92**, pp. 263106.

90. Tao, H., Landy, N. I., Bingham, C. M., Zhang, X., Averitt, R. D., and Padilla, W. J. (2008). A metamaterial absorber for the terahertz regime: Design, fabrication and characterization, *Opt. Express*, **16**, pp. 7181–7188.

91. Tao, H. et al. (2009). Highly flexible wide angle of incidence terahertz metamaterial absorber: Design, fabrication, and characterization, *Phys. Rev. B*, **78**, 241103.

92. Teperik, T. V., García de Abajo, F. J., Borisov, A. G., Abdelsalam, M., Bartlett, P. N., Sugawara, Y., and Boumberg, J. J. (2008). Omnidirectional absorption in nanostructured metal surfaces, *Nat. Photon.*, **2**, pp. 299–301.

93. Valentine, J. et al. (2008). Three-dimensional optical metamaterial with a negative refractive index, *Nature*, **455**, pp. 376–379.

94. Valentine, J., Li, J., Zentgraf, T., Bartal, G., and Zhang, X. (2009). An optical cloak made of dielectrics, *Nat. Mater.*, **8**, pp. 568–571.

95. Volakis, J. (2007). *Antenna Engineering Handbook*, 4th Ed. (McGraw-Hill Professional, New York).

96. Wan, B., Koschny, T., and Soukoulis, C. M. (2009). Wide-angle and polarization-independent chiral metamaterial absorber, *Phys. Rev. B*, **80**, 033108.

97. Wang, C., Yu, S., Chen, W., and Sun, C. (2013). Highly efficient light-trapping structure design inspired by natural evolution, *Sci. Rep.*, **3**, 1025.

98. Yin, X., Ye, Z., Rho, J., Wang, Y., and Zhang, X. (2013). Photonic spin-Hall effect at metasurfaces, *Science*, **339**, pp. 1405–1407.

99. Yu, N. et al. (2011). Light propagation with phase discontinuities: Generalized laws of reflection and refraction, *Science*, **334**, pp. 333–337.

100. Yu, N. et al. (2012). A broadband, background-free quarter-wave plate based on plasmonic metasurfaces, *Nano Lett.*, **12**, pp. 6328–6333.

101. Yu, N., and Capasso, F. (2014). Flat optics with designer metasurfaces, *Nat. Mater.*, **13**, pp. 139–150.

102. Yun, S. et al. (2012). Low-loss impedance-matched optical metamaterials with zero-phase delay, *ACS Nano*, **6**, pp. 4475–4482.

103. Yun. S. et al. (2013). Experimental verification of substrate-induced bianisotropy in optical metamaterials, *Appl. Phys. Lett.*, **103**, 233109.

104. Zhang, B. et al. (2011). Polarization-independent dual-band infrared perfect absorber based on a metal-dielectric-metal elliptical nanodisk array, *Opt. Express*, **19**, pp. 15221–15228.

105. Zhang, S., Genov, D. A., Wang, Y., Liu, M., and Zhang, X. (2005). Plasmon-induced transparency in metamaterials, *Phys. Rev. Lett.*, **101**, 047401.

106. Zhao, Y., Belkin, M. A., and Alù, A. (2012). Twisted optical metamaterials for planarized ultrathin broadband circular polarizers, *Nat. Commun.*, **3**, 870.

107. Zhao, Y., and Alù, A. (2013). Tailoring the dispersion of plasmonic nanorods to realize broadband optical meta-waveplates, *Nano Lett.*, **13**, 1086–1091.

108. Zhu, W., and Zhao, X. (2009). Metamaterial absorber with dendritic cells at infrared frequencies, *J. Opt. Soc. Am. B*, **26**, pp. 2382–2385.

109. Zhudlev, N. I. (2010). The road ahead for metamaterials, *Science*, **328**, pp. 582–583.

Index